Major Günter Goebel
Träger des Eichenlaubs
zum Ritterkreuz des Eisernen Kreuzes

THE COLLECTOR'S GUIDE TO CLOTH
THIRD REICH MILITARY HEADGEAR

GARY WILKINS
WITH A FOREWORD BY BILL SHEA

Schiffer Military History
Atglen, PA

This book is dedicated to Roy B. Merritt,
educator and poet of high caliber
who set me on the path;
And to my daughter Jen

Studio photography and book design by Robert Biondi.

Copyright © 2002 by Gary Wilkins.
Library of Congress Catalog Number: 2001094919.

Printed in China.
ISBN: 0-7643-1428-9

We are always looking for people to write books on new and related subjects. If you have an idea for a book, please contact us at the address below.

Published by Schiffer Publishing Ltd.
4880 Lower Valley Road
Atglen, PA 19310
Phone: (610) 593-1777
FAX: (610) 593-2002
E-mail: Schifferbk@aol.com.
Visit our web site at: www.schifferbooks.com
Please write for a free catalog.
This book may be purchased from the publisher.
Please include $3.95 postage.
Try your bookstore first.

In Europe, Schiffer books are distributed by:
Bushwood Books
6 Marksbury Ave.
Kew Gardens
Surrey TW9 4JF
England
Phone: 44 (0)208 392-8585
FAX: 44 (0)208 392-9876
E-mail: Bushwd@aol.com.
Free postage in the UK. Europe: air mail at cost.
Try your bookstore first.

Foreword

by Bill Shea
DBA The Ruptured Duck

When I was first approached with the concept of a book on WWII German visor caps and headgear, my reaction was something to the effect of, "time to stop flogging the dead horse".

The mere idea conjured up images in my mind of the usual several paragraphs of the same old information on the myriad of organizations who wore hats during the Third Reich.

This would be followed by the usual several images of different angles of visor and field caps from collections around the world. These would have been submitted to the author or publisher as prized possessions from collectors hoping that their hats would pass the screening test and be included in yet another "coffee table" book. Surely, previously unpublished pictures would help shed some new light on variations in style and form. They would also provide collectors with an opportunity to showcase and enhance the value of their personal trophies. At the same time, they would be sharing their findings with a worldwide collecting community.

However, I still remained reluctant and for the most part unmoved by the idea of yet another hat book.

Then I learned that the author would be a long time friend and fellow collector Gary Wilkins. I have known Gary for at least three decades and was very much aware of his military and civilian background. His passion for Third Reich military history and artifacts had motivated Gary to publish a very rudimentary soft cover "primer" on World War II military headgear back in 1986. He had combined his knowledge of the topic, his experience in handling the artifacts and his artistic talents to produce a delightful little treatise, which received local acclaim. Several years have passed since that leaflet was available and Gary has traveled many miles and lived in various interesting places such as Japan and Korea. He may have stopped collecting the artifacts themselves for quite some time, but the passion for knowledge continued to burn. During that time frame, Gary combined his computer skills with his command of the German language (among others) and continued to accumulate printed articles and information on the very intriguing topic of German haberdashcry.

The advent of electronic mail and quicker access to the files of many of the companies (even those who were in the former East) was quite enlightening. Virtually all of the companies that still existed had been renamed, and in some cases were producing a different product. Many, of course, did not survive the war but old files, old catalogues and advertisements were still accessible. Thousands of hours were spent compiling useful information heretofore never presented in any written, organized format. Furthermore, policies as they related to the cap manufacturers and how they sought to regulate themselves which have never been published or examined until now, were discovered. This "policing" was, for the most part, internal and outside the realm of the huge Third Reich bureaucracy.

Having reviewed the content of this work, I am convinced that it contains an abundance of information that will be useful for German soft headgear collectors at all levels. Enough attention has been paid to the technical aspects of construction and design to provide a learning experience for the reader. Sufficient anecdotal tidbits have been included to stir the curiosity of those interested in the personalities of that era. Abundant crisp photo images have been provided to entertain those of us who still subscribe to the philosophy that "one picture is worth 1000 words."

Lastly, the author is keenly aware of the profound impact that the Internet has, and will continue to have, over the business of buying and selling these collectibles. He has therefore provided the collecting public with many tangible observations regarding what to look for in authentic period pieces versus the altered and the outright reproductions. Certainly, this, in itself, should help all levels of collectors look more closely at what they are buying and spend a little time analyzing the characteristics of that next potential addition to their collection.

I am now convinced that this work will become an essential and helpful addition to the library of reference materials already available on the topic of Third Reich Headgear.

Contents

Introduction

This book has its origins in two simple desires: First, to list some of the points that I myself always wished had been available to me in a convenient format when I first began collecting. Second, to figure out what the devil ever happened to the Robert Lubstein company (the owner of the EREL trademark). This question has been pestering me for decades – until I decided to stop wondering and start doing! I began writing to every agency in Berlin that I could think of that might be able to shed light on the subject. Being able to write in German made things much less difficult (or so I thought). At about the same time, Bill Shea – who knew of my interest in German cap makers and maker marks in general, suggested that I keep some sort of record of the same. I was amazed to find that there were over forty makers on my list in very short order. Then, revised, over fifty makers...and the number kept growing. And thus began the mission of identifying as many makers as possible, with an associated register including historical data, if and when available. My research efforts went into full gear, but it took more than a year to secure everything required for this book – and there still remains more to be done. But, editors have established deadline dates, and so some of the information that was received too late to make this volume has to be saved for a later time. Nonetheless, what I have been able to include I am certain many collectors will find as fascinating as I did while compiling it all. Time to double-check all the caps in your collection!

The increase in online buying is another factor I wanted to address here. When I first published a basic headgear guide in 1985, the Internet was hardly a factor in the militaria market. Today it has reached such a level that militaria show attendance has gone into a steady, constant decline. This is a very unfortunate side effect of Internet shopping, because it reduces the opportunities for collectors to view a good variety of headgear examples without necessarily having to purchase an item. It allows collectors to chat with dealers and gain experience, and enjoy discussions with other collectors and dealers from all walks of life.

In comparison, items offered on many Internet sites are not always well photographed (to be polite), and there is rarely any opportunity for a 'hands on' inspection prior to purchase. Buying on a strictly 'visual' basis can be a very risky proposition, and I hope that the practical information on authentic, World War II period headgear and the many comparisons with modified contemporary caps and reproductions that I have presented here will help lessen this danger to some degree by helping collectors develop experience through practical application.

Please note that in keeping with this idea of 'practical application', many of the photographic examples used in the book are those that have appeared on the Internet (public domain), and are printed here in the same quality – good, or bad – in which they were displayed online.

Any errors, oversights or ommissions are strictly mine (I apologize in advance), and corrections or suggestions are always welcome: This hobby is supposed to be *interactive* – discussions are how we learn, and they add to the overall enjoyment of this fascinating pastime.

Gary Wilkins
Boston, MA

Acknowledgements

It is not possible to write a research work without the help of a great many individuals who are willing to share of their time and efforts, and offer all manner of assistance. All of these many people – whether individuals or representatives of some organization – are due my heartfelt thanks for their kindness. Such a work is indeed the sum of its parts – the author is merely the organizer, the person who ties it all together and provides the continuity.

As is always the case, however, there are certain individuals who went beyond the call of duty and provided critical assistance without which this book would not have been complete. These folks deserve special mention above the usual list of acknowledgments.

First and foremost in this category is Bill Shea, a leading dealer of militaria in New England. I have known Bill for a long time, and his support and encouragement over the years has been invaluable. A great many of the authentic headgear examples pictured in this volume came from Bill's 'warehouse' as they passed through on their way to personal collections across the land (not, unfortunately, my own). Bill is a militaria encyclopedia and a scrupulously honest purveyor of authentic items (who could probably outfit an entire battalion if the need ever arose!).

Another individual to whom I am indebted is Gerard Stezelberger, who, as the second largest contributor, put a selection of his first class, near (if not) mint examples of authentic caps at my disposal for the book, accompanied by some interesting discussions over the telephone (though I have yet to meet him personally).

A fellow I met and got to know somewhat late in the project, Dale Paul, was also a great help. Dale's amazing collection of autographed Knight's Cross winner photos provided much of the superb period photographic evidence used in the book, and the interesting stories about his past days writing to all of these personalities offered a pleasant diversion which was sometimes sorely needed. Other period pictures were donated by another friendly and generous fellow named Eric Mueller, and the chapter on reproduction headgear would not have been possible without the friendly assistance of Bill Bureau (Bill Bureau's Militaria), who I visited several times to discuss the subject.

My editor, Bob Biondi, deserves both my thanks and my apologies – many thanks for his patience, and many apologies for the time pressure I forced him to undergo.

Certain individuals and institutions in Germany also deserve special recognition. My warmest regards and special thanks must first go to Herr Klaus-Peter Merta, Head of the Militaria Collection at the Deutsches Historisches Museum, Berlin. I wrote to Herr Merta as one among many individuals and organizations, but he threw himself wholeheartedly into the project and sent me package after package filled with amazing information on the cap making industry during the war. He spent many hours of his own valuable time digging up all kinds of fascinating items which provided a basis for many of the details presented throughout the book, particularly Chapter 16. Herr Merta was also kind enough to take some time out of his extremely busy schedule when I requested a check into the trail of the famous maker of the EREL-Sonderklasse cap, Robert Lubstein and actually went out on a Berlin scouting mission reminiscent of the LeFevre staff's efforts in Normandy. The investigation and search paid off: the mystery that has long nagged at so many collectors over the years, especially myself: 'Whatever happened to EREL?' is answered here, for the first time (in Chapter 16). As a bonus, never before seen contemporary photographs of the Robert Lubstein building – once proudly known as the 'ERELHAUS' – are also presented. Herzlichen Dank, Herr Merta!

Frau Evelyn Bachfisch of the IHK Regensburg DV-Stelle also deserves my thanks. After receiving my request for information, Ms. Bachfisch personally sought out and visited one of the very few, still existing former cap makers (though the company has not made caps for many years) located in Regensburg, in order to request information for the book first hand 'at the source' so to speak. Wonder of wonders, she was also able to secure (from one of the company's friendly and helpful directors) an actual piece of wartime company letterhead for inclusion along with the entry on the company. My sincere gratitude is thus also due to that company for its kind assistance and willingness to share a piece of its history. Thanks also to Albert Kempf G.m.b.H. und Co. K.G., and Ludwig Vögele Mützenfabrik, two former wartime cap makers, which replied personally to my enquiry. None of these firms ignored my requests for information – and this at a time when many German companies have no desire to look at the past. For that, I thank them. Three other still-existing firms – one in St. Wendel, one in Hamburg and the last in Munich – by comparison, proved unwilling or unable to respond with even a postcard – despite several requests (all made in German).

Another person I would like to mention individually is Herr Karl-Josef Fritz of the Rastatt Historical Association. Herr Fritz was able to locate a timeline for and the origin of 'Biehler-Mütze' for me when local government offices were unable to help (no records of the firm). He also managed to secure photographs of the store front *then*, and now.

Of the many German government agencies that assisted me with the (often frustrating) research efforts into the records of individual companies, very special thanks is due the staff (particularly Frau Martina Pietsch) of the Amtsgericht Charlottenberg in Berlin which filled my mailbox with incredibly useful and fascinating Trade Register records on the seven (of fourteen) Berlin-based companies that it was able to locate. I practically ripped into the packets in anticipation of what new documents might lie within. And, for all this, the Amtsgericht asked for absolutely nothing in return. This office (and its employees) is truly a credit to the city of Berlin.

On a personal note, I also want to thank several friends who supported the many long hours that this book demanded of me, and who took no offense at my lack of attention during this time consuming project. In Germany, my friends of 25 years, Marlene and Dieter Hundrieser: Dieter assisted where he could, while Marlene took all of the Deutsch Mark Traveler's Cheques I sent her and used her bank account to disburse payments for me covering the various charges and fees that some agencies required for information. Little did she know in 1976, what answering a simple pen pal letter would lead to...

Researching a book can often have unexpected bonuses. Another fellow in Germany, Herr Heinz Beier, I learned was a comrade in arms – he was a Bundeswehr Panzer soldier for a time – while I was once an armor officer in the U.S. Army. Though we originally began working together researching cap makers in Wuppertal, Heinz eventually helped me to locate and contact a German veteran of Stalingrad (whose captured Soldbuch and hand-written war diary I had acquired and translated). This gentleman had, in fact, survived the German surrender and Russian captivity and was repatriated to Germany in 1949, where he still resides!

In the USA, my best wishes and warm thanks to my old Army buddy, former CPT Dan Flores. Dan and I served together in Seoul, Korea, and have been friends ever since. During this project, Dan has been ready and willing to provide needed fire support at critical moments, putting his services and his store, Quality Quick Photo, at my disposal any time I was in need of Quality sample photos made to order, Quick-time.

Closer to home, Melissa Newton has been a true friend who was always willing to help when needed. So also Frank Strazzulla, close friend and collecting compatriot since Junior High School when we stood together in front of our lockers – he with a Mother's Cross in Gold, myself with a Luftwaffe M43 cap (as I recall, it cost $15) – discussing militaria and World War II battles, strategy and tactics. Frank was also an immense help during the incredibly tedious job of sorting and organizing all of the photographs and advertisements used in the book. For your help with the project Frank, and many years of support and shared times, thank you very much! Finally, a thanks to my daughter, Jen, for gamely putting up with the countless militaria discussions that went on during the preparation of the manuscript.

Now for the list. Any omissions are entirely by accident, not design, and I apologize in advance for the oversight!

UNITED STATES

Individuals (not previously mentioned)
Robert Stevenson, Photographer for the US Army, Armor Magazine and the Patton Museum, Ft. Knox
Gary Baier
Eric Mueller
Richard P. Nauman
Gordon Williamson
Jeanette M. Cardamone, Research Chemist, Hides, Lipids & Wool Research, U.S. Department of Agriculture
Joe Wotka (specialist on the Polizei)

Institutions
ARMOR Magazine, Ft. Knox, Kentucky

Companies
Bancroft Cap Co.

ENGLAND

Brian Greenhalgh (Bayonet King)

FRANCE

François Saez

GERMANY

Individuals
Herr Kay Stephan
Herr Heinz Beier, Wuppertal (a Waffenkamerad), and Anke
Herr Dipl.Ing. Lothar Steins (Textiles)
Herr Dr. karl-Heinz Lehmann (Textiles)

Amtsgerichter
Amtsgericht Augsburg, Registergericht, Frau Schlee
Amtsgericht Bremen, Registergericht, (Schönwälder)
Amtsgericht Charlottenberg (Harnack), Frau Martina Pietsch, Berlin
Amtsgericht Dresden, Registergericht, Frau Kerstin Linke
Amtsgericht Erfurt, Registergericht, (Keller)
Amtsgericht Frankfurt am Main
Amtsgericht Hamburg, Herr Peter Krause
Amtsgericht Kassel (Hölscher)
Amtsgericht Koblenz
Amtsgericht Köln, Frau Kutz
Amtsgericht Offenbach am Main, Registerabteilung, Frau Kiehle
Amtsgericht Stendal, Registergericht, Herr Conrad König
Amtsgericht Stuttgart, Herr Lang
Amtsgericht Schweinfurt, Herr Geßner
Amtsgericht St. Wendel
Amtsgericht Ulm, Registergericht, Frau Krause
Amtsgericht Würzburg

Gewerbeämter
Bad Kissingen, Gewerbeamt, Frau Margarete Büchner (Verw. Angestellte)

Industrie- und Handelskammer (Chamber of Commerce and Industry) branches:
IHK Braunschweig, Herr Klaus-Peter Weidlich
IHK München
IHK Offenbach am Main, Frau H. Schlegel, Sachbearbeiterin
IHK Regensburg, DV-Stelle, Frau Evelyn Bachfisch
IHK Wuppertal-Solingen-Remscheid, Zentralrat/Information, Herr Thomas Wängler

Handwerkskammer
Handwerkskammer Berlin, Frau Marijke Lass
Handwerkskammer Dresden, Herr Jochen Noppenz, Abteilungsleiter
Handwerkskammer Düsseldorf, Herr Alexander Konrad
Handwerkskammer Koblenz, Frau Ricarda Matheus

Federal Agencies
Frau Evelyn Benke, Deutsches Patent- u. Markenamt (TIZ) Berlin

Bundeswehr
Herr Böhm, Bundesarchiv (Militärarchiv)
Herr Will and Herr Nau, Bundesamt für Wehrtechnik und Beschaffung, Koblenz
Herr Neyer, Kleiderkasse der Bundeswehr (Main Business Office), Koblenz
Dr. Kunz and Herr Lasse, Militärhistorisches Museum der Bundeswehr, Dresden

Historische Vereine (Historical Associations)
Historischer Verein Rastatt e.V., Herr Karl-Josef Fritz – many thanks!

City, and State Archives
Landesarchiv Berlin, Herr Wulf-Ekkehard Lucke
Stadt Braunschweig, Dr. Garzman, Archivdirektor
Staatsarchiv Bremen (Bollman)
Staatsarchiv Coesfeld, Herr Norbert Damberg
Stadtarchiv, Stadt Kassel, Herr Staataarchivar Klaube
Historisches Archiv – Köln (Dahl)
Hauptstaatsarchiv, Dresden
Landeshauptstadt Erfurt, Stadt Chronistin Frau Astrid Rose
Stadtarchiv, Landeshauptstadt München, Herr Löffelmeier, Archivamtsmann
Bayerisches Wirtschaftsarchiv, Dr. Eva Moser, Stv. Archivleiterin
Stadtarchiv, Landeshauptstadt Saarbrücken (Schmitt)
Stadtarchiv, Stadt and Kreis Bibliothek Münster (Schnur)

Companies
Albert Kempf G.m.b.H. + Co. K.G., ["ALKERO"], Teunz
J. Sperb G.m.b.H., Regensburg
Ludwig Vögele Mützenfabrik, Karlsruhe; Herr Dieter Vögele

Museums
Rheinische Industriemuseum Euskirchen (D. Stender)
Modetheorie und Kostümgeschichte Museum, München, Frau Prof. Ingrid Loschek
Textilmuseum Max Berk, Heidelberg, Frau Dr. Kristine Scherer
Deutsches Textilmuseum Krefeld, Frau Dr. Brigitte Tietzel, Museum Director
Bayerisches Armeemuseum, Herr Dr. Jürgen Kraus Oberkonservator
Stadtmuseum, Dresden, Frau H. Rein, Mitarbeiterin
Wehrgeschichtliches Museum Rastatt, Frau Sabina Hermes

Chapter 1

A Brief History of German Cap Making

Knight's Cross winner Kapitän zur See Alfred Schulze-Hinrichs (1899-1972) wearing the usual Kriegsmarine *"blaue Mütze"* as the Germans referred to this type of headgear – the "blue [top] cap" (see Chapter 6). Typical hand-embroidered gold bullion insignia. *Courtesy of Dale Paul.*

FROM HANDWERK TO INDUSTRIE

The art of organized cap making in Germany reaches back into the 16th century and a time when every part of a cap was made entirely by hand; when golden coins worn as ornaments were forbidden to anyone but royalty and landed gentry. Each and every cap was individually made. A handworker's guild was well established by this time and cap makers were a part of this guild. Cap making was a *Handwerk* – a handcraft. People who entered the field to become cap makers by profession had to work their way up through the three-tiered guild system of Apprentice, Journeyman, and finally, Master. After passing the Master certification test, an individual was ready (theoretically at least) to effectively manage a small to medium-sized business. In fact, a large part of the training received at the Master level involved general business principles along with the specifics of how to run a company of this type.

Most of the actual technical aspects behind the manufacture of quality caps was supposed to have been acquired already, and developed through hands-on experience at the Apprentice and Journeyman levels. A man who had passed the final certification test in these early days became a Cap-maker Master – "Kappenmachermeister" – a trained designer, overseer, and manager, all in one. A cap making business owned by a Kappenmachermeister or having such an individual to supervise operations usually employed at least one journeyman and/or apprentice as well, along with several seamstresses trained in the specialized sewing work required for this trade. Since all work was done by hand, production capacity was quite limited and most businesses were small, local affairs up until the end of the 19th century.

When the sewing machine was introduced in Germany, sometime after 1860, cap makers who could afford to do so purchased the new devices. These makers found their workload greatly reduced and production speed dramatically increased. They realized that the potential for increased sales was greater than had ever been previously possible. The potential was there – but as yet, not the demand. Business was limited to the civilian market, but times were changing along with the political climate – and national maps.

The Prussians unified the many independent German kingdoms into a single country in 1871, and in a major military conflict with France defeated that country (Franco-

Prussian War). This victory had a galvanizing effect on the new, unified Germany, and did a great deal to instill a growing fascination for the military within a German population that had always been pulled in two opposing directions – one toward creativity, the other toward things Martial. Respect for, and interest in, military service and the profession of arms in general had always found fertile ground among Germanic peoples. It took on new dimensions now, as German nationalism – still a new concept to many citizens of the country – took hold and began to grow alongside a fledgling unified military.

Turn of the Century

With the new national army (units were initially supplied by the individual German states) came new – and much more standardized – uniform requirements, and an increased demand for cloth headgear. And so in the usual way of things, many existing cap makers turned at least some of their efforts toward this potentially lucrative new customer: the military. New cap making companies opened to take part in the business. By the first few years into the twentieth century, many of the companies which would later become renowned for their Second World War military caps had already been founded: G.A. Hoffman in the 1870s, for example; Clemens Wagner and Peter Küpper in the 1890s. In 1902, the famous EREL-*Sonderklasse* maker, Robert Lubstein opened his business in Berlin. The stage was set, although new actors continued to join the cast.

A young private in a dress uniform with a standard Other Ranks visor cap. Matching aluminum insignia; the piping appears to be black for Pioniere (combat engineers). *Courtesy of Eric Mueller.*

The First World War years marked a relatively prosperous time for German cap manufacturers, who used the opportunity to expand their businesses. Though many young men went off to a war on two fronts amidst an ever-worsening military situation, and though the strain on the German economy was growing, the country itself had experienced no physical war damage. Public disillusionment with the situation was spreading, however, and the end of the war came as no surprise to many.

With Germany's eventual defeat in 1918, came the challenging conditions imposed by the Treaty of Versailles. The German economy, strained already from the war effort, could not support the financial demands required by the treaty conditions and deteriorated rapidly. Smouldering dissatisfaction grew among the German population at what it considered the cause of Germany's economic woes – unfair and insulting treaty conditions, and an ineffective government. The already teetering economy worsened during the late 1920s while the Great Depression raged in the United States. The inexperienced and poorly led democratic Weimar government could find no way out of the economic dilemma – a fact painfully clear for all to see – and it led to a growing belief among the average German that the government as it stood was powerless to effect any economic improvement. Public unrest increased, with clashes between members of opposing political parties, each with its own agenda.

With the victory of the Nazis in the political arena came the military rearmament promised by Hitler, together with a steady economic recovery – and an increase in 'national pride' (the staggering cost of which still remained to be seen). The most important factor in the economic upturn may have been simply a new, positive attitude among the population which finally replaced the near decade long sense of apathy that had previously choked out any attempts at improvement.

Knights Cross winner Generalleutnant Erwin Jolasse's Schirmmütze [visor cap] has a cover (top) in *Eskimo* (for more on Eskimo and Doeskin, see Chapter 2). Oak leaf wreath and national emblem (eagle) are in aluminum alloy [Leichtmetall]. This cap also has thick-style crown piping, which was not regulation. *Author's collection.*

During the two decades between the end of the Great War in 1918 and the start of the World War II, German cap makers struggled to keep their businesses alive. Only a very few long-established, highly successful companies had sufficient assets and/or

Luftwaffe pilot Fritz Tegtmeier (left) and an unidentified man pose in front of a Focke-Wulf Fw 190. Both are wearing the Luftwaffe M43 standard field cap. *Author's Collection.*

This rather well fed Army junior NCO wears a visor cap with standard aluminum [Leichtmetall] insignia. The cap appears to be a privately purchased piece, likely produced around 1936 or 1937. *Courtesy of Eric Mueller.*

connections to continue producing headgear for the very limited requirements of the Reichswehr military, and this business alone was not enough to survive on. Thus, alternative markets were vital. These were primarily government agencies such as the Reichsbahn, Reichspost and municipal and civic police forces – which taken together represented a larger combined market than the entire Reichswehr. The smaller makers, on the other hand, were quickly forced to shift production (if they had not done so already at the close of World War I) from military caps to civilian headwear, or to a different – if related – field, such as men's or women's clothing. Most of the companies which had their origins as furriers, reverted back to their earlier fur business.

It was during this period that the well-established Berlin firm of Robert Lubstein secured its exclusive supply contract with the Reichswehr Army Kleiderkasse and introduced its new trademark name: EREL-*Sonderklasse* (sometime around 1921). Lubstein also produced headgear (particularly Tschakos) for local police forces and any other governmental organizations that required caps. Many other manufacturers, particularly large concerns that had sprung up just before or during World War I, did not fare as well. They either went under, or flirted on the edge of financial disaster as the economy sailed on an uneven keel from one crisis to another. The Great Depression was underway in the United States. The Germans, long suffering a depressed economy and the pain of intense inflation, had already been toughing it out for so long that they did not notice much difference as the economy dropped to even lower depths.

Despite the hard times, German sewing machine manufacturers managed to stay afloat selling their current models, while putting much of their efforts into the development of new machines for specific industries – including the cap making industry.

With the rise of the NS regime, economic circumstances – for the first time in many years – turned fortuitous for the cap makers. The change did not come over night: 1931 and 1932 saw losses for many companies, though those that had survived the economic winnowing thus far managed to remain a step ahead of bankruptcy. Finally, with the year 1933, many companies began to see the first signs of a positive balance.

Rearmament had its initial impact in jump starting the economy. The business of rebuilding the armed forces required a busy economy, and money began to flow as the military became once again, a major customer. Cap makers who had been pursuing strictly civilian cap markets or fashion markets now rushed to return to military cap production. The entire textile industry was tied into the boom, and the new sewing machines developed by manufacturers like Pfaff during the down years now found a ready and welcoming market. The industry, in all of its interdependent parts, was on a roll.

'Handwerk' or 'Industrie'

In an article published in a 1936 edition of the industry trade newspaper *Uniformen-Markt* [Uniform Market], the number of 'special cap factories' existing in Germany that year was listed as 'between forty and fifty.' Although it is no longer possible to determine exactly what 'special' meant in this case, it seems safe to assume that it implied a factory which produced nothing but caps for military and government organizations, that is, large businesses quite capable of also doing contract work. It is doubtful that this *U-M* estimate accounted for any of the small to mid-sized cap makers.

Many of the larger companies were owned by merchants – not necessarily individuals with a cap making background. They ran their facilities in the true sense of an industry, not a Handwerk – efficient and fast production was the key to success at these firms. Though technically still classed as a *Handwerk* due to the nature of the final product being produced, such businesses were, in fact, cap making as an industrial enterprise.

By 1940, the number of manufacturers – all sizes included – had risen to nearly one hundred (at least fourteen makers in Berlin alone). This number may, unfortunately, also include some 'military goods' (Militäreffekten) sellers, since it is clear that some of the larger stores handling military items also sold caps under their own name, supplied, in fact, by unidentified makers. Though it is not always possible to distinguish these companies from actual manufacturers, a cap lining logo which includes the term 'Militäreffekten' in one form or another is often a good indication that the company on the logo mark did not actually manufacture the cap.

World War II

The Second World War began with no immediate impact on German cap makers, but as it progressed, this changed – particularly for the larger companies and those makers located in large cities which were targeted by allied bombing raids in the later years of the war. Manpower shortages became a serious problem for large makers, particularly the lack of skilled people with the correct qualifications. The demand for development and production of new designs (such as the M42, and M43) also contributed to strains on the industry. Government military contracts had to be satisfied when assigned, whether the price was acceptable, or not. In cities that were repeated targets of Allied air raids, bomb damage was a constant threat to production.

Nonetheless, nothing remained but to go on with business as long as it was still possible to do so – though this became increasingly difficult as the war neared its end. A few companies did become war casualties: the Kurt Triebel factory in Kassel, for example, was destroyed in late 1943 or early 1944, and the family left the area for a safer locale. A small cap maker in Dresden did not survive the firebombing of the city; a few companies in Berlin were destroyed in air raids and never reopened.

Postwar Period to the Present

At the end of hostilities in May 1945, the majority of the wartime producers – including many of those in the severely damaged Berlin – found that they had somehow survived the end of a Reich that had otherwise consumed the country along with its many victims. Amidst the ruins of the nation, these companies took stock and found that though they had survived, the prospects for the future were indeed grim. The economy would again be tough – for at least the first five postwar years, perhaps more. Their largest customer (the military) no longer existed: the market was again reduced to only government agencies – and this time the number of producers was much greater than after the end of the First World War. And so, as they had done after that conflict, so now they did again, shifting their new post war business into other, related areas. Though some companies simply switched to civilian headgear production and then moved on to men or women's apparel/fashion clothing, many of the larger companies – for reasons unknown – selected *sportswear* as their principal postwar product.

The companies located in Soviet-occupied territory at the end of the war fared poorly. Some simply closed down as the owners fled to the West. Some were located in areas that were converted to Polish territory – and with that their former existence ended. Other companies continued on for a brief time providing civilian head wear, or perhaps some caps for the Russian military; it was not long, however, before the new

Army Panzer Leutnant wearing a Panzer version officer's M43 cap. This Knight's Cross holder also sports a ribbon bar – not the usual practice with this uniform. *Courtesy of Dale Paul.*

The cap cover (top) on this young reserve infantry officer (extra, gray underlay on the shoulderboards) is in a plush, high grade *Eskimo* (for more on Eskimo and Doeskin, see Chapter 2). The cap has matching aluminum alloy [Leichtmetall] insignia, and thick-style cap crown piping [for more on this topic, see Chapter 15]. *Author's collection.*

Two Army NCOs having a break in this candid pre-war photo. Both wear visor caps, troops behind them wear the M34 field cap. *Courtesy of Eric Mueller.*

Knight's Cross winner General Eberbach in a black Panzer uniform in this picture, complete with a Panzer M38 [Officer New-Type Field Cap, see Chapter 8].The cap is piped around the crown and the top edge of the front scallop. General Eberbach was very busy during the Normandy invasion. A no-nonsense, active field commander, he managed to escape encirclement at the Falaise Gap and was not captured until sometime later. Held at the "Enfield" general's camp [England], his fellow POWs included his former subordinate, Waffen-SS Brigadeführer Kurt "Panzer" Meyer (captured earlier, around September, 6 1944, near Falaise). Meyer's division [12.SS Pz. Div. Hitlerjugend] had served under Eberbach's command during June and July of 1944. *Author's collection.*

East German government took over all private businesses and converted them into property of the state (on behalf of the people). None of the surviving wartime manufacturers in the 'East' were able to survive forty years of East German communism.

In the West, attrition was the operative word. Very few companies had chosen to continue producing caps. Clemens Wagner in Braunschweig was one that did. So, too, (Peküro) and Albert Kempf (Alkero) in Wuppertal-Ronsdorf, and Carl Halfar in Berlin (caps for the Federal Border Police). Each of these firms, one by one, eventually found themselves unable to survive as cap makers, and dropped cap production (even the postwar-founded companies) – with the sole exception of Albert Kempf. Some of the firms went out of business completely either at that time, or a few years later. The 1960s and 1970s saw the end of more makers – Clemens Wagner and Carl Halfar (Halfar, perhaps due to a competitor's political machinations), for example. Peter Küpper left military cap production behind completely and hung on as a sports cap and sportswear maker until the turn of the century when it, too, entered into insolvency proceedings and officially closed in September 2000.

Many former cap makers found themselves unable to survive postwar competition within their new business field. The new German economy was a tough place, steadily growing more integrated with the world economy, and all kinds of companies both old and new, were unable to keep up with the pace of development. One after another, they closed their doors. The serious decline in the fortunes of the cap-making industry also led to a severe reduction in the number of certified Master Cap Makers (*Mützenmachermeister* was the term used from the late 1800s). By 1965 there were not even enough Masters available to form the testing board for the very few individuals trying to achieve certification. An impromptu panel was put together, composed primarily of Master tailors. It was the only viable alternative.

Since 1990, no new Cap Makers (Journeyman level) or Master Cap Makers have been certified in Germany. Cap-making is no longer listed as a Handwerk – the field is dead.

In the year 2001, no more than ten of the more than one hundred original cap makers still exist. Of those ten, only one, Albert Kempf, still produces military visor caps as a major part of its business – but these caps are all manufactured in a foreign country where production costs are low. Of those ten, only one company, Ludwig Vögele Mützenfabrik, still manufactures military and other uniform caps in Germany – and that company will not last past the year 2005 when the current owner retires: there is no one qualified or interested in running the business. The great tradition of German military cap making *in Germany* – together with nearly all of the original makers – exists no longer.

Chapter 2

Basics of Cap Construction

INTRODUCTION

The developmental history of soft cloth headgear worn by the German armed forces during the Third Reich together with all manner of associated uniform regulations has been exhaustively documented in a number of reference works already available. This book has a completely different goal, and that is to satisfy a time-tested tenet familiar to all good soldiers and leaders in any military organization: know your enemy! The enemy we face here is inexperience and lack of knowledge, which allow unscrupulous sellers to cheat collectors out of hard-earned money.

This chapter provides information on the materials used in cap manufacture, the equipment used by cap makers, production methods and so on. Take the book along to museums and any militaria shows that are held in your vicinity, where you can actually look over and physically handle as many caps as possible. This helps you to gain experience and confidence in your ability to judge a piece. "Know your enemy!" Learn the basics of construction so that you can gauge the authenticity (or lack thereof) of a cap by individual details as well as by the overall feel of the piece.

Reproduction caps merit separate treatment, and are detailed in Chapter 14. The significant increase in the popularity and membership of reenactment groups in the 1990s has led to a valid need for such headgear – distasteful and uncomfortable as this fact may be for people who feel reproductions are a sacrilege no matter what the reason. It is a bitter pill that must be swallowed.

MANUFACTURING

Cloth headgear for the German military during the pre-war and wartime periods was manufactured in a variety of fabrics depending on climate conditions and usage, but the primary material was wool. Military uniform regulations served as a standardized base for all headgear producers, though differences naturally existed between companies. Such differences lay primarily in the training and experience of the firm's employees, the choice of material suppliers, and the selection of material grades along with other factors related to production methods. The rapid growth of the cap making industry during the early years of Germany's rearmament led to a major shortage in skilled labor, and this necessitated the retraining of people originally employed in other,

A young Army enlisted man in a pre-war photograph. Along with his Army greatcoat he is wearing a sharp-looking service visor cap. The high crown and saddle shape indicate this is a private purchase piece, which the soldier either bought himself or perhaps received from someone as a present. *Author's collection.*

Army officer's service visor cap. *Waffengattung:* Artillerie (red piping). The cap spring (shaping wire) in the crown is still in place: this is how a cap was supposed to look according to regulations. National emblem (eagle) in aluminum alloy (Leichtmetall), oakleave wreath in hand-embroidered bullion wire with an aluminum cockade. *Private collection.*

related industries. Larger producers who made not only caps but also other uniform and equipment articles were hit particularly hard from 1942 onward by shortages in skilled labor. Almi Uniformen-Mützen, a major Koblenz-based manufacturer that serviced government headgear and uniform contracts, provides a good example of this problem. An urgent request from the firm to the local economic council asking for skilled craftsmen and Russian laborers simply could not be fulfilled – there weren't enough skilled people available from the local region, nor were there facilities to house the proposed Russian workers (see Chapter 16 entry for Almi). Almi's owner most likely had to settle for producing smaller quantities than he wished, rather than increasing his production at the expense of quality.

The labor shortage situation brings up a very important factor that should be kept in mind during any evaluation of late-war caps, whether intended for historical purposes or as a potential addition to a collection. That factor is the overall workmanship of the piece under consideration. There is no question that the quality of *materials* used

to produce headgear for the German military degenerated as the war moved toward its conclusion and the availability of good quality raw materials dropped rapidly in a steadily disintegrating economy. The necessity of using lower class – even substandard – materials under such conditions was certainly both understandable and unavoidable. Rationing made procurement of high demand items increasingly difficult and consequently, the need for alternatives not subject to such strict control became ever more important. The shortage of skilled workers merely compounded the difficulties. The training period for a seamstress entering the cap-making industry, for example, was calculated at one-and-a-half years to proficiency (in 1939). Despite these conditions, the quality of actual workmanship remained high up until the end of the war. Late-war headgear, which often looks noticeably shabbier from a *material* standpoint when compared to earlier pieces, continued to be *made* well. German workers still took pride in their work – if nothing else – and right up until the end, did the best they could with the materials available. Poorly constructed caps offered for sale to collectors in today's militaria market with a seller's claim of 'late war manufacture' as justification for very shoddy workmanship, should always be suspect.

This group of Army junior officers presents an interesting array of headgear! Several are wearing the standard service visor cap; the tall fellow with outstretched left arm still has the regulation cap spring in the crown of his cap. The 2nd Lieutenant showing off his injury is wearing an officer's Old-Style Field Cap (the so-called crush cap), while the officer looking on just to the left of center is wearing a Bergmütze. *Courtesy of Eric Mueller.*

THE LAGO/AGO

Large companies under government contracts manufactured most visor caps and field headgear issued to German enlisted and non-commissioned personnel. Cap manufacturing for some of these firms was, in fact, only one part of a much larger uniform

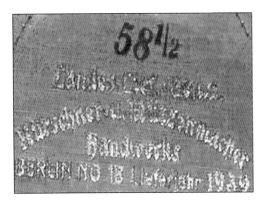

LAGO stamp on the lining of an Other Ranks service visor cap, in this case for a LAGO located in Berlin. Though difficult to read, the maker mark says:

Landes Lief. Gens.
Kürschner u. Mützenmacher
Handwerks
Berlin No. 18 Lieferjahr 1939

Translation: "Regional Supply Association furrier and cap maker's Handwerk, Berlin N O 18 (*Nord Ost* – north east) delivery year 1939."

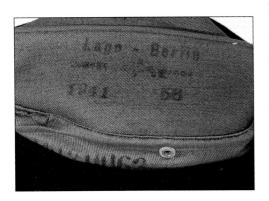

LAGO stamp on an M42 Panzer other ranks field cap, in this case: LAGO Berlin [the first three letters stood for Landes Auftragsgenossenschaft]. The acronym seems to be extinct, and does not appear in any modern German publications. *Courtesy of Bill Shea.*

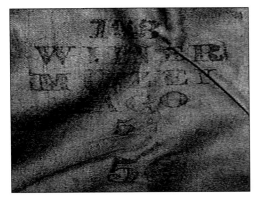

LAGO stamp on another Panzer M42 field cap. The stamp is difficult to read, but says "WIENER MÜTZEN LAGO" [Vienna cap Lago]. The date is 1943.

business. Many of the caps produced for officers, on the other hand, had their origin in one of the many medium to small-sized private makers located in cities and towns throughout the country. Quite a few of these makers were companies with long histories in the industry.

The principle advantage of the big producers lay in their ability to make large numbers of caps at a fast, sustainable pace and at a level of quality acceptable to the military or other uniformed organization (police forces, for example). At the same time, they were able to offer a bid price lower than would be possible for a small firm to meet.

The business problem was twofold for smaller makers who wished to secure a piece of the government pie: first, how to reach a high enough level of production to fill contract quantity requirements and second, how to achieve a more competitive price level (this factor was closely tied to production capacity). The solution adopted by some of these makers was to band together in a business consortium with production split amongst all the partners. Together, they had sufficient capacity to fill some of the smaller military headgear contracts and this capacity in turn allowed them to hold down production costs. The improvement in production capacity and production cost figures meant that the consortium could offer more competitive contract bid prices. In addition, it is quite possible that in some cases these associations also supported the large producers by supplying additional capacity. Such an organization was called a **LAGO** (also **AGO**), which stood for *Landes Auftragsgenossenschaft*. This title meant roughly Regional Order Association. The meaning of the "*O*" is not clear – it may simply be taken from Genossenschaft in order to give the acronym a smoother ring. The word Auftrag can mean *commission*, or *order*, in the sense of a work order.

The principal headgear product of the LAGO/AGO was field caps (M34, M42, M43) since these were well suited to mass production. Other Ranks visor caps destined for private purchase were also produced, but the number of these appears to have been extremely limited. In such a case, the LAGO information appears as a normal maker mark on the cap lining (extremely rare). On field caps, the LAGO mark consisted of the word LAGO (or AGO) with the consortium's location, e.g. Berlin, stamped in black (sometimes purple) ink on the lining. On contract (mass produced) visor caps it is usually found as a stamp on the reverse of the sweatband. Other LAGO/AGO locations that appear are Wien (Vienna) and Hessen (*Mitteldeutschland* – central Germany).

A variation on the LAGO concept was the *Landes-Lieferungs-Genossenschaft* or Regional Supply Association. LAGO members were often either cap makers, furriers, or both. The furrier business was an industry closely related to textiles, and therefore also classified as a Handwerk (craft, hand trade). Caps with a LAGO or AGO mark are uncommon items – particularly visor caps with a lining LAGO stamp, or an ink stamp on the sweatband reverse.

PERSONNEL AND EQUIPMENT

Staff

The *Mützenmachermeister* (master cap maker) was a vital part of any cap manufacturer's operation. Many companies, in fact, were founded or owned by a Mützenmachermeister. '*Herbert Grell Mützenfabrikation*', a Berlin-based cap maker, was one such company. Herr Grell, who founded the firm under his own name, was quite proud of his hard-earned title and made sure it was included in the company's commercial letterhead. If on the other hand, the company owner had originally come from a merchant background rather than from within the cap-making industry, it was nearly a certainty that he employed at least one certified Mützenmachermeister to supervise the company's cap production in terms of design, conformity with military regulations and quality control.

Every company usually employed at least one or two office personnel, perhaps also a sales assistant. Larger manufacturers often provided a service to the industry by hosting business interns. In addition, there were people who managed storerooms or material and accouterment stocks. Other employees prepped the materials which had been selected for the day's production run, stamping out cloth shapes using templates or cutting them, preparing the pasteboard cap band and so on, after which these were then released to the sewing line where the cap was put together. In large factories, much of the work proceeded in assembly line fashion. During the actual cap manufacture an assortment of sewing machines came into play, with the need for hand sewing kept to a minimum. Skilled seamstresses were an indispensable part of the business and companies regularly hired seamstresses from other fields who had been re-trained to proficiency in the particular methods and techniques used in the cap-making industry.

During visor cap production certain critical parts of the cap such as the side panels, cover/top, pasteboard cap band and the crown piping were prepared and sewn together by a *Mützenmacher* (cap maker). This individual was most likely a journeyman in the trade, with several years of experience under his belt. His responsibilities often included supervising one, two, or more seamstresses who sewed together the padding, lining cloth, visor, sweatband and the outer cap band cloth as well as the piping around the top and bottom of the cap band.

Equipment

For small to medium sized manufacturers in particular, reducing hand-sewing requirements to a minimum through the use of machines was vital in order to keep production costs down. A long-standing, leading sewing machine manufacturer very much in touch with both customer needs and industry trends was G.M. Pfaff, A.G. of Kaiserslautern. This company designed and produced a large, self-powered machine for smaller makers capable of providing nearly any sort of stitch required in cap manufacture. This 'universal' sewing machine was extremely versatile, and offered heavy or light duty zig-zag stitching (often used for sweatband attachment), stepped stitching and much more, for use with any kind of military cap. Given this capability, every part of a cap could be sewn together using this one machine. Nor did the heavy stitching required to attach Vulkanfiber or leather visors to caps present any problem: this machine, in fact, was capable of attaching both the sweatband and the visor during the same operation.

Independently driven, though not fully automated, this impressive device required an operator to oversee the sewing action and to manage the product feed line. On the other hand, it was very well designed to allow any configuration changes that might be required to be done quickly and simply without the need for any major, time-consuming readjustments or disassembly. Whether all small makers could actually afford one of these devices, is another story! A smaller machine offered by Pfaff that was also designed specifically for the cap-making industry was the Pfaff-Klasse 38-166. The 38-166 model combined the advantages of flat back stitch (or lock stitch) machines

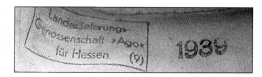

AGO stamp on the sweatband reverse of an OR visor cap. The picture unfortunately, was taken at an angle. The stamp in this case provides more information than usual on a field cap stamp: "Landeslieferungs Genossenschaft – AGO – für Hessen (9)". Translation: Regional Supply Association AGO for Hessen. The significance of the (9) is unknown.

This LAGO stamp on an Other Ranks visor cap is particularly interesting as it appeared together with the maker stamp of Clemens Wagner, a major producer. The cap is a contract piece *delivered* by Clemens Wagner; it may be that Wagner supplemented its own production with caps made by a local LAGO. Translation: "Regional – Supply – Assoc., furrier and cap maker, Central Germany". The significance of the number 27 is unknown, but these boxed numbers may have been a control code assigned to each individual LAGO organization. *Courtesy of a private collector.*

P F A F F

Spezial - Nähmaschinen für Uniformen, Mützen, Tornister, Brotbeutel usw. Leistungs-fähig, hervorragend in Kon-struktion und Material.
Verlangen Sie Angebot.

G. M. P F A F F A.-G.
Nähmaschinenfabrik
Kaiserslautern

G.M. Pfaff A.G. company advertisement. From the trade paper *Uniformen-Markt* Volume 2, January 15, 1939.

needed by every cap maker, with the versatility of a zig-zag stitch machine. The 38-166 machine could be used for sewing in the steel cap spring, sewing on cellophane sweatshields, attaching the crown piping to the cap, and for sewing visorless field caps.

A competing machine maker, Kochs Adlernähmaschinen-Werke A.G. (Kochs Eagle Sewing Machine Works) of Bielefeld, produced the Adler-Klasse 48 machine for medium difficulty sewing. One variant of this machine could also sew the visor and sweatband to the cap simultaneously.

For companies producing visor caps, another essential machine was the pasteboard cutter. This device was used to cut or stamp out the interior cap band that shaped the cap for the head while at the same time providing a stiff backing for the outer cap band cloth. Once cut to the required height and width, the pasteboard strip could then be formed into the necessary ring shape by using a steamer. Special ironing machines were also employed to dress up cloth surfaces, and stamping machines saw heavy usage as well. A cap-making company was indeed quite a busy place.

CAP MATERIALS

TUCH – CLOTH

Most standard headgear manufactured for the German military on the continent used wool as a base material. Summer weight and tropical headgear models of course, were manufactured from lighter weight textiles, particularly cotton twill/drill. (Foreign textiles, which were sometimes captured and used by the Germans, are not covered in this volume.) Some of the companies that produced cloth and lining materials specifically for caps were the firms of Julius Otto Klinke at Berlin W 8, Charlottenstraße 69, and Stenger & Becker at Berlin SW 68, Markgrafenstraße 21 (cap cloth).

In any discussion of cap materials, particularly wool, the word '*type*' is used frequently. 'Type' in this book is merely used to indicate the material originating from a particular weave of cloth.

A 'weave' is the repeating blueprint pattern (so to speak) of interlocking yarns which create a fabric when they are woven together using that pattern. Any process used to modify the surface of the raw cloth <u>after</u> weaving, on the other hand, is known as a finishing technique. Finishing techniques are what create the famous doeskin and Eskimo wool surfaces. The wool yarns and weaves and the fabric processing and finishing techniques are all part of a topic far too complex to go into here. In order for collectors to develop a good understanding of the materials German cap makers used with such impressive results in their caps, only a very basic overview of weaving and finishing is presented here.

1. *DIE GEWEBEKONSTRUKTION* (weave)

This German term refers to the pattern of interlaced yarns used to create a fabric. A choice of weaves is possible based on the ultimate look and texture desired in the finished material.

DAS GARN – YARN

The yarn used in the weave is extremely important to the final appearance of the cloth. There are two basic categories of yarn used for wool:

a. **Worsted** yarn (Kammgarn), composed only of long wool fibers (3 inches or longer) and having a fine diameter. These fibers have been *carded*, a process which straightens them and removes any remaining contaminants. They are also *combed*, which aligns them and at the same time removes any fibers shorter than 3 inches in length. These short fibers are called *'noils'* and are often used for making felt. Finally, the fibers are *drawn*. This procedure doubles them over onto themselves and then draws them out to ensure a thinner, more uniform diameter. When several worsted fibers are spun into yarn, they twist tighter and thinner than woolen fibers. As a result, cloth fabrics made from worsted yarn have a higher tensile strength. These fabrics also have a crisp, firm appearance with a tight weave. They are usually lightweight and have a surface which can be finished in a process that gives a smooth, short nap (shorter than possible with woolen fabrics). This is one form of the famous *Döskin* (doeskin) wool. A worsted fabric with such a finish combines a certain degree of the softness and smoothness attainable in a woolen fabric with the firmness and durability usual in a worsted. The most common usage for worsted fabrics is uniforms, uniform caps, suits, dresses and gabardines.

b. **Woolen** yarn (Streichgarn), composed of a mix of short and long (1 to 3 inches), coarse fibers. These fibers are only carded – there is no other processing (such as combing) prior to being spun into yarn. Due to the mixed-length fiber content and the coarser fibers themselves, the weave in woolen cloth is looser than with a worsted fabric. Woolen cloth is also bulkier and thus traps more air, which makes it a better insulator. With the application of certain finishing techniques after weaving, woolens can be given a very plush, thick nap. Cloth with such a nap, when used for caps, is doeskin. Common uses for woolens are sweaters and tweeds.

Both types of yarn can, of course, be blended with each other (and normally are), or with different kinds of yarn such as cotton or synthetics, during the weaving process. German Other Ranks Government Issue caps (NCO and enlisted grades together comprise 'Other Ranks' – that is, ranks other than officer ranks) are made from a woolen yarn blended with a small percentage of rayon. The Germans called wool blends with rayon fibers *Zellwolle*.

DAS GEWEBE – WEAVING (process)

Based on the customer's requirements for the finished fabric, a particular weave must first be chosen. When the yarn is woven in accordance with this weave pattern, the final fabric should match the customer's specifications. If the customer wants slightly diagonal, parallel ribs in the cloth, firmness but also a slight elasticity, then the target material texture in this case would be tricot (Trikot in German). The German term for the actual weave used to create Trikot cloth is the *Köperbindung* (twill weave). A twill weave produces a Trikot cloth.

To visualize the weaving process, simply imagine a borderless tic-tac-toe grid. Copy the grid several times over, one grid atop the other and connected by the ends of the vertical grid lines. Now, imagine that the horizontal grid lines actually pass over and/or under each of the vertical lines. This simple picture is the basic concept behind weaving: Yarns running in one direction are interwoven among yarns running in a perpendicular direction.

Now, replace the black tic-tac-toe lines with yarns. If each horizontal yarn were to pass over one vertical yarn and then under the next, continuing to cross under, over, under, over without change, this pattern will create the simplest weave known. In fact, this *linen* weave as it is called, was the first to be discovered and is the simplest and tightest pattern. In real weaving, the vertical yarns are actually very long; they only move in an up and down direction to allow the crossing yarn (i.e. the horizontal lines) to be worked over or under them. In English, the vertical yarns are called the *warp*, and horizontal yarns are the *weft*.

It is easy to see that there is a good deal of room for variation in the pattern of crossover, cross-under. It is possible to cross-over one warp yarn and cross-under the next two before coming up once again to repeat the process; or, crossover two or three warp yarns and then dip under the next one before repeating. Whichever weave is chosen, each will result in a final fabric with a different surface pattern and texture. The combinations are nearly endless. With the yarn factor added into the equation, we can see that either worsted or woolen yarn can be used alone in a weave, or together in a blend. The choice of Trikot cloth by the German military as its regulation standard wool obviously involved consideration of a great many factors and performance requirements.

We now know that the weave a customer selects is based on performance and appearance requirements desired in the final fabric. The coarseness or fineness of the yarns, as well as the length of the fibers that form them are all very important factors with an impact on the firmness of the final material and its ability to take a heavy or light surface nap if a finishing process is employed.

Time now to look at some of those finishing techniques, including the technique responsible for doeskin!

2. *DIE AUSRÜSTUNG* (Appretur) – FINISHING TECHNIQUES

Finishing refers to several different processes that can be applied to the surface of the material after weaving with the sole purpose of improving the wool's appearance and thus, ultimately, its appeal (that is, its marketability).

The raw cloth is often put first through a controlled shrinkage process called *fulling* or *milling*, which improves the fabric texture and tightens the weave. Once this has been completed, one or more finishing processes can be employed. One very common finishing technique for improving and softening surface texture is known in English as *brushing*. It is particularly effective with woolens since these yarns have thicker fibers. Brushing raises the ends of the fibers above the rest of the cloth surface, creating a smooth, soft nap that covers the weave beneath. The degree of nap produced can vary, ranging from a lightly brushed surface (e.g. flannel) to a deeper, plush pile. The thicker pile-like nap is created by running the surface of the fabric over cylinders covered with small, sharply pointed hooks and fine metal wires which catch and pull fiber ends to the surface. The resulting nap is smooth and relatively uniform across the entire surface of the cloth. A fabric processed in this way has a visible up side and down side.

German textile makers preferred a slightly different, two-stage finishing technique referred to as *in Strich legen* (laying in line, or aligning), and goods produced using this technique are referred to as *Strichware* (aligned goods). In the first stage, known in English as *felting* (Walken), the raw cloth is moistened in warm water and soap (or some form of detergent) and the surface is rubbed vigorously. Under this interaction, any loose fibers in the yarn begin to interlock and create a meshed layer above the underlying weave. Felting is now complete. German textile manufacturers used machines to perform this operation on a large scale.

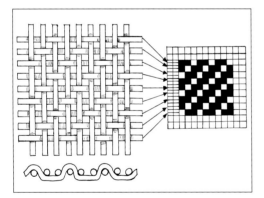

Left grid: shows the twill crossover pattern that interlocks the yarns. With each pass of the loom, the start point of the horizontal yarn is shifted. The right grid shows the pattern that this creates in the material being woven. Yarns that cross-under are shown in white, crossovers in black. The crossover points for each new horizontal row (starting at the bottom) are shifted right by one yarn. They are slightly higher than the material around them, and the points taken together form a diagonal line – a rib – in the fabric surface.

In the second stage, the felted surface of the material is run over a device with tiny studs or hooks on its surface that pass through the felted fibers. This is repeated several times and with each pass, the stud insertion depth is increased by tiny increments. In some respects, this procedure is essentially the same as *brushing*. The goal is not to completely separate the felted fibers again, but rather to stroke groups of loosened fiber ends into one and the same direction – to 'lay them in line.' Considerable care is required during the process, since the studs can not be allowed to reach deep enough that they damage the underlying weave. The result is a smooth, omni-directional nap with a very pleasing *hand* (a textile term referring to the feel). The surface has a visual appearance in which the wool seems to flow in one particular direction. This procedure is permanent – the fibers will remain fixed in the direction they are aligned.

In a final step after the alignment (laying in line) procedure, the nap can also be shorn or trimmed to a shorter, even length on a shearing machine if desired. Though this reduces the visible 'line' in the nap somewhat, it results in a soft, smooth surface texture and thus produces some of the finer grades of doeskin.

DIE WOLLSTOFFE – WOOL FABRICS

MützenTrikot – Cap Tricot

Trikot – or in this case Mützen-Trikot (cap tricot) – was the regulation fabric for German military cloth headgear made of wool. The main characteristic of this fabric is its pattern of parallel running ribs. This is the twill weave (*Köperbindung*).

There is considerable confusion in the militaria collecting community regarding which terms to use when describing or classifying the cloth used in German military headgear and uniforms. The confusion stems from two root causes:

1. Many material classification names are based on the fabric's *weave pattern*, not on the raw ingredients (yarns) used to create the material. People often get confused about exactly what they are referring to, when using a given term.

Trikot texture, standard grade. Note that the ribs in this case are of medium width and height and are quite distinct, which is indicative of fairly coarse woolen fibers used in the blend. The material shows a fuzzy surface, thus may have been *felted*. This example is from a Waffen-SS officer's visor cap, probably purchased from government stocks and thus only an average grade of Trikot.

This texture sample is from a Waffen-SS officer's Schiffchen made of very high grade Trikot. The ribs are narrow, almost delicate, the depth of the furrows between them very shallow. Such a clean texture indicates the yarn used in the weave was a worsted and woolen (the latter with a finer fiber diameter than usual) blend, or worsted yarn only. The texture of the cloth is so fine that from a slight distance, the rib pattern is nearly invisible. *Courtesy of Bill Shea.*

2. There is an innate difference in thought between American and European textile theory. The application of terms commonly used in daily speech by the public, therefore, also differs within each country. For example, regulation German military wool was Trikot (tricot). The German concept of *Trikot* and the American *tricot* differ significantly, however.

For the Germans, Trikot cloth was a *woven* fabric, based on interlocked wool yarns or blended yarns. To the American textile industry on the other hand, tricot was a *knit fabric*, with yarns interlocked by passing them through loops, i.e. through a knitting process – not by being interlaced with one another (woven). Knitted tricot fabrics in the U.S. are designed to fulfill different purposes. Americans would be most familiar with tricot as a fabric best suited for women's underwear and similar apparel, where knit fabrics are most effective or efficient.

The German Trikot is what Americans (and perhaps also British colleagues) recognize as a form of *gabardine* – a woven fabric made of wool yarn, based on a twill weave. All German regulation military wools, which the Germans termed 'Trikot', can thus be correctly labeled gabardine in English.

German Trikot is made from woolen and worsted yarns blended in specific proportions. The main characteristic of any twill weave is that at least two or more vertical (warp) yarns are crossed over at one time. With each subsequent pass of the horizontal (weft) yarn over or under the warp yarns, the starting point for the next crossover pattern offsets to the right by one yarn. This creates a diagonally running rib and furrow in the final fabric. Twill fabric also has a clear up side and down side, since the ribs are more visible on the one side of the material than on the other. If additional warp yarns are crossed over at the same time, the ribs will become wider. The two grid diagrams on page 24 show this weave. Such diagrams make it quite simple to visualize how the yarns are interwoven. The classic Trikot ribs in the finished fabric are also very easy to recognize in the right side grid diagram.

Eskimo of average, coarse grade: this fabric seen close up clearly shows the heavily felted, rough, almost hairy appearance. This sample is from an Army private purchase OR Bergmütze cap.

WOOL FINISHES: Doeskin and Eskimo

Döskin-Wolle – Doeskin Wool

Doeskin is the material that most epitomizes the German military cap from the Second World War and the Third Reich era. German officers and NCOs in sharp-looking uniforms, topped by a dashing visor cap made of doeskin – herein lies much of the fascination in the German military that such images evoke. Many cloth headgear collectors feel that a doeskin *Schirmmütze* (visor cap) made under the EREL logo is the ultimate in headgear collectibles. A rakishly shaped doeskin visor cap is usually prized above a similar cap made of a lesser material. Militaria dealers and collectors alike talk about doeskin grades in comparison with regulation (Trikot) fabrics, but the fact is that the weave used to make this material was *not* called doeskin. Like Trikot, doeskin is merely

Doeskin (*Strichtuch* – aligned cloth). This fabric example shows a particularly fine grade of doeskin, in which the *Strich*, or 'alignment line' – is fairly clear. The line would not be this distinct if the nap had been shorn. This material probably has a high percentage of woolen yarn, given its thickness.

High grade Eskimo made with a blend of woolen yarn and small amounts of mohair or silk fibers. The surface has been aligned and, later, sheared slightly in order to even off all the fiber ends. The final effect leaves almost no visible grain line. The very even, plush nap is uniform all around the cap. Army officer visor cap, Waffengattung (branch): Transportation. Note the slightly cockeyed bullion wreath – one of very few things which the Germans made mistakes on. *Courtesy of Bill Shea.*

the name of the final fabric not the weave it is made from. The soft, velvety texture typical of this material and so admired by collectors comes from the types of wool yarn used and from the finishing technique employed (see below). The actual weave pattern used for fabrics destined to become doeskin is the 'satin' weave *(German Atlasbindung)*, a pattern which produces a raw cloth with a very smooth, flat texture (no ribs). The choice of yarn used for the weave was usually a woolen, since the thicker woolen fibers produced a very plush nap after the application of certain finishing techniques. The doeskin nap is often shorn fairly short.

Eskimo-Wolle – Eskimo Wool

Eskimo was the name German cap makers gave to a fairly rugged, tough material which could be manufactured in both a *heavy* quality double weave and a *lighter* single, (but still very sturdy) weave. The double weave was exactly that – two weaves superimposed one over the other. Each weave used different component yarns. Eskimo was Strichware – meaning that the cloth was finished using the alignment process. The nap on this material also tends to be sheared fairly short.

The heavy quality, double weave Eskimo grade was a common choice for enlisted men's greatcoats and similar uniform items, while the lighter quality grade was used for officer caps. Either fine or coarse fiber yarns could be used for the weave – and the choice impacted greatly on the weight and appearance of the final cloth. Heavy duty Eskimo for field caps was usually manufactured with coarser yarns, while those used for visor cap cloth, on the other hand, were finer. The average Eskimo has a very high percentage of woolen yarn (Streichgarn). The weave pattern itself remained the same for both. The yarn for Eskimo cloth used later in the war was spun from thick, coarse woolen fibers. Materials made with a high content of such yarn often take on a somewhat hairy appearance when finished with a felting process.

Standard Eskimo was considered no more than an average wool with a slightly coarse quality finish, though quite rugged. On the other hand, there was also a very high grade form of cloth known as Chinchilla Eskimo ('chinchilla' was the industry name) which was frequently used in officer's greatcoats and caps for officers and military officials with officer rank. In fact, the *Offizier Kleiderkasse* (Officer Clothing Sales Store) catalog 1939 edition offered *only* caps made of Eskimo for officers and Army officials – no doeskin. High grade Eskimo used woolen yarn (Streichgarn), but during the yarn spinning small amounts of either short mohair fibers, or silk fibers were added. When this special yarn was used for the weave, the difference in the texture of the cloth after finishing was truly astonishing. It offered an extremely soft, lusciously smooth hand that, if sheared, produced a short, velvety smooth nap of uniform thickness. Despite its soft texture, the tension of the cloth itself remained quite firm. This version of Eskimo made superb caps and was of course, quite expensive!

LESSONS LEARNED

The kind of yarn (fine or coarse, a blend, etc.) chosen for weaving essentially *predetermines* the quality of the final cloth. Once woven, only finishing techniques can be used to make any improvements and the end effect of these is strictly limited to surface appearance. In short, if the wool yarns are of low quality, then the finished textile will be also.

Superb example of a general's Old-Style Field Cap made from top grade Trikot cloth. Extremely fine ribs with shallow, narrow furrows between them indicates worsted yarn was used for the weave. The Trikot rib pattern is so fine that it is quite difficult to distinguish from a distance. Such high grade Trikot looked different – but every bit as good – as a thin-napped doeskin. Maker: *Leparo* [Leonhard Paulig] (see Chapter 16). *Courtesy of Ruben Lopez.*

Any technique explained above, whether brushing, felting, 'aligning', shearing, or a combination of these can be used to improve the final appearance, though there were certain limitations for each method. Some of these techniques produce doeskin. For these, the yarns used in the weave play a major role in the quality of the final doeskin texture. Worsted yarns alone are unable to form as deep a nap as woolens. It is possible to nap worsted fabrics, felt them and align them, but only a thin nap can ever be achieved and this will not permit more than minimal shearing. A singeing process, on the other hand, can be used to create a smooth surface texture.

When finishing Trikot fabrics, a very fine-ribbed, high grade Trikot (woven primarily from worsted wools) will show only a slightly 'fuzzed' surface if *felted.* This actually makes the rib pattern somewhat difficult to discern. The overall appearance of this material is excellent and it was used often for Officer Old-Style Field Caps. It is also quite common with visor caps and M38 models produced by private makers, including those made by the famous Robert Lubstein firm. Officer M43 caps can also be found in fine grade finished Trikot. It was in fact, an excellent choice as a material since it satisfied military regulations while at the same time offering an impressive visual impact.

Note that when there is severe abrasive wear to a doeskin or Eskimo finish, the nap can be worn down to a point where the underlying weave becomes visible. Light moth tracking can produce a similar – if considerably more uneven – effect.

It is interesting to note that after being founded in 1955, the modern German *Bundeswehr* (Federal Armed Forces) selected Trikot for its regulation headgear and uniform material – just as its predecessor, the Wehrmacht had done. Bundeswehr wool cloth, however, includes natural or synthetic yarns in a blend. Unlike the Wehrmacht, no option exists either officially or unofficially in current Bundeswehr headgear regulations which allows a soldier to select other weaves or surface finishes such as doeskin – even if the cap is purchased at personal expense.

Tropical Materials

The most commonly used weave for tropical and warm weather headgear was a twill weave, the end result a Trikot pattern. The yarn used for tropical twills, however, was usually cotton, not wool. Far less bulky than either worsted or woolen yarn materials, cotton-based fabrics are both lighter in weight, and thinner. Tropical cloth was nothing more than Trikot woven from cotton.

Möleskin – Moleskin

Occasionally, Kriegsmarine blue or white cap covers (the upper part of the cap above the cap band) were made of Moleskin rather than wool. Moleskin was first produced late in the 17th century, a fabric based on cotton. The material has a very soft, smooth surface texture similar to fine Eskimo or doeskin, but without the heavy weight of wool.

OTHER CAP PARTS

BESATZTUCH – CAP BAND CLOTH

FILTZ – FELT

Felt was always used for the exterior *Besatzstreifen* (cap band) material on all Army visor caps and on Waffen-SS enlisted and NCO caps. Felt is not a woven fabric, but rather is produced in the same manner and under the same conditions already explained under the finishing technique known as felting. The only difference is that the process starts from scratch rather than from a previously woven wool cloth.

Felt is made with short-length wool fibers called *noils*, which come mostly from the belly of the sheep. Noils are also culled from the longer wool fibers during the combing process. With felts that are manufactured by hand, the fibers are sorted to approximately the same length and then arranged in layers atop a wicker or reed mat which serves as a flat backing for the material. The fibers are next thoroughly moistened with warm water and some form of soap or detergent. The wool is then rubbed and swished together to generate friction which, in the presence of moisture and the detergent agent, causes the fibers to intertwine and interlock with each other in a contiguous mass. Once the fibers have interlocked, the wet felt – still backed by its mat – can be rolled up and dried. The side of the wool against the mat is quite flat and thus much smoother than the upper side. When dry, the felt forms a single piece of material that can be sheared, cut and formed as desired. During the felt making process, a thin fabric sheet can be laid on top of the felt fibers, and then a new layer of fibers placed over the sheet. The final result is a sandwich material which is slightly stiffer and less prone to tearing.

Many other fibers, particularly those from animal fur, can also be 'felted' in this way. Fur from the angora rabbit, felted in such a process, was often chosen by American cap makers (Bancroft, for example) during the Second World War as a base cloth for private purchase U.S. Army officer's visor caps. In the United States and the U.K., such material went by the name of *fur felt*.

MOHÄR – MOHAIR

Cap band cloth for Luftwaffe and Navy caps was woven from black mohair. Mohair fibers are from the fur of the Angora goat, and the question of whether the Germans maintained their own stock of Angora goats to supply industry needs for mohair, or whether they imported most of their stock, remains unanswered.

SAMT – VELVET

Velvet was also used for cap bands, most notably for those of Waffen-SS officers and political leader's caps. Velvet is a medium weight *cut pile* fabric. The pile stands up very straight, hence its luster and smooth hand (feel). Velvet is woven with *two* sets of warp yarns (the vertical tic-tac-toe lines) rather than one, and it is this extra set which later produces the pile. One maker of velvet for caps (and other uses), was the firm Paul Rehmet of Berlin-Lichterfelde, Potsdamer Straße 54.

LEDER – LEATHER

The quantity of leather needed for cap construction depended of course, on the type of headgear being produced. Partial leather forehead sweatbands are often found on private purchase M38 or M43 caps, but it was the visor cap and the Officers Old-Style Field Cap which required the greatest quantity (whether sheepskin or calfskin). On Army and Waffen-SS visor caps, leather was used for the sweatband and for the chinstrap on Other Ranks (OR) caps. Luftwaffe and Kriegsmarine caps used leather for the visor and chinstraps as well as the sweatband.

SCHWEIßBAND – SWEATBAND

The demand on German leather producers was very high even in the years prior to the outbreak of the war and certainly did not decrease as the war progressed. Military footgear and load-bearing equipment needs (boots, belts and Y straps, etc.) for example, required large quantities of leather and no doubt held a considerably higher priority rating for available stocks than sweatbands or cap visors. The development of alternative materials therefore, was highly desirable. Given this logical reasoning, it is not much of a surprise that many people in the headgear collecting community concluded that the development and use of substitute leather sweatbands occurred strictly as a result of war shortages. This was not the case. Leather substitutes were available to the industry by World War I, if not earlier, and specific mention of *Ersatzleder* (substitute leather) or *Ersatz Schweißleder* can be found in patents issued to cap makers before 1930.

One common substitute material named *Alkor* was developed by the firm Alkor-Werke (owned by Karl Lissmann), however exactly which of several available types of substitute is *the* specific Alkor, remains a point of confusion. To compound the problem, it is possible that product manufacturing licenses for Alkor may have been distributed to other firms such as Dermatoid-Werke (Paul Meissner, Leipzig C1). Dermatoid-Werke advertised its material simply as "artificial leather" for caps or flag cases (the company also produced celluloid).

Often referred to as *composition* material, this type of leather substitute consisted of a very fine, mesh cloth (possibly chintz or burlap) bonded on one side with a paper thin, reddish brown leatherette material. The leatherette strip is always slightly wider than the cloth strip and the excess is folded over the top of the cloth and stitched along the reverse. The back of the cloth strip itself was also coated with a slightly yellowish,

Close up view of an artificial leather sweatband, forehead area. Age damage to the leatherette is clearly visible, with areas where it has cracked and flaked off exposing the cloth mesh backing beneath. Note the thinness of the leatherette itself: 1 mm, or less.

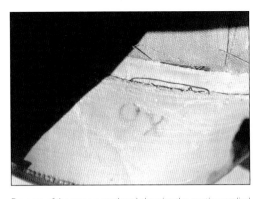

Reverse of the same sweatband, showing the coating applied to the back of the cloth as a sealer. Note at the bottom of the photo the excess leatherette is folded over the top of the band and stitched.

rubbery liquid that dried smooth, semi-transparent, and stiff. This coating was probably intended as a sealer against sweat.

Such composition sweatbands were very lightweight and quite comfortable, and they were certainly durable enough to survive the rigors of normal wartime usage. No one could foresee that they would have to last for a period of sixty post-war years or more, however, and time is often hard on this kind of material. The most common form of age damage is brittleness, which results in flaking of the leatherette and/or clear-through cracking of the band.

Many producers somehow managed to scrape up sufficient natural leather for sweatbands – albeit of low quality – right up until the end of the war. Despite the early availability of the leatherette material (and later, even a form of pressed paper substitute), these replacement sweatbands – particularly for officer caps – appear simply to have lacked popularity. Real leather remained the most common material.

A very small number of makers added a strip of extremely soft, foam-like material around the bottom of the sweatband. The strip was usually a pinkish color. This patented design was known as a *padded* (upholstered) *sweatband*, and appears most often on caps made by the Alkero firm (Albert Kempf) in Wuppertal-Ronsdorf (see Chapter 16). It was undoubtedly an expensive option. Such sweatbands usually have the letters 'DL' embossed into the band, together with **D**eutsches **L**eder (German leather).

Width: Sweatband width varied from maker to maker and this very diversity precludes width itself from being used as a factor to judge authenticity. The only *incorrect* width would be one narrower than 3.5 cm, which happens to be the norm for sweatbands on EREL-made caps, though wider bands certainly appear as well. Width may have been an optional feature.

Partial sweatbands (forehead only) on field caps taper at each end in a gentle downward curve. Such partial sweatbands on M43 caps have a very flexible thin reed rod inserted at the base of the band between two stitch lines. This rod provides a little shaping support for the front lower edge of the cap, and can easily be felt with a finger. In many cases the end of the rod can often be seen poking out of the sweatband base on one side or the other. Caution with this reed rod is important, for when dried out it is prone to snap if the front of the cap is bent.

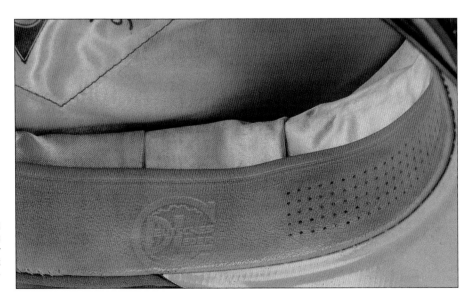

DL Mark: The sweatband on this Army officer's cap shows the DL mark quite clearly. The meaning of the encircling C is not known. The bulge behind the forehead perforation (over the visor area) is caused by the padded sponge strip that would make this cap Stirndruck frei *– free of forehead pressure. Courtesy of Gerard Stezelberger of Relic Hunter.*

Perforations: Sweatband perforations, if present, always appear in the forehead area. These are most common on officer caps, but are found on privately purchased OR caps as well. Field cap sweatbands, whether partial or full circumference, are rarely, if ever, perforated.

The patterns for perforation holes varied between makers. Some were very simple, with horizontal rows of same-diameter holes. Other makers used both large and small-sized holes (including the Robert Lubstein/EREL company), arranged in alternating rows or intermingled. A few makers used decorative patterns.

Schaumgummi (foam rubber) or Cork Strip

A problem common to visor caps in particular was pressure against the forehead caused by the edge of the visor itself. When the visor was sewn to the cap, its inside edge was located directly behind the sweatband. This caused pressure along that edge, which left a red pressure line behind on the forehead when the cap was taken off. A strip of orange foam rubber or cork was often inserted between the sweatband and the visor edge to distribute the pressure more evenly and reduce the discomfort caused by more focused pressure. Post-war visor caps still use this system, but with a strip of yellowish white polyurethane foam as the cushion material. No authentic World War II vintage cap will have such a strip since polyurethane was only developed after World War II. In addition, the life of this postwar material is limited to between fifteen and twenty years in most cases, before it dries out. Once dry it becomes completely stiff, and slowly disintegrates into a gritty powder.

Attachment: Sweatband attachment methods also varied greatly. Peküro (Peter Küpper of Wuppertal-Ronsorf) for instance, held a Reichspatent for its method of attaching sweatband to visor. This method was designed to a) alleviate skin inflammation and other discomforts caused by pressure from the visor edge on the forehead, and b) hinder the passage of sweat through to the outer cap material (a common cause of staining). The simplest method of attaching a sweatband was sewing the bottom edge directly to the lower inside edge of the cap using a zigzag or slant stitch. Due to the simplicity of these stitches, they are also the ones most commonly used by makers of reproduction caps.

Example of a somewhat ornate perforation pattern. Most were not this decorative! Cap: Gebirgsjäger (mountain troops) officer's visor. Maker: J. B. Holzinger of Berchtesgaden. *Courtesy of Gerard Stezelberger of Relic Hunter.*

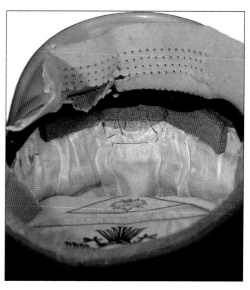

This is the usual position of the orange-colored foam rubber strip (or cork strip), positioned behind the sweatband and against the base of the cap band where the edge of the edge of the visor rests. *Courtesy of Bill Shea.*

Zigzag stitch along the bottom of the sweatband.

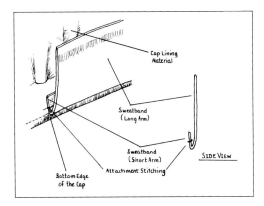

V-type attachment. When the main part of the sweatband is folded up into the cap, the stitch line is hidden behind it.

Felt strip insert (black). Notice the attachment of the sweatband.

The standard method seen especially with wider-style sweatbands on contract (Government Issue) caps was the 'V' form. The sweatband was shaped in a very lopsided 'V' form with one arm quite short compared to the other. The short arm was sewn to the cap with its finished side against the cap cloth. This left the main part of the band hanging below the cap. The sweatband was then folded upward until it was in the correct position for wear. The attachment stitch line is no longer visible when the sweatband is in place.

A strip of thick, black felt or velvet was sometimes added to the inside bottom edge of the cap at the forehead between the sweatband and the cap cloth. This felt insert provided a more comfortable, even fit on the forehead and temples. Because it raised the sweatband attachment line slightly higher than usual, the strip also served to reduce sweatband contact with the wearer's skin.

Some companies used a short strip which extended only from one end of the visor to the other, that is, covering only the forehead area; in other caps the felt insert was applied around the entire circumference of the cap. Such a full circumference strip improved the fit against the wearers head – and had the added incidental benefit of improving the interior appearance.

Sweatband ends on officer caps nearly always join at the center rear, with one end overlapping the other (usually the left side over the right). Occasionally the joining point is slightly left or right of the center line, but only rarely does it appear anywhere else. A small bow or metal brad secured the two ends together. On Government Issue (contracted) caps for Other Ranks the sweatband frequently joins at the side (still with the usual overlap), while for private purchase pieces the center rear is again the normal location.

In some instances, makers of high grade officer caps used sweatbands with vertical slots cut in the leather near the top edge of the band. These slots were similar to belt loops, and served to hold a band of different colored material that was threaded between them. The sides of the loops actually perforate the leather. Whether this was done for the design effect or to improve the fit at the top of the cap band, is not entirely clear. This system often appears on modern Bundeswehr officers caps (formerly West German, now the German Federal Armed Forces) and other caps worn by official organizations such as the police.

Right: Overlapping sweatband ends, with the usual securing bow. Shot facing toward the center rear of the cap.

Far right: The purpose of this belt-like band was to fine tune the fit. It is not under tension (i.e. not elasticized), and was perhaps added merely as an appearance enhancer. This sweatband is from a Kriegsmarine visor cap, however looped sweatbands like this appear on caps from the other armed services, as well.

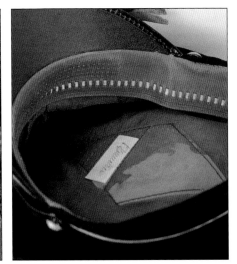

SCHIRM – VISOR

Visors used for German uniform caps were made from three primary materials: cloth covered pasteboard/cardboard, Vulkanfiber, and natural leather. Pasteboard/cardboard was used for M43 field cap visors, leather for most Luftwaffe and Kriegsmarine visor caps, while molded Vulkanfiber served for all Army and Waffen-SS visor caps (with the exception of the Officers Old-Style Field Cap).

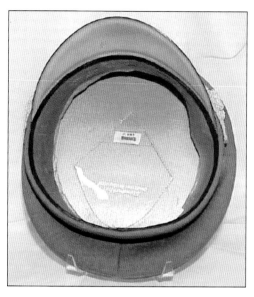

In this example, the felt strip circles the cap completely

Advertisement for the firm Gebr. Gerst G.M.B.H. The letters *GmbH* stand for *Gesellschaft mit beschränkter Haftung* – limited liability company (similar to the British L.L.C.). The text reads: "CAP VISORS for every need in all models, cap visor reinforcements, brim pasteboard, [helmet] chin straps of any kind, lacquered pasteboard, Specialty: Herkules-Russen-Visors, guaranteed hand-lacquered, Salpina cap-visor reinforcements. Approved by the RZM." From *Uniformen-Markt* Volume 2, January 15, 1940, pg. 16.

Vulkanfiber, though its exact date of introduction remains unknown, was available to German cap makers even before the First World War and can often be found in Pickelhaube helmets (spiked helmets) from that period. It was produced by treating pulp (wood pulp) with a caustic zinc-chloride or sulfuric acid solution. Mechanically, the material was very easy to shape during manufacture, yet very tough when dry, and these characteristics made it ideal for molding police tschako hats, lightweight military parade helmets, and cap visors. One of the only cap visor manufacturers whose ads regularly appeared in the trade newspaper *Uniformen-Markt* was Gebr. Gerst G.m.b.H. (Gebr. stands for the word *Gebrüder*, brothers). This company was located in the town of St. Goar.

BIESEN/PASPELN – PIPING

Piping for visor Army and Waffen-SS visor caps was produced from fine grade wool with a smooth surface (most likely a satin weave). Makers of high end, quality caps often used crown piping that was thicker than actually permitted by regulations. Piping cloth was prepared in the shape of a tube wrapped around a center cord which the Germans called a *Schnur*, which gave the piping a thicker body. Although the cord *Schnur* seems to have been most common, some makers used instead a thin, flexible reed rod; Carl Halfar Uniformmützenfabrik in Berlin is one example. Over the many years since the end of the war reed rods unfortunately tend to dry out, leaving them very brittle. Even slight bending of the piping in such a case can cause an audible snap, as the reed breaks, and such caps should be handled carefully.

Piping had a flat strip of material left on one side in a sort of tail, and when the cap was ready for the crown piping to be added, this tail strip was stitched to the cap cloth. When correctly applied, the piping was snugged tight into the seam formed by the cap cover panel and the side panels when these were sewn together. The Mützenmacher was the person who usually performed this procedure, and crown piping is so well secured to the cap that it can not be replaced without separating the cover panel completely from the cap body.

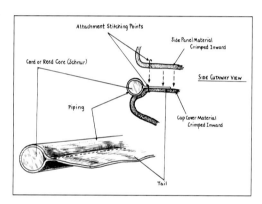

Piping was made with a round center cord used to give it shape. The cloth strip tail was used for attaching the piping to the cap cloth.

This cap, offered through the Internet, was identified as a Luftwaffe officer's visor. According to the seller, the crown piping and upper cap band piping were removed and later replaced. Note the piping cord does not sit evenly in the crown seam; instead it bulges out in a number of places. The ends of the upper cap band piping were not re-sewn end to end as they should be. Instead, they were actually inserted *into* the center seam, and then re-sewn along with it.

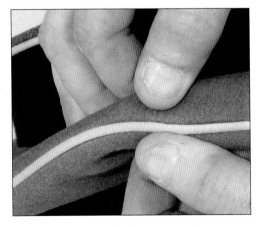

Place your thumbs on each side of a section of crown piping and gently pull the cap cloth as far away from the piping cord as possible to expose the base. Or, place one thumb directly on the piping itself and the other on the cap cloth beside it. Pull the piping one way and the cap cloth in the opposite direction to expose the piping base. While allowing for any natural fading of the piping's surface color, the base and the outer surface should match. Any obviously different colors should sound an immediate alarm.

PIPING REPLACEMENTS

Complete replacement of piping requires a very good, experienced sewer if the cap is not to be damaged, and this fact would seem to discourage any alterations beyond simple re-coloring. The cap shown at left allegedly belonged to a Luftwaffe officer who was a prisoner of the Russians. His captors required him to remove the piping (no reason given why this was necessary). The original piping was later reattached – and none too well it appears. In a situation such as this, there is simply no way to be certain of the provenance of such a piece, unless the original owner documented the story (not the case with this one).

REDYED PIPING

The crown piping was in fact, sewn so firmly into the seam between the top panel and the side panels that the tail strip can not be seen no matter how much the cap material is pulled back from around the base of the piping. Loose or noticeably lumpy piping is a major danger sign that a cap may have been tampered with. The tight seat of the piping within the seam makes access to the base difficult at best. Dyeing a light-colored or white piping to another color is no easy task, but far simpler than actually replacing the piping itself.

Should there be any question whether a cap's piping has been re-colored, place your thumbs on each side of a section of crown piping and gently but firmly pry the cap cloth away so that as much of the piping base (backside) as possible is exposed. With access to this area so difficult, re-coloring work will often fail to reach all the way through the piping cord down to the base – in which case the original color will still show. White infantry piping, for example, can be dyed pink for Panzertruppen (armor) since caps piped in this branch color command a higher market price, or black for Pioniere (combat engineers) for the same reason.

Waffen-SS caps, among the most expensive types of soft headgear in the militaria market, serve as particularly tempting targets for re-coloring since the ratio of caps piped in white is far higher than caps piped in a branch color. Caps with individual branch color piping are highly desirable and thus more expensive (for an explanation of the reason for the high number of white piped caps, see Chapter 4)

Checking the base of the piping not only serves to detect alterations; it is sometimes equally helpful in correctly identifying the branch of original caps. For example, the cap in the photo example on the left initially appeared to be an infantry visor cap with white piping slightly yellowed by age (unfortunately, in the photograph the piping looks yellower than it actually is). A check of the piping base, however, disclosed the bright lemon yellow color of Nachrichtentruppen (signals branch). Not an infantry piped cap after all, it instead turned out to be a signals visor with very faded piping.

As an interesting note, when the new Sonderführer Waffengattung (Special Service branch) was instituted in 1944 and assigned its own individual blue branch color, the military issued a policy directive regarding the new piping. Wehrmachtsbeamter (administrative officials) who had been reassigned to the new branch were required as of 1 August 1944 to have their dark green piping (forest green) removed and the cap re-piped in the new Waffenfarbe (branch color). Re-piping was preferred over purchasing an entirely new cap. In a subsequent order dated 6 September 1944, individuals in the new branch were required to have the re-piping done by a *Handwerker*, that is, a skilled seamstress or other trained person, instead of sending the cap to a manufacturer to have the work done. The intent, apparently, was to prevent any disruption in normal production that such extra work might cause.

PIPING, CELLEON PIPING

Army and Luftwaffe general's caps were piped in gold wire or gold colored celleon, in a very fine woven mesh around the central cord (Schnur). Celleon is a man-made product originating from natural materials (cellophaned rayon and artificial silk).

The piping ends were joined at the rear of the cap – never at the front – and were usually somewhere close to, or in line with the cap's rear panel seam. The most common method was to join the ends together directly. The threads used to accomplish this are usually visible, but German cap makers took pains to match the thread color as closely as possible to the color of the piping. Beware of any attachment threads of a completely different color.

Piping for field caps varied depending on the cap model. Aluminum mesh piping with an interior shaping cord served as crown piping on M43 caps for Army, Waffen-SS and Luftwaffe officers. A flatter type of mesh piping (without a tubular interior cord) was used on Army officer M38 caps, the Waffen-SS officer's *Schiffchen* (boat shaped side caps), and on Luftwaffe officer side caps. This piping was always machine-applied to the inside top edge of the cap's side skirts, with the ends of the piping usually joined at the rear.

MÜTZENKORDEL FÜR OFFIZERE – OFFICER CAP CORDS

On Army, Waffen-SS and Luftwaffe officer's caps, cords were originally made of silver wire, later aluminum wire, with gold wire or celleon wire cords for general officers. The slide knot just in front of the button bears special attention (see Appendix D). The long strand of the chin cord (on the bottom) looped under, around and over the button, then back over itself for about 2 cm or so. At this point the end of the cord, having completed the loop, was secured to the main cord beneath it with threads or metal brads, before being wrapped with a small, mesh knot covering. Just ahead of this small knot was an oval-shaped metal or pasteboard slider (one on each side of the cap) for adjusting the tightness of the chin cords. The sliders were completely wrapped in an interlocking wire filigree and together formed a large knot. Plastic, or thick metal tubes were never used for Second World War sliders. Some East German visor cap cords – otherwise virtually indistinguishable from those of the Third Reich era – do have plastic (as well as metal) sliders.

Crown piping from a Waffen-SS general's visor cap, showing the end to end joining method. The ends meet in this example just to the left of the cap's rear seam. The tiny gray horizontal threads are used to secure the ends together. The aluminum threads are interwoven to form a tubular mesh material, which was wrapped around the cotton cord core (Schnur).

Piping on an Army officer's M38 shown with the cap's side skirt turned down to expose the piping attachment. The cloth piping tail is *machine-stitched* along the inner skirt edge during manufacture and is an integral part of the cap. The stitch line is just below the base of the piping itself, which holds the piping securely against the upper edge of the skirt. The crown piping is sewn *into* the seam, at the rear of the cap. *Courtesy of Bill Shea.*

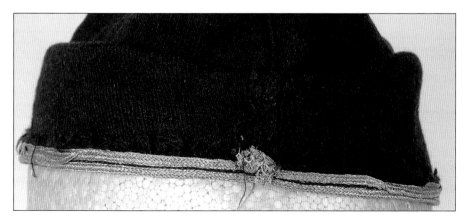

Side skirt piping on the boat-shaped *Schiffchen* style cap worn by Luftwaffe, Kriegsmarine, and Waffen-SS officers. The piping ends are arranged to overlap at the rear of the skirt. This cap is a Luftwaffe officer's *Fliegermütze. Courtesy of Bill Shea.*

Example of both thick and thin aluminum cap cords. There was no difference in the design of officer cords between the three land-based military services, Army, Waffen-SS or Luftwaffe.

There were two models of cords available to German military officers, which differed from one another only in their diameter. One version was thick, the other thin, and the choice between the two seems to have been strictly a matter of personal taste.

Active duty military officers from the rank of Leutnant (Army/Luftwaffe)/Waffen-SS Untersturmführer (2nd Lieutenant) to Oberst (Colonel) or Standartenführer in the Waffen-SS (a rank between Colonel and the lowest General rank – no U.S. equivalent) wore the same aluminum cords on their visor caps. Caps for non-military organizations used similar cords, but with diagonal stripes of another color (forest green, blue or black) alternating with the silver. Similarly, gold or celleon cords also found use in other organizations, again with diagonal stripes of a darker color at regular intervals along the cords. Active duty Army and Luftwaffe generals used gold wire or celleon cords, without any diagonal stripes of any kind.

Examples of cords for visor caps from non-military organizations. 1. Silver with light green, angled stripes 2. Silver with dark green, angled stripes 3. Gold with dark green, angled stripes. Any aluminum (or silver) chin cords with alternating stripes were not active duty military officer cords. *Courtesy Bill Shea*

FUTTER/INNENFUTTER – CAP (INTERIOR) LINING
Cap linings for all services were made of a variety of materials depending on the grade of the wearer. Lining cloth used in OR regulation issue visor caps was initially chintz,

a plain fabric usually made from cotton and glazed to give it a polished look. A water-proof oilcloth eventually took its place.

All privately purchased visor caps for officers were considered to be Sonderklasse (special grade), and therefore rated silk or rayon linings that appeared in a variety of colors. Rayon, a less expensive material often referred to as *artificial silk*, was a man made fiber that was developed in the United States by the American Viscose Company and first entered commercial use in 1910. Though man-made, rayon fibers are *not* synthetic. They are composed of regenerated cellulose, which is derived primarily from wood pulp – a natural material.

Other Ranks field caps for all services, such as the Army M34 and M42 and Bergmütze, the Kriegsmarine Bordmütze, Luftwaffe Fliegermütze and Waffen-SS Feldmütze (garrison/overseas cap with a curving, boat-like shape) were also lined with relatively simple, smooth textured fabrics, usually cotton. The weave used for lining cloth was a flat style (linen or atlas) without ribs – though a twill weave (fine ribs) was also used. Lining colors included various shades of gray or field gray. In some cases the lining can be even a slightly purplish-tinted gray, while Kriegsmarine field caps are a dark blue. Tropical cap linings, including those for the Afrika Korps, were normally in red.

The M43 Einheitsfeldmütze (standard field cap) liner was lined in the same sort of materials and also in a gray herringbone pattern that was particularly common in Luftwaffe Other Ranks M43 caps.

Army officer M38 and M43 caps, the Waffen-SS officer's Schiffchen and the Luftwaffe officer's Fliegermütze were usually lined in gray or field gray-colored cloth. Lining fabrics in privately purchased caps were often made of silk or rayon rather than cotton, however, and colors could vary.

STAHLREIFEN UND STÜTZE – STEEL CAP SPRING (SHAPING RING) AND CROWN SUPPORT

Regulations for all services required that a galvanized (non-rusting) steel cap spring (wire) be inserted in the visor cap crown to help hold the correct form. This was *not* optional on caps manufactured under government contracts; regulations required the cap spring. Soldiers were strictly forbidden to remove it, and if an individual reported to formation wearing a visor cap with the spring removed, he was subject to punishment.

Officially, the steel spring was supposed to be *sewn-in* to the cap crown, and even the German verb for this activity, *einstahlen*, implies this. Though no official reason for sewing in the ring was ever addressed specifically in any regulation, it was presumably intended to make removal difficult and therefore less likely to occur. The ring was not to be easily accessible (without breaking it, that is), and in order to remove it in one piece (i.e. as an unbroken ring) the lining had to be loosened to afford access.

More often than not, the soldier probably broke the crimped tube clip that clamped the ends of the wire together, then fed out one end through the cap interior. At that point it would be easy to pull the rest of the wire out of the cap. Cap springs – even after 60 years – still have a lot of tension, and putting the wire *back into* the crown (in the correct position) is quite difficult to do. The wire usually refuses to go back into place in the same shape that it held before being taken out.

The cap-making industry appears to have done its best to insure that all caps for all ranks, whether contract cap or high end private purchase caps, were delivered with the

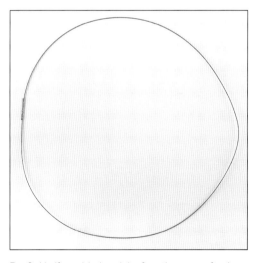

Der Stahlreife – original steel ring from the crown of an Army officer's visor cap. Note the tube clamp used to hold the ends of the ring together.

Wrapped double leather support strip. The stiffener is completely visible (and bent). This private purchase Army officer's "Extra"-cap was either manufactured or distributed by the firm *Biehler-Mütze* of Rastatt. It is quite unusual in having a very taught lining that covers *only* the top of the cap interior, but not the sides. A narrow suede leather strip is sewn over the crown seam and hides the stitching used to secure the lining. No loose cloth lining along the cap sides leaves the front of the cap and the crown support conveniently exposed to view (See Chapter 16, Rastatt, Biehler-Mütze). While this style of interior is sometimes found on Other Ranks issue (contract) caps, it is very unusual on an officer's Extramütze.

ring intact. On the other hand, it is quite possible that after the start of the war many cap manufacturers paid less attention to the sew-in requirement – simply inserting the wire into the crown once the cap was completed, instead. This would have been a time saving/cost cutting method, but of course, would also have made removal of the spring that much easier.

It is important to note that even if the wire was simply inserted into the crown rather than being sewn in, it will *never* have an attached paper or cloth tag with instructions to the wearer, "pull tag to remove steel ring before cleaning cap." Such tags are found on post-war Bundeswehr caps – but *never* on Third Reich era caps. The fact that eight out of ten German military caps have no shaping ring at all is simply a testament to the extreme popularity of the "broken cap" look among German soldiers of all ranks (see below).

Also required by regulations was a *Stütze* – a support post (strip) – which was sewn into visor caps directly behind the front seam *before* the lower edge of the lining material was tacked down. The support post extended vertically from the cap band all the way to the crown and was intended to hold the front of the cap in a straight, upright position. It was often made of a tough, stiff, padded canvas or leather strip and a steel band wrapped with a flat, wicker-like mesh cloth. The bottom of this white, open mesh cloth is sometimes visible below the lining. Another form of Stütze was a tough strip made from two pieces of leather sewn together, and a steel band covered with a fine cloth and stitched together with heavy thread along the edges. This particular type of Stütze was sometimes attached to the cap only at the bottom (at the cap band) and at the top of the strip (in the cap crown), with the central portion left unsecured.

BREAKING A CAP

Despite regulations, many German soldiers of all ranks preferred to *break* their caps – that is, to break the clean lines of the original, perfect shape. The usual way of doing this was by first pulling out the shaping ring/wire in the crown and then bending or breaking the Stütze sewn into the front of the cap. Officers and NCOs were often guilty of setting a bad example by breaking their caps in order to give them the much admired, front line look (a popular word is 'jaunty'). Once the war was fully underway, however, there was little time for much attention to or interest in enforcing these regulations, particularly since so many of the leaders themselves disregarded them. For the most part they were ignored. By that time there were, after all, far more important things to worry about.

INSIGNIA

A great variety of insignia types were used with German cloth headgear, and mixing of insignia was quite common. Some of these combinations were authorized – or at least ignored – other combinations were never authorized. Insignia alone, therefore, can not always be used as an accurate indicator of a cap's authenticity (or lack thereof), except in a few specific cases (and these must always take the *cap as a whole* into consideration).

Availability of insignia was always an issue. While the Army seemed able to maintain adequate supplies, not all types may have been available at all times. For example, an officer may have needed a replacement insignia for one of his caps. At the time it was needed, however, no officer insignia may have been available. The only alternative then would be to use whatever was on hand, even though this may have been insignia for enlisted ranks. If there was nothing else, then that is what he used. Even generals used enlisted insignia from time to time. Obviously then, it is not necessarily incorrect to find such insignia on an officers cap – in fact, this situation probably happened fairly often. Waffen-SS machine-embroidered or machine-woven officer insignia in particular seems to have often been in short supply. Once enlisted grade insignia had been applied as a replacement, it was perhaps not that easy to find an opportunity to remove it and replace it with the correct officer version. More likely, the enlisted insignia was simply left in place.

When looking at an officer's field cap with OR (other ranks/enlisted) insignia, the key consideration is *'how is the insignia applied?'* A sloppy, hand-sewn attachment of enlisted style insignia to an officer's cap for instance, is certainly not a confidence builder. The insignia – even if original – was probably added post-war and the cap has thus been modified.

Das Hoheitszeichen und der Eichenlaub Kranz mit Kokarde – National Emblem and Oak Leaf Wreath with Cockade

Headgear insignia consisted of two parts. For the:

Heer [Army], Kriegsmarine, Luftwaffe
- Das Hoheitszeichen [the national emblem] – the eagle and swastika
- Der Eichenlaub Kranz mit Reichskokarde – the oakleaf wreath with Reich cockade.

Waffen-SS
- Das Hoheitszeichen *(SS eagle)*
- Der Totenkopf – the Deaths Head badge.

For officer visor caps the insignia could be in metal, hand-embroidered bullion wire, or a combination of the two. Metal insignia was either nickel (German silver) on early caps or an aluminum alloy referred to as 'Leichtmetall.' Metal insignia in particular, was *always* very well made: well-proportioned and with crisp, high detail relief. Cheap, sloppily made insignia is rarely authentic.

Later in the war the use of low quality zinc-based alloys (often called 'pot metal') became increasingly common. Over time, this metal often oxidizes and destroys the insignia's original surface finish – which turns a uniform, dull gray. Machine-woven (BEVo) or machine-embroidered national emblems and cockades were used on field caps.

Metal Insignia Options

Leichtmetall (aluminum alloy) insignia used on visor caps was available in several options, and it is a wise to secure a reprint copy of the F.W. Assman & Sons insignia catalog in order to study the forms likely to be encountered. Early insignia was made primarily of nickel, but this soon changed to the aluminum Leichtmetall alloy. This metal was very sturdy and can not be easily bent. Some of the surface finish options included:

Army Gefreiter (lance corporal) Krohn, winner of the Knights Cross, wears an M34 field cap in this candid portrait. *Author's collection.*

Matching nickel (German silver) insignia. Nickel was used for early insignia, but by 1938 had been essentially replaced with the Leichtmetall (aluminum alloy) variety, except among a few officers who preferred the look of nicely aged nickel. The patina on this set is quite dark. This cap is piped with the *Waffenfarbe* [branch color] for *Pioniere* [combat engineers]. *Courtesy of Bill Shea.*

Matching Leichtmetall insignia on an Army artillery officer's cap. Note the metal base segment on this late-style cockade is much higher than that used with the nickel insignia. The thicker (higher profile) aluminum wreath needed a higher cockade. Also interesting in this example is the Leichtmetall Gebirgsjäger [mountain troops] Edelweiss, correctly positioned here between eagle and wreath. Any soldier of any branch serving in a Gebirgsjäger unit was authorized to wear this insignia. *Courtesy of Bill Shea.*

1. Matte aluminum
2. Polished aluminum
3. Galvanized, silver-plated (matte)
4. Galvanized, silver-plated (polished)

When evaluating the correctness of a set of insignia, the essential consideration is *whether the individual pieces match each other.* That is, do they show the same amount of tarnish, grime, wear, or shine? Or, is the shine or the wear to the eagle, for example, different than that on the oak leave wreath or Totenkopf?

After this initial and critical evaluation, take things a step further: does the insignia's condition and finish match the overall condition of the cap? Any significant discrepancy may mean that the insignia is *not* original to the cap.

Cloth Insignia Types

For field caps (M34, M38, M42, M43) regulation insignia (Hoheitszeichen and the Reichskokarde) was always cloth, and applied either:

- as two individual pieces, or
- combined together on a T-shaped piece of cloth, or
- combined together on a triangle shaped backing cloth (trapezoid)
- as one combined piece with a wreathed eagle, and cockade (combat police)

The version used depended on the model of cap and also to some degree on the period of the war. Cloth insignia was manufactured in two main varieties:

Mixed insignia on an Army officer's Panzer cap: Metal eagle, with hand-embroidered bullion wreath, and a metal cockade.

• machine-embroidered

• machine-woven (commonly called BEVo or flatwire). And in several thread colors. The backing cloth could also vary considerably, even on the same type of headgear.

Despite these variations, it's important to remember that most original cloth insignia for field caps was either machine-sewn, or professionally hand-sewn to the cap *at the time of manufacture or purchase.* If the insignia was applied at a later date, the sewing work was still normally done by some individual in the unit with experience and a neat hand. Poorly sewn, sloppily applied insignia always requires very careful scrutiny, and should be considered guilty until proven innocent, rather than the other way around.

The most problematical cloth insignia is tropical, and the only effective way to become familiar with tropical insignia is simply to view as many authenticated original examples as possible.

Luftwaffe Other Ranks M43 cap with trapezoid insignia in machine embroidered white cotton thread. The white stitching is authentic. *Courtesy of Bill Shea.*

Insignia/cap Relationships

In certain cases where a specific type of cap is closely associated with a specific type of insignia (such as the Army Bergmütze with the 'T' style machine woven insignia), divergence from the norm is something that should always be carefully checked before making any decision about purchasing the piece. Though exceptions did occur they require, as always, close scrutiny.

VISOR CAP CONSTRUCTION METHOD (All Services Overview)

An article in the industry trade newspaper *Uniformen-Markt* for 17/15 October 1936, described the actual procedure for making a visor cap as follows:

1. *Zuschnitt des Oberstoffes* – Cutting the upper material;

2. *Zuschnitt des Futters und der Einlagestoffe* – Cutting of the lining and filling (or padding) material;

3. *Druck des Futterbodens mit Klischee* – Printing the maker mark into the base lining;

4. *Einrichten durch den Mützenmacher* – Preparation by the cap maker;

5. *Maschinenarbeit = Teile und Ränder nähen, paspelieren, Paspel einnähen, Steifgaze, Watte und Stahlreifen einnähen* – Machine work: sew side panels and edges, sew around seams, sew in the piping; sew in [interior] stiffener cloth, padding, and the steel cap ring [spring];

6. *Stützen befestigen und Mütze hochstützen* – Secure the [front] support and prop up the crown;

7. *Futter aufschlagen* – Stitch the lining;

8. *Ventile und Abzeichen annähen* – Sew on vents [if any] and insignia;

9. *Schirm und Schweißleder annähen* – Sew on the visor and sweatband;

10. *Bügeln* – Press; [this included steaming to fix the form]. Caps were pressed, apparently to neaten up the cloth finish.

11. *Auslegen und Fertigmachen zum Versand* – Set out (the piece) and prepare for shipping

Knights Cross Winner General Meinrad von Lauchert's service visor cap has a very high crown. With a superb saddle-shape in this side view, this cap is truly what the Germans meant by the term Extramütze (Special-cap) – a term applied to any high-quality, private purchase cap. Von Lauchert was a Panzer general. *Courtesy of Dale Paul.*

(the *U-M* source was probably the G.A. Hoffman factory, in Berlin. See Chapter 16, G.A. Hoffman)

It is unlikely that every manufacturer followed the exact same order – there was some amount of variation.

Materials required to produce a visor cap:

1. Base cloth
2. Cap band cloth [*this was the outer cloth – felt, mohair or velvet band*]
3. Piping cloth and cord [Schnur]
4. Lining [*cloth*]
5. Horsehair, cheesecloth, felt and wadding
6. Insert for inside cap edge [*felt strip, for example, between the edge and the sweatband*]
7. Steel, wide and curved
8. Post [*crown support*]
9. Visor
10. Leather
11. Strap [*or cords*]
12. Buttons
13. Vents
14. Emblems and cockade
15. Celluloid angles [*stiffeners sewn into the upper part of the cap to help hold the form*]
16. Printing materials [*either for stamping a maker mark in ink or a transfer*]
17. Sponge [foam]strip [*as an option, for reducing pressure on the forehead*]
18. Yarn and silk

The interior (structural) cap band for visor caps was made of lacquered pasteboard, sometimes with a thin cloth glued over it. Kriegsmarine caps occasionally turn up with a cap band made of a thick celluloid material rather than the normal pasteboard. Metal was never used for bands, nor was plastic (highly flexible plastics, in fact, did not yet exist in the 1940s). Many of the reproductions available today *do* use plastic, rather than a pasteboard band.

LESSONS LEARNED

You are now armed with a basic knowledge of how German military caps were made, and the materials used in their construction. This knowledge is particularly important when considering a purchase of a cap through the internet, where your decision will be based almost exclusively on visual information. This is the Internet's greatest drawback, despite its convenience. At a militaria show, you can inspect an item physically – gauge the colors, stitching, weight; study the wear, and so on. With Internet shopping this is not available to you – you must depend solely on what you can identify *visually*, unless the seller allows you an inspection period after purchase. With this in mind, it is extremely important that you know what you are looking at in order to avoid costly mistakes. If the pictures of an item do not provide all the answers you need, you can use your new knowledge to determine which questions you ought to ask.

Time now, to look at the specific cap models for each military service.

Chapter 3

Die Heeres Schirmmütze - The Army [Service] Visor Cap

GENERAL

All visor caps issued to enlisted (Other Ranks) troops from government stocks were required to be manufactured in Teller (circular plate) form. Officially, this specification *remained unchanged throughout the war.* Most Other Ranks visor caps manufactured per government contract remained in this form until the destruction of the Third Reich in 1945, though there were no doubt some makers who supplied caps with the increasingly popular saddle shape. Once the war was fully underway, frontline soldiers holding ranks lower than sergeant actually had very few opportunities to wear a visor cap. While on leave from the front, the normal headgear was the M34 field cap (garrison cap), later, the standard field cap (M43).

Army officer's visor cap, *Waffengattung* [branch of service]: *Pioniere* [combat engineer]. Doeskin Strichtuch (cloth finished using the alignment process), with a lightly shorn nap. *Courtesy of Bill Shea.*

Army officer's visor cap, Waffengattung (branch of service): General Stab (General Staff). Very fine grade, long-napped doeskin Strichtuch. Aluminum insignia, with silver bullion cords (thus they are tarnished) rather than the usual aluminum variety. *Courtesy of Gerard Stezelberger of Relic Hunter.*

Enlisted grade issue caps frequently had the unit marking stamped in ink on the raw (reverse) side of the sweatband (left rear).

This was often (but not always) accompanied by another stamp showing the manufacturer's name in a rectangular box and sometimes the year of production (or a LAGO stamp).

A Government Issue Other Ranks cap should be Teller shaped, with a reddish brown colored lining often described as *rust* or *havana brown* (crème on earlier caps), a reddish brown sweatband, and a visor reverse (underside) painted in a reddish brown shade to match the sweatband (see below).

The more popular *Sattelform* (saddle shape) increasingly supplanted the Teller form in caps manufactured for *private purchase sales* from 1935 onward, but it was never *officially* adopted by any branch of the Wehrmacht. As a means of ensuring that any cap purchased privately on the market was in accord with regulations, soldiers were required to present such a cap at the unit for inspection. If passed as acceptable, the sweatband reverse was stamped in ink with the word 'Geprüft' (inspected).

Construction methods were essentially the same regardless of the branch of the armed services the cap was intended for – only the materials and measurements differed. The upper part of any visor cap consisted of four side panels (two front and two rear) and a top panel, all of the same base cloth – in this case feldgrau (fieldgray) for the Army. Another *Uniformen-Markt* article provides more detail on construction methods.

At the start of construction, the top panel was positioned with the bottom (unfinished) side up – this would later become the interior surface. The panel edges were carefully crimped inward and the crown piping tail strip was stitched down with the crimp. This left the piping itself running along the outer edge of the panel. Next, the side panels were sewn together. At all seams, the edges of the panels were first turned inward and then the edges sewn together on the inside. The top edges of the side panels were sewn to the cover panel (and piping tail) in the same fashion. The cap top was now complete. At this point a layer of cheesecloth could be added, cap padding (if desired) could be attached, and the lining sewn in.

The cap band was prepared by sewing the felt cap band cloth together around the previously prepared pasteboard band. The tail strip of the upper cap band piping was also sewn to the top edge of the felt band.

Just above the pasteboard's bottom edge, an additional strip of base cloth was attached at this point, with the lower cap band piping inserted into the seam during the sewing. The strip of base cloth extended below the pasteboard edge. This excess was folded around the bottom of the pasteboard, forming a cloth rim that served as the cap's lower edge. The previously assembled cap top section was now sewn to the cap band. Visor and sweatband were attached, and with that the cap was essentially completed but for the trimmings.

Once again, a key point to remember regarding construction is that every panel seam on German visor caps (of any branch of military service or organization) is stitched on the *inside* of the cap – no threads should be visible along the exterior seam lines.

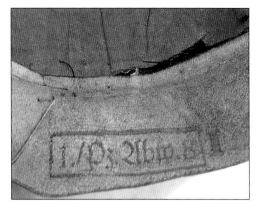

Unit marking stamp. I/Pz Abw. 8 II on an Other Ranks issue visor cap.

"Inspected" [and approved] stamp. Regulation issue Other Ranks caps did not require an inspection, of course, and thus did not have this stamp. Only privately purchased caps were inspected to ensure that they did not significantly diverge from regulations. Note the greenish blue tinted lining, which would be incorrect on an issue cap, but on a private purchase cap could be overlooked. *Courtesy of Private Collection.*

Fine example of an Army officer's cap, Waffengattung (branch): *Panzertruppen* [armor]. *Leichtmetall* [aluminum] national emblem, hand-embroidered bullion wreath. *Courtesy of Gerard Stezelberger of Relic Hunter.*

Other Ranks private purchase visor cap, Panzer branch. This cap came in the original shopping bag received at purchase during the war. Mint, unworn. The insignia finish has oxidized over time, which means the base metal was not aluminum, but rather a cheaper substitute. The piping color is the same overall, and exactly the correct shade of pink. The maker's mark is EREL-*Sonderklasse* of Berlin. *Courtesy of Bill Shea.*

Right: Medical officer's cap in Eskimo wool. Leichtmetall eagle and bullion wreath, with high profile aluminum cockade. This cap is in nearly flawless condition. *Courtesy of Gerard Stezelberger of Relic Hunter.*

Extremely rare example of a visor with the owner's name and a date stamped on the reverse. Note the crosshatch pattern molded into the surface common to German military Vulkanfiber visors. *Courtesy of Brian Greenhalgh (England).*

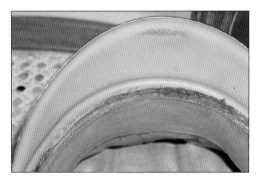

Green oxide colored visor reverse. Not a common color, but quite authentic. *Courtesy of a Private Collection.*

VISORS for Army Caps

Visors for all Army and Army administrative official visor caps were made of Vulkanfiber, lacquered on the upper, visible surface (commonly done by hand), and with a raised ridge along the brim. Vulkanfiber Army (or Waffen-SS) cap visors *without* the standard ridge along the brim were never authorized. A fine crosshatched mesh pattern is usually (but not always) molded or scored into the underside of the visor.

A rarity is for an individual's name to be imprinted in black ink on the visor reverse (this usually on tan colored visors). This is sometimes accompanied by a date. Though names do occasionally appear there are no regulations concerning this practice and it does not seem to have been very common. It is also unknown whether the name was added privately at the owner's expense (most likely), or was applied by the individual's unit.

Visor Reverse (underside) Colors:

The usual color for the visor reverse on Army caps was either:

1. Reddish brown (the most common color on Government Issue OR visors; also common on officer visors)
2. Tan (also appears on OR visors, very common on officer visors)
3. Crème (pale)

A common misconception in the collecting community is that for a visor reverse to be correct, it must be painted one of these colors when in fact, other colors were also used and are not incorrect (though they appear much less often). Among these are:

1. Matte green oxide
2. Gray
3. Black (rare)

The only official word on this matter was a requirement that the color of the visor reverse match the color of the sweatband leather on all *Government Issue* (contract made) caps. Makers of "Extra"-caps for private purchase generally followed this practice also, though they were not obligated to do so.

SWEATBAND COLORS:

While *Ersatz* (subsitute) leather sweatbands were nearly always reddish brown, natural leather sweatbands for visor caps and for Officer Old-Style Field Caps were produced in several colors:

1. Reddish-Brown (most common on Government Issue caps for Other Ranks)
2. Tan
3. Brown
4. Crème (pale color)
5. Maroon
6. Various shades of gray

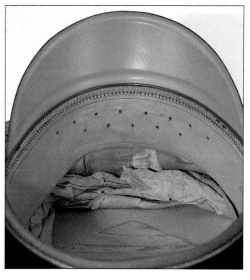

Pale, creamy gray sweatband. Note that the visor reverse is the usual tan color, and thus does not match the sweatband. The cap is an Army general's visor cap. *Courtesy of Bill Shea.*

Of these, the first three are most common. Army issue caps for Other Ranks usually appear in reddish brown, especially those of pre-war manufacture. There was no hard and fast rule governing sweatband color for officer caps other than the general caveat that the visor reverse and the sweatband color should match. This was often ignored, however, on private purchase grade (Extra) caps.

Army visor caps with a very dark brown or even black sweatband are rare, but correct. Dark brown, for example, is quite common on caps made by the J. Sperb company of Regensburg, which also used the more usual tan. Another rare color is a very pale, creamy gray which most often appears with sweatbands on private purchase Army (or Luftwaffe) general's caps.

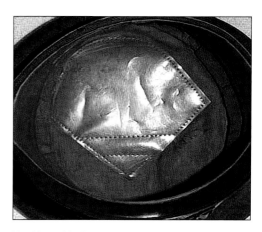

Matching reddish brown cotton moire lining, reddish brown sweatband and visor reverse. This is the standard color combination for a Government Issue enlisted or NCO cap (i.e. Other Ranks).

LINING

Lining Colors/Patterns:

Cap lining colors most common on Army visor caps:

1. Reddish brown [also called havana brown, or rust] (regulation color for OR caps)
2. Pale gold (champagne)
3. Gold

Other colors (less common)
4. Gray (plain, or in a chevron pattern material similar to Herringbone)
5. Bright lime green
6. Greenish blue (pale)
7. White
8. Black

Gray lining.

As mentioned earlier in this chapter, linings were initially cotton (sometimes chintz) for Other Ranks (enlisted) issue caps, later also a rust colored, water proof oil cloth. With the addition of a maker's mark, private purchase OR caps often used these materials as well. The cap making industry was well aware that enlisted soldiers were required to have a private purchase cap inspected before they could be worn, and so cap makers tended to conform to regulations as much as possible in order to reduce the chance of rejection. The quality difference between the private OR cap and the issue OR cap was therefore mostly external – in the wool used for the cap material, and perhaps in a better quality sweatband. The combination of rust colored lining, reddish brown sweatband and reddish brown visor underside is the standard color pattern even in private purchase Other Ranks caps.

Most officer grade "Extra"- caps were lined in either silk, or rayon. The difference between the two is very difficult to determine by feel alone. Visually however, real (natural) silk has slightly more sheen to it than rayon. Beyond this, there is precious little else that can be used to distinguish between them. Both are of natural origin. No completely synthetic fiber, such as nylon, polyester, or acrylic, was used by German manufacturers until well into the post-war period.

One of the rarer color combinations, and particularly unusual with this manufacturer: Lime green lining, reddish brown sweatband, tan visor reverse. Army artillery officer's cap; maker: EREL-*Sonderklasse* (logo appears only embossed on the sweatband). *Courtesy of Bill Shea.*

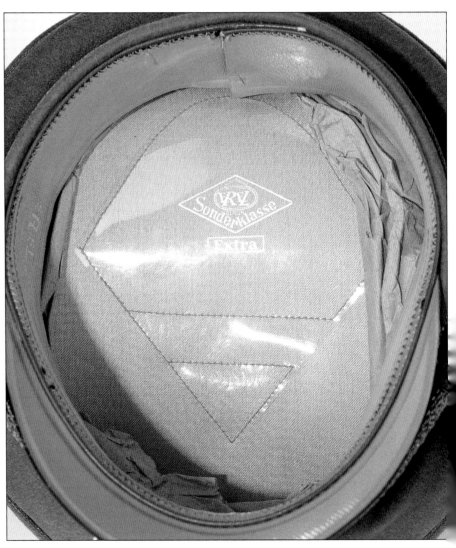

Right: Gold lining, tan sweatband, tan visor reverse.

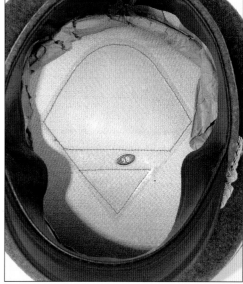

Gold lining, dark brown sweatband, creme visor reverse. Maker: J. Sperb, (Regensburg).

Pale greenish blue lining, maroon sweatband.

A cheesecloth sheet was often inserted between the base cloth and the lining material on visor caps, which normally is never visible unless the lining itself has been damaged.

The upper part of the cap lining was usually stitched along the crown seam to prevent it from otherwise hanging down on the wearer's head.

The *bottom* edge on the other hand, was usually hand-tacked to the cap cloth just above the sweatband sewing. This attachment was done using large, loose, saw-tooth stitches. The bottom of the lining was usually trimmed at a level below the top edge of the sweatband so that it remained out of sight. Nearly every maker used this method, in both high grade and contract caps. An exception, however, was the firm of J. Sperb, which seems to have preferred to machine-stitch the lining clear around the cap (following the top of the cap band). Every Sperb-made Army example seen to date has exhibited this peculiarity of construction. Excess lining material below the stitch line on a J. Sperb cap hangs down loose behind the sweatband and is never tacked down.

Internal Padding:

If any padding was used in a cap to improve its form (usually a cotton batting known as 'vlies'), it was secured to the cap cloth under the lining to prevent slippage, and should never be visible when the lining is correctly in place.

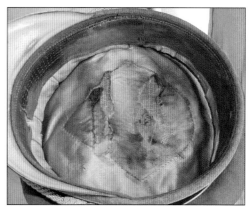

The cheesecloth backing (between the cap cloth and the lining material) is easily seen through the holes in this severely damaged lining. The cap is an EREL-*Sonderklasse*, and the extra material (used to create an "air cushion") added to the bottom of the sweatband at the forehead area was a concept patented by the maker of this cap, Robert Lubstein (Berlin). *Private Collection.*

Visor cap for an officer in the military police (Gendarmerie), piped in orange. High grade Eskimo wool, the cloth is not regulation field gray. Aluminum alloy eagle, bullion wreath and cockade. *Courtesy of Gerard Stezelberger of Relic Hunter.*

CHIN CORD BUTTONS

Chincord buttons for Army officer caps were available in two attachment styles:

1. Ring type
2. Split pin type

Both button styles had a silver colored, lightly pebbled outer surface. The back edge of the button was crimped inward and fit over the edge of a convex plate made of brass colored metal. The base of the prongs in the split pin button type was secured behind the plate, with the prongs passing through two holes arranged side by side. The

Ring and split pin-type officer chin cord buttons for Army, Waffen-SS or Luftwaffe caps, viewed from the front and rear, and a set of Other Ranks chinstrap buttons (front and rear). Note the inwardly crimped edges of the buttons, which serves to hold the back plate in position. *Courtesy of Bill Shea.*

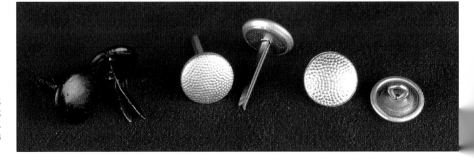

Far left: A dashing young Army Panzer Oberleutnant (1LT) wearing a visor cap made of Eskimo wool (less the cap spring). Matching hand-embroidered bullion insignia with an aluminum cockade. He is also wearing an early version black Panzer wrapper, indicated by the piping present along the upper collar edge (later discontinued).

Left: This handsome young enlisted man has already seen some action, as he wears an infantry assault badge and a wound badge – both in silver. His cap is decidedly non-regulation (he has removed the cap spring), and he has given it a real 'comfortable' look. Matching Leichtmetall insignia, with a high-profile cockade. *Author's collection.*

ring-type button had in the convex plate, with a ring attached at the center. Other Ranks buttons were similar, but painted black on the exterior. Chin cord buttons, like nearly every other part of World War II caps today, are readily available as reproductions.

INSIGNIA

Army visor cap insignia is actually a complicated subject that requires good attention to detail, and a thorough study of insignia references. Basic Army visor cap insignia consisted of two main types:

a. Metal
b. Hand-embroidered bullion wire

Of the two, the metal insignia was available to all ranks, while the embroidered bullion insignia was restricted to use only by officers.

Officers wore the same metal insignia available to enlisted soldiers or NCOs, but with a better finish. Such a finish on a set of insignia, however, a polished version for example, was an option that could only be had at extra cost. The average soldier could, of course, buy better insignia provided he was willing to go out and spend the Reichmarks needed for a set with a higher-class finish than the one his cap had been originally issued with – few soldiers probably did so. NCOs, on the other hand, were far more likely candidates for such a purchase.

Officers, blessed (more or less) with greater financial flexibility, had more opportunity to select alternate finishes for their insignia – or they could opt for hand-embroidered [*handgestickt*] bullion wire versions instead.

Fine example of a matching hand-embroidered bullion insignia including the cockade, on this officer's cap made by Clemens Wagner of Braunschweig. High-grade Eskimo wool top. The branch piping is the cornflower blue of the medical service.

Other Ranks private purchase visor cap with the rare branch piping color of motorized transport troops. This cap is an EREL *Extra* model. Aluminum national emblem and oak leaf wreath, good quality Doeskin wool with a fairly visible *Strich* grain – meaning that the nap has not been shorn. This cap probably belonged to an NCO and includes the EREL-*Sonderklasse* trademark cockade vent system; note the red painted, wire mesh at the center of the cockade. For additional examples, see Appendix E. *Courtesy of Brian Greenhalgh (England).*

Leichtmetall vs. Nickel

Early insignia was made of a nickel alloy commonly known as German Silver. This metal develops a distinctive, dark patina as it ages and is quite easy to recognize. The nickel insignia was later replaced by an aluminum alloy which the Germans referred to as 'Leichtmetall' (actually, there are several different aluminum alloys which fall within the technical classification 'Leichtmetall'). This material was strong, not easily conductive or prone to tarnish, and held a good shine. In fact, it proved to be an ideal choice for sharp-looking insignia, and most caps with a delivery date of 1938 or later will have insignia of Leichtmetall. The point to remember is that German soldiers or officers would never have mixed nickel and aluminum (Leichtmetall) insignia on the same cap. Such an action would have seemed, well, unnatural.

Vented Insignia Models

Certain models of oak leaf wreath insignia were designed with holes along the base edge of the leaves that allowed air to enter the cap. This system of perforations, with its associated vent system that allowed air to pass through the cap band into the cap, was

The *Frischluft* wreath as seen from the side. Small, rectangular holes at the base of the wreath where it rests against the cap band material allow fresh air to enter the cap and cool the wearer's head. Army officer visor cap, maker: J.B. Holzinger of Berchtesgaden. Waffengattung (branch): Mountain Troops. *Courtesy of Gerard Stezelberger of Relic Hunter.*

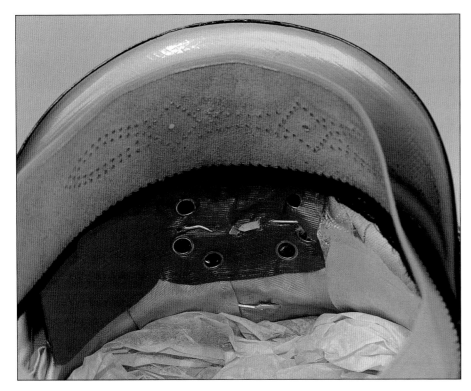

The *Frischluft* air system (same cap) from the inside. This is the forehead area of the sweatband, with the sweatband leather turned down. The small panel can be lifted and latched up to allow fresh air in, or closed, as desired. How much of a cost to the buyer this option represented, is not known. The Frischluft system was developed by the company HPC (see the maker Hermann Potthoff, Chapter 16). *Courtesy of Gerard Stezelberger of Relic Hunter.*

usually referred to with the term 'Frischluft' (fresh air). It appears on caps made by HPC of Coesfeld, and the word 'Frischluft'' in fact, appears somewhere on the cap's lining or printed on the sweatband. HPC held the rights to this invention.

Only a cap with aluminum alloy (Leichtmetall) insignia will have this system, which is rare in any case. Inside the cap, the corresponding holes through the cap band should be present in the forehead area.

The EREL ventilation system, which was designed to fulfill a similar purpose, used a different exterior set-up altogether that did not involve the oak leaf wreath. ERELs instead had a hollow cockade, in the center of which was a red-painted, wire mesh vent rather than the usual red felt or painted metal center. Inside the cap, a hole passed through the cap band to allow air to enter the interior (For a picture of the cap band hole, see Chapter 16, Robert Lubstein).

The EREL vented cockade system was available for Luftwaffe, Army and Waffen-SS visor caps, and such caps came with a special foil or paper tag attached to the sweatshield that explained the benefits of this system. The Frischluft system was not available for Luftwaffe, Waffen-SS or Kriegsmarine officer visor caps due to the lack of a *metal* wreath (the wreath for two of these services was usually hand-embroidered).

Both of these ventilation systems were no doubt expensive, private-choice options usually affordable only by officers and some NCOs (unless the cap was bought as a present). They are only found on private purchase caps, usually with a maker mark on the lining – on EREL caps, or for the Frischluft system on caps made by HPC and Holzinger. It may have been possible for other companies to manufacture caps with one of these vent systems if the firm first:

1. Obtained permission from the company that actually held the Patent or Deutsches Reichs Gebrauchsmuster (a sort of junior patent);

2. Paid a licensing fee;

Knight's Cross Winner (September 30, 1943) Oberleutnant Josef Herbst, a member of the Panzer Grenadier Regiment *Grossdeutschland*, has given his visor cap the ideal broken look. Matching Leichtmetall insignia. *Author's collection.*

3. Printed the D.R.G.M. notice or number on the cap lining or sweatband (any patent or D.R.G.M. application accepted by and registered with a number at the Reichspatentamt – the National Patent Office – was legally protected from unauthorized use). Several examples of both patents and Gebrauchsmuster are shown in Appendix C.

Field/Company Grade Officer Insignia

The German Army made no distinction between field grade (Major to Colonel) and company grade officers (2nd Lieutenant to Captain) on any of its officer cap models, including the visor cap. The same insignia was used for all officer grades below the rank of general. Contrast this with the American Army where, since World War II, the ranks of Major through Colonel have a row of so-called *scrambled eggs* – that is, golden oak leaves – embroidered on the top of the visor on a black felt covering. Company grade ('junior') officers, on the other hand, have a plain, black visor.

Indeed, there were no design differences in *metal* visor cap insignia from the rank of private all the way up through colonel in the German Army. There *were* differences, however, in mixing *combinations*. NCO and lower ranks basically had no options – all insignia was metal. Officers, with the added possibility of bullion insignia, had several combination options. All are correct, and all can appear at one time or another on caps available for purchase in the militaria market. The usual combinations are:

1. *Bullion hand-embroidered insignia only* – both national emblem (das Hoheitszeichen) – the eagle – and the oak leaf wreath (Eichenlaub Kranz) with cockade in metal or bullion.
2. *Aluminum or nickel (on earlier caps) insignia only* – both national emblem and oak leaf wreath and cockade.
3. *Bullion hand-embroidered national emblem with a metal oak leave wreath and cockade.*
4. *Metal national emblem, with a bullion oak leaf wreath (metal or bullion cockade).*

A key point to note when viewing metal insignia is how well the two pieces match each other in finish and overall appearance. Issued or purchased as a matched set, they should still look like a matched set with no obvious disparities between national emblem and wreath. Remember also that aluminum and nickel insignia were never combined.

General Officer Insignia

It is nearly impossible to modify a field/company grade officer's cap into a general's cap. The piping itself would have to be replaced, and this is simply not worth the effort. Reproductions are more of a threat here and are discussed later in Chapter 14.

Most notable with general's caps is the German military's insignia 'golding' policy – the intent of which was to change all insignia for generals from aluminum to gold. Until January 1, 1943, general's caps had used a national emblem and oak leave wreath insignia in the same aluminum metal or silver bullion as lower ranking officers, but with gold chin cords and gold mesh piping. After January 1st however, the new regulations required aluminum insignia to be changed to gold. This is significant, because with the changeover came an increase in the popularity of gold-colored celleon thread

A general's cap in perfect condition from the period before 1943, with gold piping – but aluminum insignia. The cap never had chincords, but if present they would be gold colored. *Courtesy of Gerard Stezelberger of Relic Hunter.*

for insignia. By now it had become clear to many generals that the gold celleon not only held up perfectly well (and was less expensive?), but also had the added benefit and convenience of not tarnishing like bullion. The only reasons then, to choose bullion over celleon, were prestige among peers, or a true liking for the color of the bullion over the celleon.

With general's insignia after January of 1943 there are three variations; all three appear on authentic caps:

1. All gold bullion insignia (the cockade, however, always remained silver, whether bullion or aluminum).
2. All celleon insignia (except cockade).
3. Gold bullion insignia – but with highlights and details picked out in celleon thread for contrast (there *was* a difference in both color and luster).

Insignia Mounting Prongs
If it is possible to access the back of the insignia, check the shape of the mounting prongs. Insignia with wedge-shaped, tapered prongs are usually reproduction. View a

reprinted F.W. Assman insignia catalog and study the forms of the insignia reverse, and the detail of the insignia front – especially the eagle's head and wings.

Heavily plated or chromed insignia are nearly always all reproductions. Note that original eagles could have two or three prongs.

Machine-Woven Insignia

The machine-woven (BEVo-style) insignia was never authorized for wear on the service visor cap and finding a cap with such insignia is a cause for suspicion. Machine-woven two piece insignia was intended for use on the officer's old-style field cap (see Chapter 7), the M38 Officer's 'New-Type' Field Cap and the M34 and M43 field caps.

MODIFIED POST WAR AND CONTEMPORARY CAPS

Bundeswehr Army Caps

The Bundeswehr is the Federal Armed Forces of post-war Germany, the successor to the Wehrmacht. It is divided into three independent services: the Army, Navy and the Air Force, all of which use visor caps. In addition to the Bundeswehr, the German *Bundes Grenzschutz* or BGS [Federal Border Defense Force (border police)] – and some civilian organizations also made use of visor caps.

Though there is no official confirming documentation, it is clear that the new caps designed for the Army of the Federal Republic were intended to look as different from Third Reich era headgear as possible. No doubt, the fledgling West German government in 1955 had no desire, as it set about re-militarizing West Germany, to provide the new army with a cap that would in any way evoke association with the old Wehrmacht. If indeed this was the intent, it most certainly succeeded. The model of visor cap now used by the modern German Army is quite different from both its own first model, and

Oberst (Colonel) Heinz-Georg Lemm wears his cap at a jaunty angle, tipped down to the ear. For actions subsequent to the Ardennes Offensive, Oberst Lemm received the Swords to his Knights Cross on March 15, 1945, while commanding officer of the 27th Fusilier Regiment. *Author's Collection.*

Close up of the pink Panzer branch piping on an original Heer (Army) Other Ranks visor cap. A private purchase piece, unused. There is no fading whatsoever to the piping, which is the correct, pre-1945 warm, pale pink color. *Courtesy of Bill Shea.*

Bundeswehr (1960s vintage) Other Ranks cap. The branch is Panzer. This cap, interestingly enough, was manufactured in 1965 by Clemens Wagner – a famous wartime producer. Dark gray cap cloth, with a brownish gray, ribbed cap band. The visor is molded plastic. The chinstrap (which is the same for *all* ranks) is a paper thin, shiny vinyl.

its Wehrmacht predecessor. It can not be modified into a convincing Wehrmacht cap without an inordinate amount of effort. Bundeswehr Army or Luftwaffe visor caps do not, therefore, represent any *significant* danger to any but the most inexperienced collector who has done no research at all. A comparison is provided here in any case, for convenience.

Although the base cloth for Bundeswehr visor caps is Trikot, all comparison stops there. The exterior color of the Army cap is a medium gray on officer caps (a darker, slate gray on first model, early caps), and darker gray on Other Ranks caps. The Wehrmacht-style twisted aluminum cap cords for officers were eliminated and replaced by an extremely cheap-looking, paper-thin, shiny vinyl chin strap. The chin strap is the same for all ranks. The pebbled aluminum buttons on officer caps were retained, and are nearly identical to the Wehrmacht version.

A warm gray-colored, *ribbed* cap band on winter version officer caps and a black ribbed band on summer caps (the cap cloth on the latter being a very light gray) is a radical departure from the smooth textured, dark blue green felt cap band on pre-1945 Army caps.

Although the earliest model of Bundeswehr officer visor caps had plain black visors (similar to the WWII form, but made of plastic), the obverse (top) of later and all current model officer cap visors is covered with cloth and has a row of silver scalloping along the front edge similar to that used on World War II Kriegsmarine visor caps. Field grade officers (Major through Colonel) have a single row of aluminum wire oak leaves. A 'U'-shaped strip of black leather (or vinyl) is wrapped around the edge of the brim and stitched through the visor. Bundeswehr Army *Other Ranks* caps have a plain, thin black plastic visor (no cloth covering), with a very slight molded ridge at the brim and a noticeable vertical ridge where the visor joins the bottom of the cap (this presumably to support the chin strap). Vulkanfiber visors were completely replaced by plastic versions on all contemporary German caps (since the early 1970s).

Sweatbands are made of polyurethane (similar to a soft vinyl), and on all caps, irregardless of branch of service (Army, Air Force, etc.), or the wearer's rank (officer

Close up of the pink Panzer branch piping on a *post war* Bundeswehr Army Other Ranks visor cap. Notice that the tint is shifted slightly to the violet – this is the standard modern color for Panzer branch pink. It is too violet in comparison with the wartime color: the Second World War tint no longer exists in post-1945 production.

Bundeswehr Army officer cap, 1980s model: major divergences from Wehrmacht design put these caps beyond convincing modification: cloth color, warm gray ribbed cap band, cloth-covered visor; scalloped visor ornamentation, and vinyl chinstrap.

Early Bundesgrenzschutz visor cap. Fine police green Trikot wool. Dark green felt cap band. Pronounced saddle shape; note the alternating aluminum and dark green cap cords (these were *not* used by regular army officers during the war). Superficially, early BGS officer caps are very similar to World War II Army caps. Maker: Peküro (Peter Küpper) of Wuppertal-Ronsdorf.

eather sweatbands included), they are gray. The visor underside color on all enlisted caps is also the same shade of gray. Indeed, the sweatbands for nearly all modern German caps from any federal state or city police force as well as caps for paramilitary organizations like the BGS, are *all* gray. Cap linings for Army caps are also gray for all ranks, though materials differ.

The black felt or satin insert strip between sweatband and cap cloth still appears on many Bundeswehr officer caps. Caps for Other Ranks, however, frequently have a very flexible, light gray vinyl strip in place of the felt.

Piping colors also show some divergence. Contemporary pink Panzer branch piping, for example, has a more violet tint than the World War II shade. Nearly all contemporary reproduction caps make use of the modern Bundeswehr piping, and so the color will be slightly off.

In the Bundeswehr, only Other Ranks (enlisted) visor caps are piped in branch colors, and all cap insignia is metal. Most of the major branch colors (Waffenfarbe) remain unchanged except for *Infanterie* (infantry). During the Second World War, Infanterie branch color was white, while that of the Panzer *Grenadiere* (motorized infantry) was grass green. In the Bundeswehr, on the other hand, Infanterie branch color has been changed completely from white to grass green – which the Germans call 'Jäger grün'. The Artillerie branch color, on the other hand, remains bright red (hochrot), Panzer is still pink (rosa) – though the tint is different – Pioniere black (schwarz), and so on.

All Bundeswehr officer caps, whether Army or Air Force, are uniformly piped in aluminum mesh cord with silver bullion wire hand-embroidered insignia. The insignia is actually glued to the cap – not sewn. Bundeswehr Army officer caps therefore no longer identify the wearer's branch.

BUNDESGRENZSCHUTZ (BGS) – FEDERAL BORDER PATROL
(Combatant Status)

Early BGS visor caps (up to the 1970s) offer more opportunity for modification. The sharp appearance of these early model caps actually has more in common with Third Reich period headgear than any contemporary Bundeswehr cap. Early BGS visor caps have a high peak (almost excessively so), a dark green felt cap band with aluminum and green twisted aluminum wire cords (for mid-career level officers).

The insignia on early BGS caps was hand-embroidered bullion (aluminum) wire. The oak leaf wreath looked quite similar to the Wehrmacht Kriegsmarine silver wreath worn by administrative officials in officer's rank, since it extends well above the top of the cap band (the leaf arrangement at the bottom center of the BGS wreath, however, is quite different).

Unfamiliar as these caps are to many novice collectors, however, a BGS bullion wreath can easily turn up on a World War II vintage visor cap that had no original insignia, and then be passed off as an authentic wartime Army officer's visor.

The flaws to note:

a. Height of the embroidered silver bullion wreath, which projects well above the cap band, is incorrect (for a pre-1945) Army cap. An original wreath should fit between the top and bottom of the cap band (within the piping rings).

Early post-war police cap, altered to appear as a World War II Army administrative official's visor cap (in officer's rank). The two obvious air vent grommets expose its true nature – not pre-1945 vintage. Fine grade material in police green, with blue green colored felt cap band. The visor is Vulkanfiber rather than the later plastic. Officer chin cords and insignia are original. The chin cord button is the correct distance from the end of the visor, but set slightly high on the cap band.

Interior. This early postwar cap still uses the wartime-style stiff celluloid sweatshield. The maker is Carl Isken of Köln. Carl Isken police caps and Bundeswehr Other Ranks caps are quite common (all post 1945) – including many early examples like this one. Note the postwar pattern matching gray lining, gray sweatband and gray visor reverse. The air vent grommets are also quite hard to overlook in this view.

Contemporary model police cap, this one for the police forces of the federal state of Nordrhein-Westfalen. The BGS, and nearly all state and city police forces now use this moss green color. The police star for this particular police department is positioned on the crown, instead of on the cap band. Mid-career level officer's cap made by Albert Kempf (Alkero) – a former wartime maker.

Early model (postwar) police cap for the police force of one of the German Bundesländer (federal states). Wartime-style police green Trikot cloth. Police-Star on the very dark green, felt cap band, with grass green piping. Mid-career level aluminum and green striped (wartime-style) chincords. Note the wrapping along the visor edge. *Courtesy of Kay Stephan, Germany.*

b. No cap band piping – also incorrect (only the crown is piped).

c. Cap's base cloth is police green, not field gray.

d. Visor is made of very thin Vulkanfiber, similar to that used on some Luftwaffe caps during the war.

During the early 1970s, the BGS cap and uniform underwent major design modifications – as did those of all German police forces. These modifications corrected the oversight which had left uncomfortable similarities to pre-1945 Third Reich police patterns. The most notable was a change in color – from the World War II shade of police green to a very deep, dark, moss green (*Moos-grün*), which became standard for all German federal, state and city police forces.

The height of the crown was reduced slightly, the twisted aluminum (or aluminum and green) chin cords were dropped completely and replaced with a flat, silver strap with slides for officers (a design borrowed from US military and police officer caps). The bullion federal eagle insignia was completely replaced by a gold colored, multi-pointed, metal *Polizeistern* (police star) with a crest mounted in its center. Old model BGS caps still exist, however, and can be easily modified to pass for World War II headgear.

MODIFIED POST-WAR GERMAN POLICE CAPS

In the years immediately following the end of World War II the redesign of municipal and federal police headgear and uniforms had yet to take place. This means that these early post-war caps are a potential danger for any inexperienced collector who has not done a little research.

Sweeping changes were eventually made that definitively removed any similarity to World War II headgear, but until then a good number of these early caps (1950s through the 1960s) were produced, many of which are still in existence.

Cover (cap top) materials on early postwar police caps are often wool, and in the correct World War II shade of police green shade. Sweatbands were often real leather (tan colored), and the twisted aluminum cords were still in use for career officer grade police ranks. Visors were also still made of the wartime Luftwaffe-style very thin Vulkanfiber, though the thin plastic version was just beginning to appear. In fact, these early post war caps differ very little from pre-1945 original headgear, with the occasional exception of two minor – *but very significant* – points:

1. Many makers began introducing a new, much more supple plastic film for the sweatshield material which eventually replaced celluloid in all postwar caps. The new material wrinkles quite easily and will not normally crack or remain bent as does the older celluloid.
2. Postwar police caps often have two metal rimmed, circular air vents installed beneath the side overhangs, one on each side of the side seam for a total of four vents per cap.

Early postwar caps like these can be easily altered/modified to look like Wehrmacht pieces – *from a distance*. The air vents can not be hidden, however, and *no Third Reich era military visor caps had air vents*. A look inside the cap merely confirms the diagnosis if the sweatshield turns out to be the soft, post-war plastic.

Chapter 4

Part 1: The Waffen-SS Visor Cap
Part 2: The Police Visor Cap

Part 1:
DIE WAFFEN-SS SCHIRMMÜTZE -
THE WAFFEN-SS [SERVICE] VISOR CAP

Waffen-SS visor caps have always been among the most sought after of headgear collectibles. An advanced collection is simply not complete without at least one Waffen-SS officer's visor cap, and one Other Ranks version – more than one, if possible! The availability of SS items, particularly officer's caps, is much more limited than Army headgear for a number of reasons. Chief among these:

1. The Waffen-SS was a smaller service than the Army, and so the number of caps, both for Officers and Other Ranks, would be comparatively fewer.
2. The aggressive nature and leadership style of Waffen-SS officers led to a very high rate of these leaders being killed in action. Any cap with them in the field at that time was likely lost as well.
3. At the end of the war, it was very undesirable and in some cases downright unhealthy, for a German soldier to be captured wearing anything with an SS association. A good many artifacts were therefore destroyed or otherwise 'ditched' by their owners just prior to capture.

The price of Waffen-SS headgear – of any type – has reached such high levels that potential buyers simply can not afford to purchase a cap without thoroughly inspecting it first.

And yet, people do just that all the time – even buying *obviously* bad pieces through Internet auction sites at prices lower than market norms for the genuine article (but still quite expensive). They seem to always believe they just *may* get lucky this time around and the cap will turn out to be some kind of a bargain, when in fact, all they are really getting is junk and a reputation for being either a sucker, or a fool. It is hard to conceive that people would actually choose to be cheated, yet it happens all the time.

In nine out of ten cases, cheap prices mean fakes or reproductions – you do, indeed, usually get what you pay for. In all fairness, that tenth case does come down the road once in a rare while (don't hold your breath), when a lower price does not neces-

Excellent example of a private purchase Waffen-SS officer's visor cap in regulation Trikot wool (very fine, private purchase grade). The piping Waffenfarbe is either the white of Infanterie, or the universal white officer piping in effect before or after 1940. There is no mothing or other external damage. A cap in this condition and quality was valued at upwards of $4,000 in the year 2001.

sarily mean a given cap is no good. Occasionally, an excellent, authentic item may be offered at (considerably) less than market value simply because the seller is trying to raise money quickly, perhaps in order to make some other purchase, and is willing to take a loss (or not, depending on what he *originally* paid for the item). While it is wise to always be skeptical of caps offered at below market prices, do inspect an otherwise promising piece before dismissing it – it is possible (if very rare) to get lucky if in the right place at the right time. A cap that is obvious junk, however, will never get any better.

CONSTRUCTION

Internally, Waffen-SS visor caps were essentially the same as Army caps. The lining colors and materials, sweatband and visor underside colors were no different. Exterior construction points are also the same, with only material differences such as the black felt (OR) or black velvet (officer) cap band, and on OR caps the SS version of the black leather chin strap with its distinctive center buckle. Waffen-SS cap piping also differed from the Army system. Despite these superficial points, however, construction was otherwise the same with, for example, the cap band seam always at the cap rear. Material measurements remained the same.

PIPING

Piping on German Army caps had traditionally been in the appropriate branch color (Waffenfarbe) for all enlisted men and officers up to the rank of General, with general's caps piped in gold. The Waffen-SS on the other hand, was somewhat different:

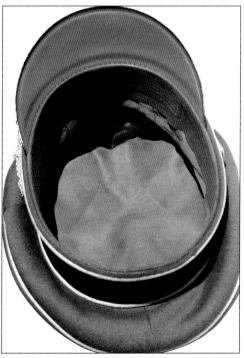

Interior view. Reddish gold- colored lining, lightly soiled but in fine shape. Wide, reddish- brown leather sweatband, with ends overlapping at the rear. The cap once had a sweatshield, but this did not survive the war – nothing remains but the impression. There is no sign of any maker mark

Another superb example of a mint Waffen-SS officer's cap. Nicely matching Leichtmetall insignia. High quality, plush Eskimo wool cover. White infantry (or universal) officer piping. *Courtesy of Dave Curtis.*

1. Other Ranks caps (for NCO and enlisted grades) were piped in the wearer's Waffenfarbe.
2. Officer caps up to the rank of general were piped in a universal white, irregardless of the wearer's branch.
3. Waffen-SS general rank officer caps were piped in silver (aluminum) mesh woven around a white cotton or wool central cord (Schnur).

There was an exception to the universal piping rule, however: In 1940 the regulation was changed to allow Waffen-SS officers to pipe their caps in the wearer's appropriate branch of service color. This practice, exactly the same as that used by the Army, was very popular. And, it was apparently for just that reason – because it was too much like the Army – that Reichsführer-SS Himmler rescinded the order at the end of that same year and returned all officer piping (below general rank) to the original, universal white.

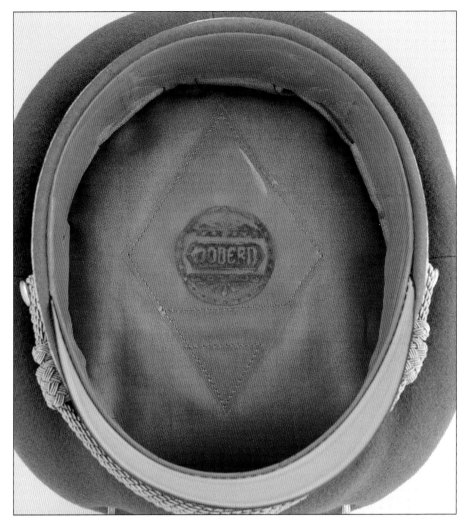

Generic maker mark "Modern". The diamond-shaped sweatshield is typical of smaller-sized (less well-known) cap makers. The very thin leather sweatband is surprising, given the high quality cap exterior. The golden orange lining is not a common color. Note also the sweatband overlap – in this case it falls on the left side rather than at the rear. *Courtesy of Dave Curtis.*

It would appear that many officers rushed to get a branch piped cap while this was allowed and wore them throughout the war, regardless of official regulations. Because of the short time period in which Waffenfarbe piping was authorized on Waffen-SS officer caps, their availability was limited. Branch piped caps were not to be made after the regulation was rescinded. Logic dictates that the number of white piped caps will, as a matter of course far exceed the number with branch piping. This causes confusion since a white piped Waffen-SS officer's visor cap may be an infantry (white) branch piped cap, or simply the universal white SS officer piping – there is no way to know which, unless the cap happens to be dated 1940.

The prevalence of caps with white piping also makes them good candidates for modification – primarily that of dyeing the white to *Panzer* (armor) pink, or *Pionier* (combat engineer) black, for example.

VISORS for WAFFEN-SS Caps

Visors for all Waffen-SS visor caps were made of Vulkanfiber, lacquered on the upper, visible surface, with a *molded* raised ridge along the brim in exactly the same pattern as Army models. Waffen-SS cap visors *without* a standard brim ridge were never authorized, nor were there any wrapped and stitched brim edges similar to those found on Kriegsmarine or Luftwaffe caps. The officer's visor cap in the following picture was

Waffen-SS Knights Cross winner Sturmbannführer (Major) Hans Hausser wears his service visor cap at a jaunty (and off-center) angle. As with many SS officers, he has the Totenkopf insignia centered but very high on the cap band, touching the upper cap band piping. *Author's collection.*

Waffen-SS visor cap offered publicly at a starting price of $1500 (on the Internet). Quite a few good items, in fact, can be found with careful Internet searching, but this was *not* one of them. For a reenactor, a fine substitute for the real thing – but certainly not at a price of $1500. Though offered for auction three times, no one (fortunately) ever won the cap.

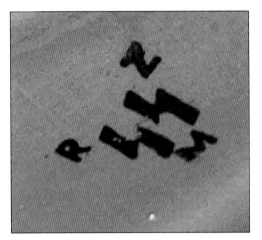

RZM SS ink mark on the underside of an *Allgemeine-SS* Other Ranks cap visor. Note the lack of the usual surface crosshatching. Waffen-SS caps did not have any RZM mark. *Courtesy of Bill Shea.*

offered in an Internet venue as an authentic cap, piped in lemon yellow for Signals troops. The seller who offered it, at a starting bid of $1500, did not in any way whatsoever indicate that the cap was not authentic. The visor however, is incorrect, with a glossy edge showing an underlying mesh pattern while the visor surface itself – which should be lacquered – is completely dull. Authentic Vulkanfiber visors never have such a completely matte appearance regardless of how well worn or damaged they may be – and considering the fine condition of the cap *top* on this piece, wear is certainly not an issue here. Note also, that the Totenkopf has been moved at least once (if the chin-shaped indentation visible beneath the insignia is any indication) – meaning that the cap has been played with and is suspect as a result. The cap cloth color is also off, though this may have been due to the lighting used for the picture. Additional interior discrepancies (colors and construction) confirmed that this cap was a definite piece to stay away from.

As with Army caps, a fine crosshatch pattern is usually (but not always) molded or scored into the visor reverse. SS RZM marks in black ink which sometimes appear on the underside of Allgemeine-SS visors *are not found on Waffen-SS caps*. The color of an RZM-marked visor underside is almost always reddish tan. Marked visors are quite rare.

Visor Reverse Colors:

The most common visor reverse color on Waffen-SS caps was either:

1. Reddish brown (most common color on OR visors, especially in the pre-war years; also common on officer visors)
2. Tan (also appears on OR visors, very common on officer visors)
3. Crème (pale tan)

As with Army visors, other colors do appear though far less frequently.

Whether the color matching scheme employed by the Army also applied officially to Waffen-SS caps is not certain; there does not seem to have been any official color coordination regulation issued specifically for the SS. Most Waffen-SS examples in general, do seem to follow the Army pattern: sweatband and visor reverse color in matching tones, and as close a match as possible with the lining color (this for Other Ranks caps). Private purchase *Extramütze* models for Other Ranks showed a tendency toward overall compliance with the matching colors system, but makers were not officially bound to do so.

LINING

Lining Colors/Patterns:

Cap lining colors most common on Waffen-SS visor caps:

1. Reddish brown ['havana brown'/ rust] (regulation on OR visor caps)
2. Pale Gold (champagne)
3. Gold
4. Pale greenish-blue
5. Gray (plain, or in a chevron pattern material similar to Herringbone)
6. Brown (light)

Militaria dealers sometimes refer to a particular lining color or other item on a cap as being textbook, but this term can be misleading. *Textbook* implies a standard – but with the variety possible in a private purchase visor cap, there was no standard since each maker could use any colors it wished. Some companies certainly followed a fairly regular, self-standardized color arrangement that they only altered due to changes in material availability. Robert Lubstein for instance, the maker of EREL Sonderklasse caps, nearly always used gold colored linings for his *Extra* model (the shade sometimes varied slightly), while the higher grade EREL *Privat* model lining was typically a pale bluish green.

Regulation caps of course, are all made of Trikot wool – irregardless of whether a cap is a private purchase *Extramütze* (that is, a cap made with finest grade Trikot and other materials), or simply a piece made under contract for the government. Any cap made of any material other than Trikot is, by definition, automatically *non-regulation*. A cap made of doeskin, for example, is non-regulation, as is Eskimo wool. Caps made of either of these wools were intended for private purchase and as such, can appear with any interior color scheme that the maker chose. Thus, using the term 'textbook' with a private purchase cap is not really possible.

For Waffen-SS officer visor caps *without a maker mark* but which are nonetheless clearly private purchase (i.e. the cap material is a very high grade Trikot, doeskin, or Eskimo), there is a particular color combination that is considered as close to an ideal

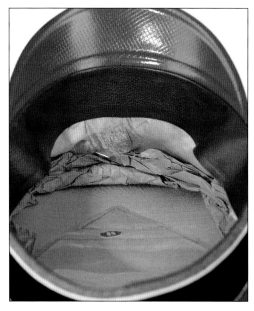

This interior color combination represents the ideal pattern for Waffen-SS officer's service visor caps without any maker mark (but in a private purchase quality wool). It is the so-called "textbook" combination. This mint, unissued cap even came with the original stuffing paper in place that had been used to pad the cap sides and help maintain the form while in storage. *Courtesy of Bill Shea.*

The interior of Foreign Minister Joachim von Ribbentrop's Waffen-SS visor cap varies greatly from the norm. Von Ribbentrop himself most likely requested this all dark combination lining, sweatband and visor reverse. The Holters Uniformen logo mark is printed in white – reversed from the company's usual logo (in black ink). *Courtesy of Bill Shea.*

(i.e. textbook) as it is possible to get. For Waffen-SS officer caps, this is a pale, bluish green unmarked lining, with a maroon sweatband and a reddish tan visor reverse, as in the example at the top of the previous page.

A very few special caps, on the other hand, were decidedly non-standard even given the differences between makers – that is, they have an interior color scheme that diverges significantly from the usual patterns. Presumably, such divergences were all personally requested by the buyer, and no doubt a hefty premium was paid for this extravagance. The all dark interior of the cap belonging to Reichs Außenminister von Ribbentrop is a perfect example. Such complete exceptions are extremely unusual, however, and generally appear only on named items (caps belonging to a high level, named individual).

CAP BAND CLOTH: *FELT AND VELVET*
The only material difference between Waffen-SS and Army visor caps was the cap band cloth:

- Waffen-SS officer visor caps were made with *black velvet* cap bands
- Waffen-SS Other Ranks visor caps were made with *black felt* cap bands.

Army caps, irregardless of the wearer's rank, had dark bluish-green felt (badge cloth) cap bands.

These differences in cap bands are significant, and require careful scrutiny. German-made felts used for the cap bands of Waffen-SS Other Ranks visor caps (or on Army caps) were *short napped*, with a very even hand (feel), as it is called in textile terminology. The nap, though short, is still quite noticeable.

Felt is *not* a woven material, and so there *should never be any kind of weave pattern* visible in the cloth – no weave exists. The felt cap band cloth was often backed with a sheet of cheesecloth-like material for added firmness.

Similarly, German-made velvet as used on Waffen-SS officer's caps and political leaders' visor caps was extremely soft and had a beautiful luster. More importantly, it was made with a *very low pile* – that is, the pile is cut short. Velvets with thick (long) piles *were not used* by German cap makers on their caps, whether Waffen-SS, political, or any other type. Velvets with a long pile indicate a reproduction.

The cap band, whether felt or velvet, was applied to the cap very securely and in such a way that there is no sagging – even when exposed (briefly) to wet conditions. Of course, repeated and/or prolonged exposure to water could eventually cause sagging or other damage, but signs of water exposure in this case should apply to the entire cap – not just the cap band. In other words, such damage should be general, not limited to the band alone.

Modified Army Caps
The Gebirgsjäger (mountain troops) NCO cap displayed in the picture at left was offered by an Internet seller during 1999 through a popular auction venue, and is highly suspect for the following reasons:

Interesting example claimed to be a Waffen-SS NCO Gebirgsjäger cap (mountain troops). Note the severe wrinkling of the cap band, the lumpy spots on each side of the crossbones, and the strange shade of grass green branch piping.

1. The cap band is not the correct kind of felt – if it is felt at all (it looks more like black colored linen cloth). German felts had a very short *but definitely visible* nap. This band, however, is very thin and has no nap at all. No amount of wear could account for such an overall even lack of a nap.

2. Visibly loose play of the cap band cloth. Note in particular the major wrinkle – the ridge and furrow visible on the lower right under the Totenkopf. The cloth should not be so loose that it is able to bunch up – unless it has suffered repeated and prolonged exposure to rain and dampness. The cap wool itself, however, does not show any equivalent signs of distress and thus water damage as a cause for the cap band condition, simply does not seem to fit the equation. This cap band appears to be a replacement or has been applied over the original band material (probably an Army band). This would also explain why the cloth seems so thin – so as not to cause a noticeable increase in thickness of the cap band.

3. Indications of other modifications involve the cap band area around the skull, judging from the lumpy areas on each side of the bones (both right and left sides). There is the distinct impression of another set of holes beneath the present cap band cloth.

4. Finally, though the original camera film, the lighting, or both may have been partly to blame, the piping color as it appears in the photo is simply not the correct shade of grass green for the *Gebirgsjäger* (mountain troops) branch.

This cap is extremely questionable, and would not be a wise purchase.

CAP BAND MODIFICATIONS

Authentic Army visor caps may be modified to look like Waffen-SS caps by replacing or covering the blue green Army cap band with a black felt or velvet band, by replacing Army with SS insignia, and by switching the leather chin strap from the Army version to the SS version (with its center buckle) on an Other Ranks visor cap. Replacing cap bands is no easy trick, and any such complicated effort will almost certainly show signs of tampering. An alternative is to put a new band over the original band – this, however, changes the surface height of the cap band area in comparison with the cap band piping, and is usually noticeable.

A physical check of the cap band can certainly help in making a determination. If the dealer refuses to allow a physical inspection, best to take your business elsewhere. Checking the band is simple:

• Hold the cap firmly and place a finger on the cap band surface. Gently push up and down. There should be little to no looseness – that is, there should be very little play in the cap band cloth. The band on original caps was often backed with a sheet of material to stiffen it, so that it would not shift around. If the cap band cloth is easily moveable under mild finger pressure, leave the cap alone.

• Use your thumbs on each side of the Totenkopf insignia to feel for any extra holes that may be hidden under the cap band cloth. The dimensions of the Totenkopf cap insignia are narrower than those of the Army oak leaf wreath. Consequently, any holes in the underlying pasteboard band made by the prongs of an Army wreath will be *outside* the edges of the SS Totenkopf and can still be felt underneath any new cap band material.

CHIN STRAPS/CORDS

Officer chin cords on Waffen-SS visor caps were the same twisted aluminum cords and pebbled aluminum buttons used for Army officer caps. Waffen-SS officers in general's rank also used aluminum cords – never gold. The black leather chin strap for Other

To test for unseen holes in a questionable Waffen-SS cap band, hold the cap with the visor facing you. Put a thumb on each side of the Totenkopf crossbones and press gently. Covered holes from any previous insignia (e.g. an Army wreath) should be quite easy to feel.

Authentic Army visor cap piped in black for the Pioniere (combat engineer) branch, which has been deliberately ruined by adding SS insignia. Vulkanfiber visor; the chincords have been removed.

SS eagle offered on the Internet as a "guaranteed original". Contrast this incredibly flawed piece with another reproduction in the following picture.

Reproduction SS eagle. Detail and dimensions of this are better, but the surface shows severe bubbling and pitting irregularities common to poor quality castings. There is also edge "flash" remaining on the neck, on the right wing (viewer's right) and the rear edge of the wreath. (Flash is metal that overruns the edges of the casting mold). Compare the detail on this reproduction with the authentic piece in the picture on the next page.

Ranks Waffen-SS visors differed from the Army version by having a connecting ring in the center. It was made from two doubled-over straps, with the loose ends of each strap secured to the cap band with a black pebbled button. The strap section ran to the center of the cap where it passed through a rectangular ring, doubled back over on itself, and ran back toward the side of the cap where it inserted into a buckle. An example of this arrangement appears in the picture on page 73. Lack of stock availability of course, might lead to a replacement chinstrap with an Army or Luftwaffe version – whatever was available at the time.

INSIGNIA

Waffen-SS insignia consisted of the SS version of the national emblem and the Totenkopf. There was no oak leave wreath and cockade of course, and so the only method possible for ventilating an SS cap involved using the eye sockets of the Totenkopf insignia as the means of air entry. The only maker that appears to have employed such a system on an SS cap is Robert Lubstein (EREL). Examples of such ventilated caps are extremely scarce, which indicates they were not at all common and most likely the option was available only at a considerable extra expense.

Since the SS was not considered an official branch of Germany's armed services, it was not entitled to wear the national emblem on the right breast of the uniform, or the *Reichskokarde* (Reich cockade) on the cap. The *Totenkopf* (death's head) was the SS alternative to the oak leaf wreath and cockade of the national armed services, and it certainly made a definitive image statement. The SS national emblem was also designed to be noticeably different from the Army, Luftwaffe and Kriegsmarine forms through the use of a distinctive double-beveled wingtip pattern.

As a general rule of thumb, the prongs of metal SS insignia normally do not pierce the cap's lining material. *This factor in itself should not be used as an exclusive evaluator of authenticity*. When originally applied either before, or at the time of sale, the insignia prongs were carefully bent over beneath the lining, and were not allowed to pierce it. Over the years since the war many headgear collectors and dealers alike have had a lot to say about the authenticity of a cap which has prongs through the lining, the general attitude being negative. From time to time a cap needed to be cleaned, however, and for this to be done the insignia first had to be removed. It does not seem reasonable to expect that every single soldier who replaced his insignia on a laundered cap would go to the trouble each and every time of making sure that the prongs did not go through the lining. Now and again, some must surely have done so.

Types:
Insignia for Waffen-SS visor caps included:

1. Leichtmetall for Other Ranks caps;
2. Leichtmetall for officer ranks including generals, or
3. Hand-embroidered bullion wire for officer cap eagles with a Leichtmetall Totenkopf.

There was no choice of insignia options for Other Ranks visor caps – only metal insignia was permitted.

Officers, on the other hand, had more possibilities. They could choose the standard metal insignia (perhaps with a polished surface), or opt for a hand-embroidered bullion officer eagle combined with a metal Totenkopf.

In rare cases, some officers wore the smaller, early model jawless skull (very similar to the Hussar skull used on Army Panzer uniform collar tabs).

Machine-woven (flat wire, BEVo) insignia such as that used for the officer's old-style field cap was *not* authorized for wear on the service visor cap, and thus rarely appears in photographs. Nonetheless, as with every rule there *were* always exceptions. Waffen-SS Standartenführer Wilhelm Mohnke of the 1.SS Pz Div. Leibstandarte, for example, wears a service visor cap with (a somewhat crudely bordered) machine-woven eagle in the picture appearing on page 117 of Mark C. Yerger's fascinating book *Waffen-SS Commanders: Krüger to Zimmerman* (1999, Schiffer Publishing Ltd.). The extreme rarity of such exceptions, however, makes the appearance of such insignia on a vintage cap extremely questionable at best.

In the previous picture, the Army cap band on the pionier cap remains in its original blue green color, with wreath removed and an SS Totenkopf added. The eagle is an enlisted ranks machine-woven sleeve eagle; the cap is intended to look like an officer's old-style field cap (no chin cords), and to be fair, the seller clearly advertised it as a makeover. The tampering destroyed the value of the original Army service visor cap that it was. This item sold for over $170 on the Internet in the year 2000. A clean, unplayed with original Army Pionier officer visor cap would actually have commanded a much higher price.

Quality:

The insignia detail on the SS eagle and Totenkopf should always be crisp and clear.

Despite production by a number of different makers, there is a degree of consistency in the detail quality of authentic insignia that is generally lacking (despite years of practice!) in the majority of reproduction versions available on the market today. Some reproductions are excellent, others terrible – so incredibly crude that it is hard to believe anyone with eyes might think they are real, but so it is.

Fact: The Germans did not make sloppy metal insignia! As an example, consider the "guaranteed original" SS cap eagle on the previous page. The eagle can certainly be guaranteed original – that is, guaranteed to be originally made in someone's garage, by lamplight. Note the inconsistent width of the inner feathers between the right and left wings; the uneven height of left and right legs; the bulbous head (looks like it's wearing a falconer's hood) and scrawny body with an indentation on the right side. The wreath sides are flattened, the leaves are inconsistent and the swastika very unevenly shaped. The upper edges of the wings are so thin as to be almost non-existent.

The metal used for most reproduction insignia is usually soft and easily bent, while authentic insignia is much sturdier.

The use of non-SS insignia on Waffen-SS caps was not authorized, but *was* tolerated, since supply problems sometimes left no other alternative particularly for enlisted men's caps. With officers, this practice seems to have been more a matter of personal preference. It was certainly possible to find a bullion (rarely metal) Army eagle on a Waffen-SS officer's cap. In some instances, officers might even use the machine-woven (BEVo) Army eagle on their service visor caps. Such rare exceptions to the rule, however, require very careful scrutiny to insure that the Army insignia has not been applied post-war. The Army eagle on the Waffen-SS officer's cap in the following example was probably added post-1945. It is not correctly positioned, being too close to the cap band, and is actually an Army breast eagle insignia rather than the slightly smaller cap eagle. The hand stitching is not very neatly done and the backing cloth

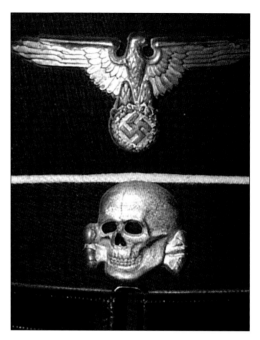

Authentic insignia on an Allgemeine-SS Other Ranks (OR) cap. Triangular legs, very exact, straight body edges. The thick upper wing edges are very clear, as is the excellent feather detailing. There is no bubbling or pitting to the metal surface. Of the two rows of curving feathers, the top three feathers on each outer row are shorter than the ones below them – not the case in the first reproduction. This particular photograph also provides a good shot of the rectangular center ring for an OR chinstrap. *Courtesy of Bill Shea.*

Waffen-SS Knights Cross winner Hauptsturmführer (captain) Ludwig *Lutz* Hoffmann, who received his award on May 9, 1945 while CO of 111th Bn, 23rd SS-Panzer Grenadier Regiment "Norge" (Russian Front). Hoffmann wears his service visor cap at a dashingly rakish (and non-regulation) angle in this autographed photograph. Note the Army pattern cap eagle. *Author's collection.*

Although this cap was seen only as a jpeg picture, careful viewing, and a good idea about what to look for, can sometimes provide enough information to make a preliminary judgement regarding an item's authenticity. The cap cover (Deckel) in this example is made of a fine grade (private purchase quality) Trikot. The Army eagle may be a postwar application. There is no explanation for the noticeable (and uneven) discoloration on the lower cap band piping.

Offered on the Internet (fall of 2000) at a price of over $1000, the front seam of this Waffen-SS officer visor cap drifts noticeably off-center towards the crown, and the insignia is thus thoroughly out of line. Such sloppiness *simply didn't happen* in German production or usage. Note also the incorrectly rounded ends of the swastika arms on the eagle – although good, it is not original. The Totenkopf also appears to be a reproduction.

border varies considerably in width, being almost non-existent below the wings. Hand-embroidered bullion insignia was the only area in which German manufacturers were not always consistent in either form or neatness. Quite a few authentic bullion Army eagles, for examples, appear in forms that are not entirely satisfying from an aesthetic point of view, yet are nonetheless correct. The same is true of bullion SS eagles. Study every authentic example of embroidered insignia that you can locate, in order to get a feel for the many variations in manufacture.

Sepp Dietrich, commander of the 1st SS Panzer Division Leibstandarte Adolf Hitler, fancied the Army pattern eagle and appears in many period photographs wearing one on his cap – but only in bullion (the aluminum eagle does not seem to have held the same allure for him). Obersturmbannführer (Lieutenant Colonel) Johannes Mühlen-kamp of the 2. SS Panzer Division Das Reich, also appears in quite a few photographs wearing an Army eagle on his visor cap.

With Other Ranks caps, the use of non-SS insignia was more likely a response to unit shortages of the correct SS version. If an item was not available for one reason or another, well – there was always an Army unit right around the corner from which to borrow when in need.

OTHER MODIFICATION TELLTALES

While the visor cap in the example at left shows enough external flaws to raise concerns, a study of the interior reveals evidence that damns this piece beyond any chance of redemption:

1. **RZM tag**: *Neither Waffen-SS visor caps nor Waffen-SS field caps had RZM tags.* As a military rather than a political force, the Waffen-SS escaped from under the RZM umbrella, though certain specific items, like insignia and belt buckles, remained under RZM control. Waffen-SS clothing, however, did not fall into this category. Only caps worn by members of political entities such as the *Allgemeine-SS* (General-SS), political leaders and organizations such as the DAF (Deutsche Arbeitsfront), etc. will have RZM tags.
2. **Circled SS runes on the lining**: *Waffen-SS caps do not have the circular, gold SS rune mark on the cap lining* (Allgemeine-SS caps usually do).
3. **Thread color**: *The black thread used for the sweatshield stitching (and not straight, either) is highly suspect.*

Left: Interior, cap shown at left. The cap lining and sweatband reverse show no wear at all – a dangerous sign, especially on an already questionable piece. This cap is a poor reproduction seeking to dazzle the inexperienced collector with an impressive-looking RZM tag – which itself, is the fatal flaw. Right: A correct RZM tag, from an *Allgemeine-SS* black visor cap where they do appear (in several forms). Note that the tag is secured to the cap band behind the sweatband using the lining tack down threads (which pass directly through the tag). The information on the tag is also *quite specific*, clearly stating *Tuchmütze* [cloth cap], with the manufacturer's RZM number stamped in purple ink, in this case, 143.

OFFICER UPGRADES

Did SS NCOs ever upgrade their felt visor cap sweatband to officer velvet, after being promoted to Untersturmführer (2LT)? That is, upgraded by attaching one band over the other? An interesting question indeed and the answer is *Yes* – it was done, though authentic examples are exceedingly rare. By far the easiest choice was simply to go out and buy a brand new cap. On the other hand, if a cap was a true favorite with the owner, broken in to a point where it was such a comfortable fit that nothing could be better, then the evidence is out there: upgrades *were* done. The scarcity of authentic examples, however, means that a painstaking inspection of the entire cap must be carried out if any purchase is being considered.

Below: Front view of a black, Allgemeine- SS officer's cap. This positively stunning, very high quality piece is an authentic upgrade from an NCO cap to an officer version. The side view shows the situation more clearly. Right: The unusual thickness of the velvet cap band becomes quite obvious in this side profile picture. Applied directly over the original *felt* band, the new velvet cap band surface is raised higher than the leading edge of the cap band piping. The attachment sewing is unnoticeable and indicative of a very experienced professional. An extremely rare answer to the upgrade question. *Courtesy of Bill Shea.*

POST-WAR [WEST] GERMAN CONVERSIONS

In most cases, West German (before the reunification) military and civilian caps are too different from the World War II models to be easily modified – it would simply require more work than would be worthwhile, particularly when there are far better targets available for alteration, such as East German caps. This does not mean that no [West] German caps were or are ever modified. In fact, one Internet seller with a reputation as an excellent source of fine DDR (East German) militaria offered an all black West German (non-military) visor cap with pink piping for auction on an Internet site. With a reproduction SS-style eagle and Totenkopf applied, he offered it as a Waffen-SS panzer cap. Amazingly enough, seven people bid on this item, which reached a final price of $72 (year 2000 dollars). In his description, the seller honestly stated that the cap was a modified West German piece. But more than this, it was a complete fantasy cap – no pink piped, black service visor cap ever existed (the black SS visor cap was always piped in white for all ranks below general). The seller immediately took three similar caps to a militaria show and sold them all within two hours of putting them on display. It is all the more astounding since, with a little effort, the buyers could have instead purchased a first class quality reproduction Waffen-SS field gray visor cap for approximately $150. This example simply proves that some people are willing – or foolish enough – to spend hard-earned money on almost anything given an opportunity, and it is exactly this that the true shady dealers count on.

DDR CONVERSIONS

The most common cap used for modification into Waffen-SS model visor caps is the East German Army officer's cap. East German Communists never felt, assumed or admitted to any association with Germany's National Socialist past. This being the case, East German Army uniform designers did not need to concern themselves with any of the image constraints (that is, similarities to World War II uniforms) that the West German Bundeswehr had to deal with. East German (DDR) Army uniforms are therefore, quite similar in cut and overall design, if not in color, to those of the pre-1945 German Army. There were some practical changes made to the East German visor cap design, however, including some style adjustments attributable to Soviet Army influence.

The East German officer's visor cap has a ridge, but no actual cord piping around the bottom of the cap band. The cap band top, and cap crown piping are both white cord.

The similarities with vintage World War II caps were as follows:

1. Trikot cloth material
2. Black visor similar to the Vulkanfiber visor, including the rounded front brim ridge
3. Waffenfarbe piping; white for officers, in branch colors for Other Ranks (remember that the Bundeswehr chose silver mesh piping as a universal piping for all officers)
4. Felt cap band

East German visor caps for all ranks are made of a fine grade, warm gray colored Trikot. No other material was authorized, and the East German Trikot is wool, blended with synthetic fibers. The East German gray color is simply not close enough to field gray, however, for it to ever pass as a pre-1945 German Army visor cap. Even for reenacting, the color differential is too great, and thus these caps are rarely substituted for pre-1945 Army caps.

The color is similar enough to the pre-war Waffen-SS mouse gray color, on the other hand, to pass for a Waffen-SS cap copy. The smoky gray colored *felt* cap band used on NVA caps (*National Volksarmee* – National People's Army) lends itself quite easily to being hand-dyed black. As an Other Ranks (enlisted grade) cap, the piece needs only the addition of a correct World War II-style leather chinstrap and buttons and the appropriate insignia to be complete.

East German branch piping colors for Other Ranks caps are essentially the same as those of the pre-1945 German Army, and so do not require any adjustment.

DDR Army officer caps are modified in the same fashion as those for enlisted grades: the cap band is hand-dyed black. Black or not, however, the material remains felt – not the correct velvet – so the modification is already fundamentally flawed. The stiffening ring in the cap crown is often removed in order to soften up the top edges. East German visor caps normally have a very rigid, flat top with a low, somewhat rounded crown (this being the Soviet Army influence). Softening the edges helps make the crown look a little higher. The East German insignia is replaced with reproduction or original SS cap insignia, but the twisted aluminum cap cords are left as they are since East German cords are nearly identical to the World War II variety. (Note that original East German caps have no insignia on the crown.) The only thing remaining to change is the cap piping, which can either be left as is (white) for Infanterie, or can be hand-dyed to another color. Such dyeing is common because the East German Army did away with individual branch colors on its officer visor caps, instituting a universal white cap piping for all officers regardless of branch – similar to Waffen-SS regulations during most of the Second World War. With East German Army officer uniforms, in fact, the only color indication of the wearer's branch is the underlay material for the shoulderboards. Thus, to create any branch other than infantry, the white piping must be hand dyed.

All things considered, it is quite easy to recognize modified East German Army headgear, and thus few sellers bother trying to pass them off as authentic caps. People do buy them, nonetheless. The key differences from original World War II period Waffen-SS caps, summed up:

East German Army officer's cap, modified to appear as Waffen-SS (Internet offering). The cap spring in the crown has not been removed, and the top retains its original Soviet influenced, rounded shape. The cap band felt has been dyed black, the chin cords are original East German manufacture (visually, no difference from World War II cords). The SS insignia is reproduction. The piping has not been dyed – this is the East German universal white piping used on all officer's caps.

Interior of an East German Army officer visor cap. The diamond-shaped sweatshield with embossed NVA is one of several easy giveaways on such caps. It is not possible to remove the heat sealed sweatshield (no threads) without damaging the very tight lining beneath. Note the reddish-brown vinyl sweatband, and the glossy smooth, black visor reverse.

1. **Plastic visor**: The visor, which superficially resembles the old Vulkanfiber visor, is made of a smooth, easily scratched, glossy black plastic with a black underside (all ranks.)

2. **Missing Piping**: Unlike World War II vintage Waffen-SS originals, East German caps have no piping around the bottom of the cap band.

3. **Lining material and color**: The lining cloth on East German caps is always very tightly attached to the cap top, and the color is always light gray.

4. **Vinyl Sweatband**: Always made of a maroon colored vinyl on NVA caps.

5. **Translucent Plastic Sweatshield**: East German sweatshields are a translucent plastic diamond (not a rhomboid), which are heat sealed (not sewn) to the lining. Embossed or molded into the thick surface is the cap size (e.g. 56) and the letters 'NVA'.

These modified caps are quite common, and sell within a price range of $35 to $45 in year 2000 dollars.

THE JOACHIM VON RIBBENTROP WAR BOOTY CAP

By May of 1945, American troops in Italy had doggedly fought their way north through the Italian Alps, onward through crumbling German resistance into southern Germany itself. Though still offering the beautiful landscapes of Heidi fame to anyone with the time for viewing, Bavaria was in fact, awash with confusion and apprehension. What would the American conquerors do, now that Germany was defeated? A few high-level Nazi party members and government officials had a good idea of the reckoning that was in store for them, and had decided to desert the ship while it was still – if just barely – afloat. Of the very few routes out of Germany still open, the easiest – or so it appeared to many of those who sought to disappear – was through Switzerland, through western Austria, or Spain.

Company E, 71st Infantry, 44th Division, 6th Army Corps, 7th US Army was among the U.S. forces that had advanced into Bavaria. Commanded by CPT Howard Goldsmith Jr., the company received new orders to swing back to the south and slightly eastward, a route which would take it into the northwestern part of the Tyrolian Alps in Austria. The mission: to patrol, track and interdict any Nazi officials who might seek to escape in that direction.

Company E pulled into the Austrian village of Umhausen, which CPT Goldsmith had selected as his base of operations. The village and surrounding area was quickly secured. The men of Company E had had no opportunity to relax in several months. With only half the company needed at any one time to fulfill mission requirements during the night (patrolling, security and other garrison duties), CPT Goldsmith wanted to make sure that the other half had an opportunity to sleep again in warm beds, and to eat good food. A picturesque town with stunning scenic views, and a tourist spot before the war, Umhausen was well equipped to serve this need with several hotels. CPT Goldsmith commandeered five of these, and ordered that beds be made available for every off-duty soldier.

The operation proceeded smoothly except at the Hotel Krone. Though one of Umhausen's larger hotels, this establishment could only offer one bed for every two men. Annoyed at the seeming refusal of the hotel owner to make more beds available despite the hotel's size, CPT Goldsmith decided to inspect the place. The English-

Pièce de résistance! Short of one of Sepp Dietrich's caps, this example of a Waffen-SS general's visor cap is about as good as it gets! It was once the personal property of Reichs Außenminister (Foreign Minister) Joachim von Ribbentrop. Fine grade doeskin cover. Note the secondary ridgeline just inside the normal brim ridge on the Vulkanfiber visor. *Courtesy of Bill Shea.*

speaking hotel owner seemed quite nervous and agitated as he followed the captain on his inspection. When Goldsmith tried to go up to the third floor, the sweating owner sought to block his way. The captain, sensing that something was sorely amiss and not a small man, firmly shoved the owner aside and continued on his tour.

The third floor rooms were all filled with a vast assortment of trunks, suitcases, boxes and even file cabinets. CPT Goldsmith instructed the owner to have everything moved out to a different location so that beds could be set up for his men. To his surprise the owner refused, claiming that he was storing the goods for people who would be quite upset if they were moved or disturbed. At that point, a lustrous yellow glint caught the captain's eye from farther within the room. He moved to take a closer look and found a gold plate inscribed 'Le Louvre, Paris.' Curiosity now thoroughly piqued, he asked the hotel owner what the trunks and boxes contained, but the man refused to give any answer. The hard muzzle of Goldsmith's weapon at the man's neck induced a sudden change of mind, and brought the answer to the question: Everything belonged to the German Foreign Minister.

In fact, the boxes, suitcases and trunks contained the biggest cache of war booty ever to be found by American soldiers: the personal possessions (quite a few appeared

WAC Staff Sergeant Lucille Mullens (Goldsmith's sister-in-law) with an assortment of von Ribbentrop's uniforms. Note the Waffen-SS general's visor (with the Totenkopf insignia set high on the cap band) resting on the back of the sofa just to the left of SSG Mullen's head: 'tis the very one! *Courtesy of Bill Shea.*

Waffen-SS Knights Cross winner Sturmbannführer (Major) Waldemar Fegelein wears his service visor cap as most officers – with a noticeable cant – in this autographed photo. Ritterkreuz, awarded December 16, 1943 while CO of 2nd SS-Cavalry Regiment, 8th SS-Cavalry Division "Florian Geyer", 8th Army, Army Group South, Russian front. Waldemar was the younger brother of Waffen-SS Oberführer Hermann Fegelein (who was executed in Berlin in 1945). *Author's collection.*

to have been stolen) of the former German Foreign Minister (Reichs Außenminister) Joachim von Ribbentrop. All of these possessions had been moved from Germany to the Hotel Krone in Umhausen a short time earlier, with Frau von Ribbentrop settling into the hotel along with the goods. What she planned to do from there – to slip out of the country perhaps, or to stay for an indefinite period – is not known.

Company E's arrival came completely unexpected. It may be that, having heard news of the American commandeering of hotels, Mrs. von Ribbentrop had simply panicked and fled unnoticed from the Krone just a few hours before Captain Goldsmith's arrival to investigate the bed shortage situation at the hotel. She took with her all she could carry.

Goldsmith immediately notified his higher headquarters of the find, then began a complete inspection of everything in the rooms. Meticulously organized records were discovered, together with a variety of documents stored in filing cabinets, which detailed all manner of state activities. Many of these files were later used against von Ribbentrop during his trial for war crimes at Nuremberg a year later, in 1946.

An extensive collection of jewelry, sabers, daggers, and pistols was found together with etchings and sculptures, many of which had been taken from various museums. Wines, liquors and gin alongside paintings by the old masters Van Gogh, and Renoir, among others – all packed away in the third floor rooms. Flags, banners, diplomatic sashes, and a collection of foreign awards presented to von Ribbentrop were there, and a twenty-four piece silver service for 200 people. There was also a collection of British made civilian suits and thirty four Foreign Ministry and Waffen-SS general's uniforms (von Ribbentrop had also held dual rank as an SS-Gruppenführer und Generalleutnant der Waffen-SS; U.S. equivalent = Lieutenant General). Amongst this incredible display was also a Waffen-SS general's visor cap – the very one shown here.

Unfortunately for her husband, who was arrested a year later in Hamburg, the items Frau Ribbentrop took with her when she fled did not include any of the damning records stored among the Umhausen goods. These were later used against the former Foreign Minister during the trials in Nuremberg; he was executed on October 16, 1946. Mrs. Ribbentrop left behind a wardrobe of her own when she fled the village. But the story does not end with the discovery of the Ribbentrop booty. Captain Goldsmith was permitted to keep all of the awards, daggers, pistols and uniforms that he found, and these he sent back to his sister-in-law's family in the United States. His sister-in-law, Lucille Mullens, was herself an Army Staff Sergeant present when the uniforms were displayed for photographs at the hotel – she even posed with them in one shot.

The majority of the von Ribbentrop goods were confiscated by military authorities, who returned the artwork to the original owners, and spirited away the documents. Captain Goldsmith was allowed to keep Ribbentrop's awards, pistols, uniforms, banners and other regalia, which he sent back to the Mullen's home in Omaha, Nebraska.

After completing his postwar occupation duties, Goldsmith returned to the U.S. where he settled in Texas and continued serving the military in the National Guard, retiring with the rank of Lieutenant Colonel. On special occasions, Colonel Goldsmith put his trophies on display at the National Guard Armory and they often received notice in local news. One year, however, a local write-up attracted greater attention and as a result Colonel Goldsmith eventually sold his entire Ribbentrop collection to a persuasive fellow who led the Colonel to believe the items would be displayed in a private museum. Instead, many of the pieces were immediately offered up for auction, where they headed off in different directions. After many years, the Ribbentrop cap has finally settled in among other named pieces in a notable New England collection.

Part 2:
DIE POLIZEI SCHIRMMÜTZE - THE POLICE VISOR CAP

Organizationally, all police forces [Ordnungspolizei] were under the authority of the SS. The combat police units were also part of the SS and not one of Germany's constitutional armed services – which status was jealously and zealously guarded by those same official forces: das Heer (Army), die Kriegsmarine (Navy) and die Luftwaffe (Air Force). The World War II combat police troops were, however, classed as combatants – just as Germany's modern border patrol [police] forces are today (known as the Bundes Grenzschutz or BGS). The Schutzpolizei provided the largest contingent of troops for combat forces.

Like its SS parent organization, the Schutzpolizei (literally protective or defense police) were prohibited from wearing the national emblem above the right breast pocket

Schutzpolizei (combat police) officer's visor cap in Trikot. Green piping around the cap crown and dark brown felt cap band with aluminum chincords. *Courtesy of Gerard Stezelberger of Relic Hunter.*

of their uniforms, this being the sole right of the legally recognized military services. The police eagle, superimposed on its distinctive, full oval wreath, was located instead on the wearer's left sleeve and together with the national colors in the form of the Reichskokarde (cockade), on the police visor cap as well. Police service visor caps are worth special attention owing to the fact that they differ significantly in several ways from visor cap models used by the military services, and because early post war caps are fairly similar to wartime originals.

CONSTRUCTION

The first visual difference from Army and Waffen-SS caps is the color of the base cloth: police green. The police green color was traditional to German police forces, a color both distinctly lighter and distinctly greener than the military's field gray.

Caps were available in the same variety of wool grades and types as for Army/ Waffen-SS caps – from Trikot to doeskin. The distinction between issue and private purchase ("Extra"-Mütze) caps, however, is somewhat more muddled with police headgear, since the various police organizations (combat police, national police, municipal police, etc.) were subject to different uniform regulations.

Police visor cap construction on the other hand, did not differ from the manufacturing pattern used for Army and Waffen-SS caps. The cap band seam was always located at the center rear of the cap, and the overlapping ends of the sweatband usually met at the cap rear as well. Interior construction in other respects (lining, galvanized steel cap spring, etc.) also remained the same.

PIPING

Schutzpolizei caps were piped in green. There was no difference between officer and Other Ranks piping – it was the same for all. Administrative Police [Verwaltungspolizei] officials had caps piped in light gray (hellgrau). Administrative officers could be assigned to duty with Army units as well as with police units.

Authentic hellgrau piped administrative official's caps like the superb example shown in the next picture are fairly rare, and command high prices.

LINING

Linings for officers were of either natural silk or rayon, and lining colors were essentially the same as with the caps of other services with the most common color being gold – though this depended on the maker. A large majority of police officer caps were private purchase, and so there is a good deal of variety possible in interior colors.

Other ranks visor caps generally have tan cotton linings, often with a matching tan sweatband.

VISOR

Visors were exactly the same as those common to all Army and Waffen-SS caps. Vulkanfiber, with a gently curving ridge along the front of the brim. No pre-1945 visors for any German headgear were ever manufactured in plastic. The visor reverse is found in several colors depending on the cap maker, but are most commonly:

1. Green for Other Ranks caps (slightly deeper and not as bright a green as that used on Luftwaffe visors);
2. Creme, dark tan, or a reddish brown for officers (depending on the maker).

CAP BAND

The exterior cap band is the second distinct difference between police caps and those of other organizations, including the Waffen-SS. Police cap bands are always brown-colored felt – no other form of cap band is correct. Early police cap bands were a medium brown; this was later changed to a dark chocolate brown – but the material was still felt, for both Officer and Other Ranks caps.

INSIGNIA

Insignia for combat police visor caps is always in metal – no known example with hand-embroidered bullion insignia has been seen. The cockade, centered at the cap crown in the spot where the eagle normally appears on Army or Waffen-SS caps, is metal and frequently the polished aluminum type with a high base roundel. The police eagle/wreath is centered on the front of the cap band directly below the cockade. Alignment is such that the top edge of the wings and the top ring of cap band piping form an

Superb example of a police administrative official's light gray piped visor cap (in officer's rank). Fine quality police green Strichtuch (doeskin) and plush brown felt cap band (short nap), with aluminum cap cords. The eagle is in the preferred position on the cap band, with the upper edge of the eagle wings even with the piping around the top of the band. The visor shows light crazing (fine cracks). *Courtesy of Bill Shea.*

unbroken horizontal line though this is not always the case. The police eagle/wreath insignia was manufactured in several patterns for officer grades and enlisted:

1. Two-piece construction (usually for officers). The eagle is a separate piece, and raised above the wreath. The detail is average quality.

2. One-piece eagle with high relief (for officers). The eagle is raised well above the wreath in 3d-style – quite distinct – and the overall quality of the detail is superb.

3. One-piece, stamped Leichtmetall (aluminum) insignia for Other Ranks caps.

In addition, the insignia surface finish options available to the military were also available to the police forces, e.g. matte frosted, silver plated (polished), etc.

VENTILATION

The third major difference, and perhaps the most significant (though not consistently present), is cap ventilation: many national police visor caps were equipped with ventilation grommets, which do not appear on any other type of visor cap for any other service. The grommets were arranged two per side, aligned with each other so that one grommet falls on each side of the cap's side seams. In addition, the grommets on police caps (except combat police caps) were relatively large, with the air hole covered from the inside with a fine wire screen laid cross the hole. Unfortunately, as is so often the case, this can not be considered a hard and fast rule since *combat police* caps sometimes have small grommets with a baked on police green enamel, and no mesh screen.

MODIFIED ORIGINALS AND POST-WAR CAPS

Until the late 1990s, police caps never attracted much attention among collectors in comparison with other types of headgear. This situation is changing rapidly. It may be that police headgear, long relatively ignored, is seen as an affordable alternative to other types of headgear that have increased beyond the price limits many collectors are willing to pay. Unfortunately, this growth in interest has also pushed the market value of these caps higher as well. From 1999 through 2000, the value for original police visor caps in good condition increased by approximately $300. Steadily growing demand has continued to fuel the price rise.

Combat police helmets experienced a similar jump in price as interest in them grew – perhaps as an alternative to far more expensive Waffen-SS helmets. Demand increased rapidly, together with prices.

Not only higher prices are a cause for concern, however, for with this increased popularity comes a corresponding increase in the number of modified originals entering the militaria market. Such modifications to police visor caps fall primarily within one of two areas:

1. *Upgrades from Other Ranks to officer level.* Since there is no difference in piping between officer and enlisted personnel, this presents a significant problem officer upgrades can be made with little more than a change of insignia and the replacement of the enlisted-style chin strap with aluminum officer cap cords and buttons. Since different insignia makers sometimes located the insignia prongs in different positions, carefully check the cap band to see if there are any previou holes which might indicate the present insignia is a replacement. Also check:

Mint, unissued Waffen-SS officer's service visor cap in average grade wool with a felted finish. This type of finish is *not* doeskin. Nicely matching aluminum insignia and aluminum chin cords. The cap has a bluish green lining with a maroon sweatband. *Courtesy of Bill Shea.*

a. Visor underside color: green is indicative of enlisted caps.

b. Lining material: if the lining is not silk or rayon, then the cap was probably not originally an officer cap.

c. Overall material quality: officer caps are not as likely to be made of average or lower grade materials.

While enlisted to officer upgrades surely occurred, there is enough uncertainty about them that any such example is difficult to prove.

2. *Modifications to early postwar police visor caps.* Modification of postwar police caps is not especially difficult, but these caps include several physical details that can not be [easily] changed and which help to identify them for what they really are:

a. The ventilation grommets, while present, are too small, and there is no wire mesh covering the holes.

b. The cap band is *green* felt on post-war caps, not the correct *brown* felt.

c. Post-war police caps *do not* have Waffenfarbe piping around the top and bottom of the cap band (though the cap crown is piped in a dark blue green).

d. The sweatshield may be a soft, supple, vinyl-like plastic rather than the pre-1945 stiff celluloid, and the maker mark may also include a post-war date – always a dead giveaway!

These points should be enough to help even beginning collectors spot a modified cap and avoid a costly error in the search for a high quality original.

Waffen-SS officer's private purchase service visor cap, in fine doeskin. This piece has a very rare Waffenfarbe (branch color): Copper brown for Aufklärungstruppen (Reconnaissance troops). Author's early collection.

Chapter 5

Die Luftwaffe Schirmmütze - Air Force [Service] Visor Cap

Luftwaffe visor caps are quite distinctive from field gray Army and Waffen-SS headgear, and hold a good deal of attraction for many collectors. Luftwaffe officer visor caps were made of gray blue colored Trikot cloth, or an aligned finish (*Strichware*) material such as doeskin. Caps for Non-Commissioned Officers (sergeants) and enlisted ranks were a gray blue impregnated (water-proofed) Trikot made from worsted yarn with a mix of 30% rayon (Zellwolle). In other words, somewhat lesser quality than the wool used for most officer caps.

The Luftwaffe gray blue *Fliegerblau* (flier's blue) color was used for the:

1. Standard service visor cap
2. White top summer visor cap *bottom edge*

Luftwaffe officer's service visor cap. Fine doeskin, with a very short nap. Maker: EREL-*Sonderklasse*. Ink stamped June 1942 on the sweatband reverse. The piping on this piece is subdued — that is, matte rather than shiny. Aluminum cockade, bullion insignia. The visor is the thin Vulkanfiber type with simulated stitching on the upper brim and no pique edge wrapping.

Full front view. Superb bullion insignia.

3. M43 field cap
4. *Fliegermütze* (the boat-shaped garrison cap)

Degrees of difference in color value and tint between textile producers did exist, of course, but such differences were only permitted within a narrow range. Luftwaffe cloth was *never* a deep blue color, nor was it ever completely gray.

CONSTRUCTION

The blue Luftwaffe visor cap, whether for officers or for Other Ranks personnel, was manufactured with a non-removable top. During the pre-war years instances arose where blue cap covers were made that could be interchanged with the white top summer cap, but this practice was expressly forbidden by regulations. The cap-making industry seems to have policed itself on this issue by putting pressure on any companies that did not comply. Cap covers on all blue top visor caps should, therefore, be an integral part of the cap.

Luftwaffe blue service visor caps tend to be much more heavily padded than caps made for the other military services. As a result, they are also heavier, on average weighing in at around 230 grams. The sides look very well-rounded where they overhang the cap band and this is due to a thick, felt insert which is sewn along the sides of the cover between the lining and cap cloth, and which follows the line of the crown piping. The purpose of such inserts is strictly to give the sides added firmness – and thereby enhance the overall shape of the cap. This interior padding is quite easy to feel if the overhanging cap sides are given a squeeze. It is never visible to the eye, however, on authentic caps.

The required galvanized (rust-free), steel cap spring was sewn into and around the crown of the cap directly behind the crown piping and the padded crown stiffening support was positioned into the front of the cap directly behind the vertical center seam. The crown support was 4.5 cm wide and was intended to prop the cap crown in

Fine example of a Luftwaffe officer's service visor cap in regulation Trikot wool. The Trikot is very fine grade (narrow, delicate ridgelines), identifying this as a private purchase cap rather than one from government stocks. Hand-embroidered, bullion insignia, with the cockade also bullion. Early model national emblem (eagle) with the characteristic droop tail. *Courtesy of Gerard Stezelberger of Relic Hunter.*

Luftwaffe NCO or enlisted grade visor cap. Waffenfarbe: Flight personnel or paratrooper. Private purchase piece made of a medium-nap doeskin. Second model national emblem. The three-piece chinstrap has squared buckles (Army-style). Lacquered leather visor with edge wrapping. The cap's sides are well padded, giving them a thick, rounded appearance at the side overhang. *Private collection.*

a vertical position. This it did, unless purposely broken (by being bent). Again, none of this internal structure is visible when looking into the cap since it was all installed behind the lining.

SIDE PANELS

Four panels of equal length, with a height at the front of 6.5 cm, and on the sides at 4.75 cm. As with all German military visor caps, the seams are turned in and stitched on the interior – no sewing is normally visible along the exterior seam.

CAP BAND – *External*

Luftwaffe cap bands were made of ribbed black mohair until 1938. After that year, production began shifting to black, ribbed rayon as the principal material. The band itself was 4.3 cm wide, with 2 mm wide piping around the top and the bottom edges. In addition, there was also a 2-3 mm wide ridge (*Vorstoß*) at the bottom of the cap band made of the standard gray blue cloth, which served as a support for the cap band itself and prevented slippage from the correct position. The width of the support ridge, the cap band and the top and bottom cap band piping all together was 4.9 cm.

A significant construction difference from Army and Waffen-SS models that goes well beyond color and type of material is the location of the closing seam for the *mohair* cap band. On Luftwaffe visor caps, this is located at the *front center* of the cap, directly behind the wreath insignia. On all Army and Waffen-SS visor caps, in contrast, the cap band's closing seam is located at the center rear. Luftwaffe visor caps, in this sense, are more in line with Kriegsmarine cap design than they are with the Army/ Waffen-SS pattern. Any Luftwaffe visor cap with the cap band seam located at the rear of the cap is indisputably *incorrect*.

PIPING

Luftwaffe visor cap piping also differed from the Army system in that:

1. Waffenfarbe (branch color) piping was only used on NCO and enlisted rank caps (i.e. Other Ranks).
2. The piping for officer grades below general officer was universal – that is, the same for all: aluminum mesh wrapped around a central cord (Schnur).
3. Piping for generals was gold wire mesh or gold celleon mesh wrapped around a central cord.

This piping system, in fact, was in many respects quite similar to that used by the Waffen-SS, which, except for one half a year when Army-style branch piping was allowed for officers, used a universal white piping for all field and company grade officer visor caps irregardless of branch. Luftwaffe NCO and enlisted ranks, as with their counterparts in the Waffen-SS, were piped in branch color.

Luftwaffe caps having colored branch piping but aluminum chin cords (rather than the normal black leather chin strap for other ranks), can be only one of two things:

1. A badly researched modification
2. A cap for an officer cadet, since *Offizier Anwarter* were entitled to wear officer grade twisted aluminum chin cords on their enlisted/NCO caps (thus the cap would still have branch piping), until the soldier received his actual promotion to Leutnant (2nd Lieutenant).

Officer's subdued silver piping (non-shiny). The piping is formed from fine bluish gray colored threads interwoven into a mesh surrounding a tubular, white cotton cord core (Schnur).

A beautiful example of an Other Ranks cap broken to a more casual look. The cap spring has been removed and the cap has sagged in all the right places. Thin, lacquered Vulkanfiber visor with simulated sewing line along the rim. The lacquer is lightly crazed; that is, it shows a fine tracery of cracks. Waffenfarbe (branch color) is for the Luftnachrichtenschule. Note the narrow ridges on the top and bottom of the chinstrap, and the correct rounded Luftwaffe buckles. *Internet offering.*

Matte Piping

A notable and fairly rare variation on officer caps is matte aluminum piping. That is, instead of the normal shiny aluminum the piping is subdued – almost a light blue gray color, in fact. The photograph above offers an example of a cap with exactly this type piping. It is still unclear whether this variation represented anything more than an option available to customers later in the war (in combat areas, subdued piping would certainly be safer from a tactical standpoint). Specific references to subdued piping have yet to be found in any uniform regulations.

VISOR

Luftwaffe officer cap visors from the years before 1942 – that is, from the early war period – were usually made of leather, with a thick leather or patent leather strip wrapped around the visor edge and sewn into place with a single row of stitching. This factor is another point Luftwaffe caps had in common with Kriegsmarine headgear. Caps for Other Ranks (NCOs and enlisted grades) usually had visors of Vulkanfiber on issue

Luftwaffe general's cap, rear. This picture shows the location where the ends of the crown piping join together. The general's piping on this cap is gold-colored celleon mesh wrapped around a central cord of white wool, or cotton. *Courtesy of Bill Shea.*

Standard green visor reverse. A section of the leather brim wrapping on this visor has come off, affording an interesting view of the thread holes (not normally visible). The brim wrapping on a Second World War era Luftwaffe visor cap *should never be of vinyl, or plastic.* This cap is a "double" EREL (maker mark on the sweatband, as well as on the cap lining). *Private collection.*

caps, again with wrapped edges (there should not be a molded, curving front edge ridge on Luftwaffe Vulkanfiber visors).

Visor Reverse (Underside) Color

The normal color for the reverse side of the visor on Luftwaffe caps for all ranks including generals, was an apple green. The shade of green varied within a narrow range, depending on the manufacturer. Luftwaffe visors usually (but not always) had a fine, checkered crosshatch pattern scored or molded into the visor material.

Mid-war period visors tend to be of the thin Vulkanfiber form, including caps for officers. Instead of the edge wrapping, this style has a fine, simulated stitch line scored into the surface near the front edge. Many collectors prefer this type of thin visor to the heavier leather type. The thin Vulkanfiber visors were also colored green on the reverse.

Under no circumstances should any Luftwaffe visor be made of plastic. Post-war visors look very similar to the thin, wartime Vulkanfiber type, but are made of a smooth black plastic with either unpainted black, or gray (painted) reverse sides. Modern visors *do not* have any crosshatch pattern.

SWEATBAND

Luftwaffe visor caps sweatbands show no significant differences from those used on Army caps. Regulation standard color for government issue enlisted rank cap sweatbands was reddish-brown, to match the havana brown (rust) colored cotton liner material. As with the Army, enlisted men were required to have the unit identifications stamped on the reverse of the sweatband on caps dated prior to the outbreak of hostilities. With the start of the Second World War this practice was halted for security reasons.

LINING

On other ranks issue caps, linings were nearly always rust colored (the so-called havana brown) chintz, or cotton moire. Chintz, which was often waterproofed, has a characteristic slightly shiny surface. With Luftwaffe caps, the rhomboid sweatshield was positioned on the lining in such a way that its upper, rounded end always extended *under* the back of the cap band, so that the top of the curve is not visible.

Superb example of a Luftwaffe general's visor cap. Felted doeskin material (slightly furry appearance), gold bullion insignia with highlights picked out in gold celleon thread. The piping is gold celleon. The color difference between slightly tarnished gold bullion, and gold celleon (which doesn't tarnish), is very clear on examples like this. *Courtesy of Bill Shea.*

INSIGNIA

Studying reference works on German insignia is well advised for any collector. Reproduction insignia is readily available, and ranges in quality from excellent to poor. With the upper-end, high quality insignia range, the differences between reproduction and original are sometimes very few. On Luftwaffe service visor caps for Other Ranks (NCO and enlisted grades) the insignia was always metal and consisted of two parts:

1. Luftwaffe-style national symbol (eagle with upturned wings)
2. Oak leaf wreath with stylized wings and cockade, all in one piece.

The eagle could be either the early variety with droop tail and short wingspan, or the later model with wider wings and a more horizontal tail.

The oak leaf wreath and wings were curved back slightly to fit around the rounded cap band. On most wreaths the wings projected straight out (horizontally) from the oak leaves, but another, less frequently encountered version has the upper edge of the wings with a slight downward taper.

Other Ranks cap, showing two different unit markings stamped in ink on the reverse of the sweatband (left rear). The attachment of the lining to the base of the cap is poorly done, and may have been post-war re-stitched.

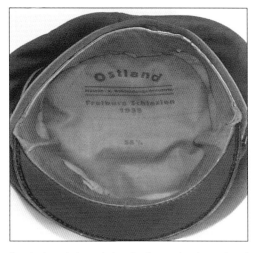

Standard regulation-style interior for a privately purchased Luftwaffe Other Ranks visor cap. Rust colored lining (reddish brown), reddish-brown sweatband, and the standard, green visor reverse. No sweatshield. This maker, Ostland, was located in Freiburg-Schlesien – an area that became part of Poland after 1945.

Nicely matched Luftwaffe insignia set in Leichtmetall (aluminum alloy) for an Other Ranks cap. Note the attachment points of the cockade with the surrounding wreath. Early (first) model eagle (short wingspan, large swastika and a droop tail).

The majority of Luftwaffe Other Ranks insignia is made of the Leichtmetall aluminum alloy, which does not oxidize easily and so remains bright. Late war, lower quality insignia made of pot metal (as it is called), on the other hand, contained far less aluminum and much more zinc. This alloy is not as stable as Leichtmetall, however, and is far more susceptible to oxidation. Over time, the surface finish tends to be lost completely leaving the insignia a dull, uniform gray color.

NOTE: The hand-embroidered bullion national emblem (eagle) on Luftwaffe officer's caps should be given special attention. Ideally, it will be applied to the cap crown rather loosely, with more space between each individual stitch than common on a breast eagle, for example. The reason for this is not clear, but the majority of authentic caps have the eagle attached in this manner. Although it *is* characteristic, the world is not an ideal place. Caps do appear with tighter stitching, but these should be carefully checked – particularly when this applies to the oak leaf wreath. The wreath, when correctly applied, should have a bit of play under finger pressure.

Metal insignia options available for private purchase caps included a polished finish, unpolished (matte) finish, and so on. Officer insignia was normally in hand-embroidered silver bullion wire, and this was especially true of the national emblem. Of special note:

• An officer's cap with both national emblem and the wreath insignia in metal is not regulation – though this did, in fact, occur. A combination is far more likely (and common) however, with the wreath usually in bullion and the eagle in Leichtmetall.

CHIN CORDS/CHIN STRAPS

Luftwaffe officer twisted aluminum wire chin cords were exactly the same as those used on Army and Waffen-SS officer caps. The usual two versions of silver button were also used to hold the cords to the cap: split pin and ring type. Luftwaffe officers for some reason, seem to have preferred the ring type button over the split pin, since the majority of Luftwaffe officer visor caps use the former. For general officers, gold wire or gold celleon twisted cords were used. For other ranks, black leather chin straps (black pebbled buttons) with two short pieces and a longer central band is correct. There was a considerable mix of buckle types and both the Luftwaffe oval form and the Army square form buckles are correct. The buckles should match, however, with both oval, or both square.

BUNDESWEHR [LUFTWAFFE] VISOR CAPS

As with Army caps, modern Bundeswehr Luftwaffe officer visor caps differ so greatly from pre-1945 caps that they are of no real concern to even beginning collectors with basic experience. The single most immediate and noticeable difference is the color, which is a much deeper and darker blue than the wartime Fliegerblau. Next is the cloth covered visor with either an aluminum wire embroidery scalloped edge for junior officers, or a row of oak leaves for field grade officers. This is the same system employed during World War II on Kriegsmarine visors, but never on original wartime Luftwaffe headgear.

Other Ranks (enlisted rank) visor caps on the other hand, are a bit more problematical. Dyeing the sweatband black and switching the buttons and the chin strap to the

Luftwaffe general's insignia, close-up. A beautiful example done completely in non-tarnishing celleon thread. The height of the bullion cockade above the surface of the wreath (i.e. the profile) is clear from this angle. The cockade's central red dot is actually vented (wire). Note also, the angular spiral pattern of the wire strands that make up the silver and black cockade roundels. Celleon chin cords. The cockade vent would indicate that this cap is an EREL-*Sonderklasse* manufactured by Robert Lubstein, Berlin.

World War II version, adding the pre-1945 style of metal insignia is not beyond the capability of a talented individual! Fortunately, the physical design and materials used in the manufacture of modern caps prevent them from being passed off to anyone with any experience, however, as anything more than a piece for use in reenacting. This is because:

1. The visor is plastic.
2. There will be a pair of holes set fairly close together in the cloth beneath the eagle. The holes are from the German federal cockade. The metal gold, red, and black federal cockade on Bundeswehr Other Ranks and officer caps is positioned at the crown in the location formerly occupied by the Luftwaffe eagle during the war.
3. The interior lining, the sweatband and the visor underside will all be gray – another dead giveaway of a postwar cap.
4. Insignia: on officer caps, the aluminum bullion federal wreath with superimposed wings is actually glued to the cap band rather than sewn, and remnants of this glue may remain.
5. Sweatshield: soft, postwar plastic or vinyl, not the stiff wartime celluloid.

A makers mark on the lining may also serve as a telltale, since these normally give the year of production (i.e. postwar dated), and possibly a Patent or Gebrauchsmuster notice, in which case the mark will say:

D.**B**.G.M. – Deutsches ***Bundes*** (Federal) Gebrauchsmuster instead of
D.**R**.G.M. – Deutsches ***Reichs*** Gebrauchsmuster.

Finally, the fact that the Bundeswehr allows a mix of yarn – including a synthetic blend – in its Trikot cloth, means that the entire cap contains post-war material.

Bundeswehr Luftwaffe field grade officer's cap. The visor is cloth-covered, with aluminum wire bullion oak leaves (scalloped edge for Junior officers). World War II Luftwaffe caps have neither of these features. The cap color is much too dark and deep blue, and the cap band is also blue instead of black like wartime originals. These caps are far too different to be conveniently, or convincingly modified.

DDR CAPS

East German Luftwaffe caps also differ too greatly from the pre-1945 Luftwaffe cap in construction, color and overall shape to pose any kind of modification threat to knowledgeable collectors. They are in effect, the same mouse gray cap used by the East German Army (see the example in Chapter 3), but with different insignia.

REPRODUCTIONS

Given the fact that:

1. Modifications to *authentic* Luftwaffe Other Ranks caps are difficult at best - and with (as yet) little reason *to* modify them, why bother? Original caps are still a reasonable price value – with the possible exception of a very rare branch color.
2. Conversion of an authentic Luftwaffe officer's cap to a general's cap is practically impossible;
3. Modified *Bundeswehr* Other Ranks caps are very easy to identify as such;
4. DDR (East German) caps are too different to modify.

It is easy to see that the greatest danger to the collecting community today comes from reproductions, and these will be discussed in more detail in Chapter 14. Some of the better quality reproductions on the market even use original Luftwaffe visors in their construction. This is particularly troublesome when it comes to reproductions of Hermann Göring [HG] Luft-Land (field) Division visor caps (white piping). Original HG caps command the highest prices for a Luftwaffe Other Ranks visor currently acceptable in the militaria market (and will always remain higher than caps with other piping colors).

DIE LUFTWAFFE (WEIßE) SOMMERMÜTZE – (WHITE-TOP) SUMMER CAP

Examples of this cap are much in demand, and not easy to find – particularly white top officer grade caps.

CONSTRUCTION

The body of the cap is the same as that for the blue visor cap (though the bottom edge of the cap remained the normal fliegerblau color), except for the addition of a special, thick cloth rim around the top edge of the cap band. This rim served as a backing for the bottom edge of the white cover when this was in place. The summer cap had a permanent interior lining attached to the top of the cap band behind the mounting rim. *The lining should not be removable, or separated from the cap.* The usual crown support was installed at the front center of the cap band. This support (Stütze) maintained the height and form of the front crown. The support system varied between makers – there were a number of them in use. One form of support for the white summer cap was patented by the cap maker Felix Weissbach of Glauchau (see Chapter 16).

Sweatbands were usually reddish tan, or crème colored as on the blue top cap.

Interior of a Bundeswehr cap. In common with all *modern* German military caps, the interior is gray: sweatband, flexible vinyl sweatband cushion, and the lining. The visor reverse is black. Of particular note is the modern soft vinyl sweatshield, and the makers mark: *Bamberger Mützenindustrie*. This firm is a postwar manufacturer, which did not exist prior to 1945.

Original Luftwaffe administrative (green piping) official's cap for an enlisted man or NCO. A rare piece of headgear! The cap spring is still in place in the crown. A heavily felted finish gives the wool a slightly fuzzy surface. The rectangular connecting ring between the short and long chinstrap sections is visible from this angle. Note the placement of the chinstrap button – slightly low on this cap. *Courtesy of Bill Shea.*

Luftwaffe officer's summer cap. Standard cotton fabric for the cap top. The removable, hand-embroidered bullion national emblem, on a white wool base (which often yellows with age) is attached to the top of the cap using small snaps on the back of each wingtip and the bottom of the swastika. The cap is an EREL-*Sonderklasse* (Robert Lubstein), offered through the Offizier Kleiderkasse. *Courtesy of Bill Shea.*

Private purchase summer cap for Other Ranks. Waffenfarbe piping is red for Luftwaffe Flak Artillery. The national emblem and oak leaf wreath are the usual Leichtmetall. Standard leather chinstrap with Luftwaffe rounded buckles. This cap came in its original bag, and storage box. The bag sports the same maker name as appears on the cap lining. *Courtesy of Bill Shea.*

VISOR

The visor is usually leather, with a green reverse. It should be noted that uncolored (black) visor reverses sometimes appear on summer caps and are not necessarily incorrect.

LINING

Permanently attached, usually cotton for Other Ranks caps with very fine grade cotton rayon, or silk for officers. Once the war began the military ceased issuing the white top cap to enlisted or NCO personnel, though for a time they remained available for private purchase. Many of the existing examples therefore, are private purchase caps and have a sweatshield, often with a makers mark as well.

The lining color was usually white, though some makers including Robert Lubstein used a pale gold (champagne), or a very pale creme. The sweatshield was attached in the usual fashion, with the makers mark, if present, appearing either on the celluloid, or on the lining beneath it. The EREL (Robert Lubstein) logo mark on the white top cap appears in gold color.

PIPING

The summer visor cap was piped around the top and bottom of the cap band in aluminum mesh cord for officers, in Waffenfarbe cord for Other Ranks. The white cover wa always piped in the same white as the base material, regardless of rank. *A cap with colored piping on the white cover is incorrect.*

MODIFICATIONS/REPRODUCTIONS

The most likely enemy on the white top cap front will be reproduction caps. There are a number of reasons for this. Original white top visor caps do not lend themselves well to convincing modifications, and since even the NCO cap commands a respectable price in today's militaria market, there is not much point in modifying one.

It would certainly be simple enough to replace the enlisted metal cap eagle with an officer's original bullion eagle (the only obstacle here being finding one with the correct white wool backing). The bullion eagle would have to include the necessary snaps on the back. A bullion eagle with a *fliegerblau* wool backing on the other hand, is highly unusual and suspect. The main problem with an officer upgrade modification lies in the lack of aluminum officer cap band piping.

Unlike modified pieces, a reproduction officer's white top summer visor cap can be made to order. A few firms can even provide caps made to the same design and nearly the same materials as those used in originals (up to and including original visors). A cap that appears too new always deserves careful inspection and a bit of skepticism – after all, examples like the boxed and bagged NCO cap shown earlier are extremely rare! Makers marks are easily forged, and do not in themselves provide any guarantee of authenticity. The benefit of makers marks is that they present the reproduction maker (or forger) with one more item that has to be correctly copied – and thus another opportunity to make an error that can be spotted. Make certain that the evidence of aging and/or wear shown by the cap is equal over all of its components.

Authentic Luftwaffe visor caps are slowly but steadily moving into the high-end price range for headgear collectibles. They are consistently in demand, and add some spice to almost any collection – or serve as a dashing starter piece that's hard to beat. Make sure that when the right one comes along, you are ready, and able to spot it.

Interior view. Gold colored rayon lining, with a rhomboid sweatshield. The maker mark says "Methmanns Nordmark Mütze, bürgt für Qualität [stands for quality], Friedrich Methmann Flensburg." The company's brand name seems to have been *Nordmark Mütze* [Northmark Cap]. *Courtesy of Bill Shea.*

The EREL-*Sonderklasse* logo from the officer's summer visor cap seen earlier. Note the EREL diamond filled in white, as is the interior of the rectangle that identifies the cap grade as Extra under the EREL-*Sonderklasse* grading system. *Courtesy of Bill Shea.*

The complete set of cap, bag and box. The cap owner, whose name and unit appear on a tag inside the cap, was only a *Gefreiter* [Lance Corporal] and most likely received the cap as a present from his family. A find such as this is extremely rare. *Courtesy of Bill Shea.*

Chapter 6

Die Kriegsmarine Schirmmütze (Model 1938) - Navy Visor Cap

The *Kriegsmarine* (Navy) cap design was quite distinctive from that of any sister service. Navy caps were bigger, and had a look all their own when the cap spring was removed and the cover outline broken in the style popular among German military personnel. As with all German military visor caps regardless of branch of service, Kriegsmarine caps were required to have a galvanized cap spring sewn into the cap cover and positioned behind the cover piping. Without the spring in place, the excess material on these caps gave them something of a billowing appearance – even on the more oval-shaped caps which became available from 1938 onward. The preferred look among sailors was for a cap crown bent slightly aft (rearward), with the rest of the cap cover tapering generally downward and toward the rear, like the back slope of a hill.

CONSTRUCTION

The Kriegsmarine M1938 model visor cap differed slightly from its 1935 model predecessor in being more saddle-shaped, and a bit less bulky, though the construction was otherwise the same. The 1935 model blue top KM (Kriegsmarine) visor caps were wider on the sides.

The model 1938 cap in contrast, had slightly narrower side panels and a higher front crown. Despite the move to a more oval top, the cap *still* looked bulky in comparison with visor caps from other services.

KM visor caps were available in two basic models:

1. Blue cap (non-removable top/cover) [die Blaue Schirmmütze]
2. Cap with removable white and blue top [Schirmmütze mit abnehmbarem blauen und weißen Bezug]

Blue Top (non-removable)

The blue-topped visor cap (Model 1935) was constructed in the same fashion as the visor caps of other services, however the width of the side panel measurements was 6.5 cm at the side seams and 5.5 cm at the rear. In comparison, the Luftwaffe cap was only 4.75 cm wide at both the side and rear seams. For officers, the material used for the blue cover was usually doeskin, though a lighter weight, navy blue moleskin (a cotton

Picture perfect blue top senior officer's visor cap (private purchase), in navy blue doeskin. Gold bullion national emblem and wreath, bullion oak leaves, and bullion cockade. This cap has a rare Edelweiss esperit badge mounted on the left side cap band (against regulations). Below: Front view. *Courtesy of Bill Shea.*

cloth) was also available. For NCOs, the navy blue cloth was normally a medium weight, finished wool material known as Feldwebeltuch.

The cap cover was piped around the crown in the same navy blue color as the base cloth itself. The correct cap band for NCO or officer Kriegsmarine visors was ribbed, black mohair – the same as that found on early Luftwaffe caps, and again like the Luftwaffe model, the closing seam for the cap band was located dead center at the front of the cap. The seam was hidden from view behind the Kriegsmarine version of the oak leaf wreath, the only version used by any German armed service that extended above the top of the cap band.

Raised, rounded ridges around the bottom and top of the cap band served to keep the mohair band from slipping out of its proper position.

As with all German military visor caps, the cap front had a vertical crown support sewn into the cap behind the front seam. This padded, burlap-wrapped wire helped support the crown in an upright position. Cap wearers commonly bent this out of its original shape in order to create the popular veteran appearance.

The front area of each of the side panels was heavily padded with a thick layer of wadding and a double layer of thick cheesecloth on Kriegsmarine NCO and officer caps. Although later caps were, in fact, more saddle-shaped than the 1935 version, the side panel measurements still remained slightly wider than those of their Army or Waffen-SS counterparts.

The blue cap. The junior officer (left) has his cap spring correctly in place. The officer on the right wears a perfect example of a nicely broken cap. He is probably Rheinhard Suhren, commander of U-564 and he wears the cap badge for that boat. Both caps are in either navy blue doeskin, or navy blue cotton moleskin. *Courtesy of Dale Paul.*

White Top (Removable)

The body of the white top summer cap was manufactured in the same manner as the non-removable blue top version, but without any cover (top) attached. The top of the cap band was made with a strip of stiff material designed to fit against the bottom inner edge of the white cap cover, creating a narrow shelf. The raw, bottom edge of the cover was reinforced with a narrow strip of material shaped in a *U*. This rimmed bottom edge could now be inserted into the cap band top where it rested on the shelf. Though attachment methods varied depending on the manufacturer, the usual method was a wide tongue of material affixed to the bottom front edge of the cover. This tongue was fitted with a female snap, the male end being affixed to the cap body, behind the top edge of the cap band. Though the cover shown in the following picture is the removable blue top version, the tongue/snap system was also used on the white cover.

Though Kriegsmarine visors of (any kind) are always in demand, the white top officer's summer visor is far and away one of the most desirable models. This is prima-

Blue top visor cap (regular Navy officer) in navy blue doeskin. Gold bullion visor embroidery, wreath and eagle with celleon highlights (the bullion is tarnished, thus darker). Very nice specimen, evenly aged. *Courtesy of Gerard Stezelberger of Relic Hunter.*

rily due to the use of this cap as a symbol of the U-Boat commander, where it served as a means of quick recognition for crewmembers in a vessel running in red light (night/combat lights).

U-Boat commanders were permitted to wear the white cap all year round, however, during the summer months the white top cap did not really offer any major distinction since many other officers not associated with submarines also wore the cap during the summer season. Unless a given cap is named to an individual traceable to a U-Boat command, a white top visor cap is not a U-Boat cap *simply by virtue of its being white* – it could just as well have belonged to a desk jockey.

Both the thick, reinforced bottom edge of the cap cover and the tongue with its snap, are visible in this picture. The cap top (white or blue) was unlined. *Courtesy of a Private Collection.*

In this case a real U-Boat captain: Knights Cross winner and CO of U-159, Kapitän-Leutnant *Helmut Witte* wearing his white top summer cap. *Author's collection.*

Senior administrative officer's white top visor cap (made by EREL, and purchased from the Navy officer's clothing store, the O.K.K.). White moleskin cover; silver bullion wreath, aluminum eagle. The chincords are the usual, Army-type twisted aluminum cord (thus, they remain untarnished). *Courtesy of Gerard Stezelberger of Relic Hunter.*

White top administrative officer's visor cap, possibly from government stocks as it has an unmarked lining. The chincords on this example are twisted silver cord (thus tarnished). *Courtesy of Gerard Stezelberger of Relic Hunter.*

VISORS

1935 model caps had Vulkanfiber visors with the forward brim edge wrapped in a shiny leather piqué reinforcement, attached with either a single row or double row of stitching depending on the maker. The Vulkanfiber visor was later replaced with natural leather, though the leading edge brim reinforcement remained unchanged.

Navy officer caps from 1936 onward had a fine black felt covering on the visor obverse (top) surface underlying either gold bullion scalloping or bullion oak leaves. It is interesting to note that post-Second World War American military caps for Army field grade officers (Germans would say staff officer) also share this feature, as do modern Bundeswehr Army officer caps; the practice is no longer confined to the Navy.

A single row of gold wire scalloping along the inner line of the brim's leading edge denoted junior officer caps from Leutnant zur See (*Ensign*) through the rank of Kapitän-Leutnant (*Lieutenant*) – commonly referred to as Herr Kaleun, by German sailors. Senior grades below Admiral had a single row of oak leaves in gold bullion or celleon wire, while Admirals sported two rows of oak leaves. Visor caps for Other Ranks (in the Navy, this was limited to NCOs) had a plain, glossy (lacquered) visor obverse.

Senior officer grades wore one row of oak leaves, here embroidered in celleon thread. Admirals had two rows. The visor surface is covered with fine, dark navy blue felt. Note the single line of stitching used to secure the leather edge strip (piqué).

Visor Reverse (Underside) Color

The reverse of the Kriegsmarine cap visor was left in black, without the crosshatch pattern common on Vulkanfiber visors. Any texturing was in the form of a lightly pebbled surface. Colored visor undersides on regular Navy caps are rare, and should be very carefully proofed before buying. Beware, if you are offered a KM white top visor with a *green* visor underside – it may be a Luftwaffe summer white-top visor modified to appear Kriegsmarine, since the Navy cap has a slightly higher value.

Officially at least, Luftwaffe regulations specifically forbid blue top visors to have removable tops (Bezug – top/cover). Removable tops were *only* permitted on the white summer visor. The bottom edge of the Luftwaffe summer visor cap, however, was always the standard Luftwaffe fliegerblau base cloth, which of course, would not appear on a Kriegsmarine cap.

SWEATBAND

Most KM visor cap sweatbands are made from real leather. Substitute (Ersatz) leather bands appear far less often on KM visor caps than they do on those of the sister services. One possible explanation is that the Navy, the cap makers – or both – may have determined that natural leather withstood the harsh conditions of sea duty better than composition bands, although there is no official documentation to confirm this hypothesis. The most commonly encountered KM sweatband colors are:

1. Reddish brown
2. Tan
3. Dark brown
4. Gray (least frequent)

In general, no attempt was made to match the sweatband with the lining color (common practice on Army and Waffen-SS caps), nor with the visor reverse. A number of manufacturers added the narrow interlacing strip near the top of the sweatband as explained earlier in Chapter 2. While fairly rare on Army, Luftwaffe and Waffen-SS visor caps, these strips are often found on the sweatbands of privately purchased KM

Close up of the interlaced sweatband strip on a Kriegsmarine private purchase visor cap.

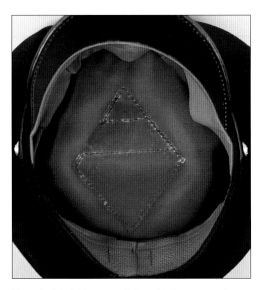

Unmarked dark blue rayon lining, with the commonly seen diamond-shaped sweatshield. *Courtesy of Gerard Stezelberger of Relic Hunter.*

Knights Cross (awarded August 21, 1941) winner Konteradmiral Kurt Weyher, Commander of the *Hilfskreuzer* [raider] Orion wearing a post-1936 white-topped visor. Bullion wreath and metal national emblem (eagle). The cap spring is in place, if crooked. The eagle is positioned so low on the crown that the swastika overhangs the top of the wreath. Admiral's double-row oak leaf visor embroidery. *Courtesy of Dale Paul.*

Right: Officer's hand-embroidered insignia with (tarnished) bullion cockade, on a black felt backing. Embroidered insignia do show considerable variation between makers, but though very well done, this one is a reproduction. It is nearly perfect, except that original wreaths *do not* show complete acorn stems running in a line across the base of the vertical leaves.

caps. Such strips are also a very common option on postwar Bundeswehr officer visor caps (all branches).

LINING

Navy blue top caps, whether from government stock, or private purchase examples made in doeskin or cotton moleskin, were commonly lined in dark blue rayon. There was considerable variation in the shade of blue, from a very dark navy to purplish blue, to a lighter cornflower blue. All are correct. Other colors appear less frequently and include gray and gold.

White top visor caps are often lined in gold colored, crème, and more rarely, white rayon.

Makers marks do not generally appear on Kriegsmarine caps, though why this is so remains a mystery. Navy visor caps may have a sweatshield, or not. Diamond shaped sweatshields (which are indicative to some degree of smaller makers) are common, and are usually unmarked. One maker often seen, however, was Steinmetz & Hehl, a company based in Hamburg. Caps made by this firm have the company name in unusually large print in white ink, with a diamond shaped sweatshield. Another common maker is EREL (Robert Lubstein), whose logo – when it appears – is usually accompanied by the O.K.K. etiquette beneath the EREL-Sonderklasse mark.

INSIGNIA

Regulation Kriegsmarine NCO visor cap insignia was the national emblem (eagle) and wreath in gold colored Leichtmetall (aluminum) with an aluminum cockade.

For officer caps, the insignia was gold bullion wire or gold celleon, with the national emblem in either bullion (or celleon) or gilt aluminum. It is interesting to note that among naval officers (admirals in particular), celleon insignia never completely replaced bullion – despite the fact that celleon did not tarnish.

Army officer caps with the optional hand embroidered bullion wreath always had a bit of wadding inserted beneath the wreath's backing cloth, which served to give the wreath a slight vaulted curve with the highest point being the cockade at the center. Navy bullion wreaths, including the silver bullion wreath worn by Navy administrative officials with officer rank, were similarly vaulted.

CHIN STRAP/CORDS

The Kriegsmarine was the only active military service whose regular officers did not wear twisted aluminum chin cords. KM officer caps used instead a leather chin strap with either a leather slide loop or a black metal buckle at the right temple, and a snap (on the end of the strap loop) at the left temple. Contemporary Bundesmarine (Federal Navy) officer visor caps *do not have this type of chinstrap.*

The naval administrative officer's cap with silver bullion insignia used the Army-style twisted aluminum chin cords (not the regular Navy leather chin straps). Another Navy idiosyncrasy is evident here, since administrative officers preferred cords made of real silver wire rather than the aluminum version used by the other services. An ideal administrative officer's cap, therefore, is correct with very dark (tarnished) twisted silver, chin cords – though the aluminum variety was also used.

BUNDESMARINE (BM) – FEDERAL NAVY VISOR CAPS

Current German Bundesmarine visor cap regulations specify only one color for visor cap tops: *white*. There is no longer any blue version for unscrupulous sellers to modify. Bundesmarine caps are becoming increasingly difficult to find. Though superficially similar to the pre-1945 model, they differ markedly in quality and materials from their Wehrmacht predecessors, and should not be that difficult to recognize even if modified.

In Germany, collecting Second World War vintage headgear is (for obvious reasons) difficult at best due to strict federal laws against any display of the *Hakenkreuz* (swastika). In the past, this even reached into the hobby of plastic modeling where imported Japanese-made German World War II aircraft packages were unsealed and opened, and the swastikas (for the aircraft tail) cut from the decal sheet. Any swastikas on the box cover painting were also blackened out with magic marker.

Regulations outlined in paragraph 86 of the federal penal code (*Strafgesetzbuch*) governing the purchase and sale of Third Reich militaria items in Germany are very strict. Many German collectors therefore, follow the path of least resistance and instead collect international headgear from a variety of historical periods and nations (British, U.S., Soviet, and East German, for example). They also collect Bundeswehr items. In the United States, Bundeswehr Heer (army) and Luftwaffe (air force) caps are quite easy to come by and are still very inexpensive, since interest in these caps among American collectors (or British, Australian, New Zealand and Japanese collectors – yes, Japanese) is relatively low. This will most likely remain the case for several years to come since in these countries, it is only high prices that serves to dampen interest in German World War II historical headgear – *not* official government policies. Bundeswehr Heer and Luftwaffe officer caps in the U.S. militaria market fall within an average pricing range of from $15 to $25 in 2001 dollars depending on the seller, with Other Ranks caps running between $8 and $12. These low price levels do not hold true for German Bundesmarine caps, however.

A rare, color photograph of Gross Admiral Karl Dönitz (autographed), who was appointed acting Führer and Reichskanzler by Hitler at his death. Dönitz wears his doe-skin-topped blue cap with cap spring in place, in keeping with official regulations. *Courtesy of Dale Paul.*

Bundeswehr naval officer's cap, old form. The cover on the older form (1955-1990) is oval, the newer design more rounded. The white ribbed (twill weave) cloth is a cotton/*synthetic* blend, with the ribs quite pronounced – much more so than on original wartime caps. The metal federal cockade is attached either with prongs or by a screw post on the back (variations exist). *Courtesy of Kay Stephan.*

The Bundeswehr was never a very large military force to begin with, and the 1990s saw further reductions in its manpower levels. With the end of the Cold War and the dissolution of the Soviet Union and its allies, the federal armed forces were downsized again in 1994 to a total personnel strength of 370,000 officers and enlisted personnel for all services *combined*. German defense ministry figures clearly identify the Bundesmarine (BM) as the smallest of its three armed services, with a personnel strength of only 27,200. The small size of the pre-1945 Kriegsmarine (relative to the Army, SS and Luftwaffe), and much more so its modern counterpart, means that German naval caps of *any* period are scarce.

Few, if any, Internet militaria sites in the U.S. offer modern Bundesmarine caps, in fact. Even in Germany they are difficult to locate. In terms of U.S. prices, these caps are quite expensive in comparison with Bundeswehr Army or Luftwaffe caps. Prices range from about $50 for a used officer's cap (when available), to $120 for a mint, unissued piece. According to a number of German militaria dealers, the popularity (and resulting scarcity) of Bundesmarine caps among collectors is due to two main factors:

1. The small size of the service.
2. The intent of many buyers to modify these caps into wartime Kriegsmarine look-alikes.

Bundesmarine New Form (junior officer) cap with the more circular top. The interior front has a thick, padded white front support and the band-like (flat), adjustable shaping ring – none of which is ever found on authentic wartime caps. The cockade is bullion (on a metal disk), sewn to the top of the black wreath backing (i.e. positioned lower than on the Old Form cap). The cockade has a screw post, which passes first through the cover, then through the interior padded support and is secured inside the cap with a disk-shaped nut.

Of most concern here, is the second reason. In fact, of all contemporary models of German military cap, the only one that really poses a significant threat to inexperienced collectors – as a modified piece – is the Bundesmarine cap.

Identification

The following are key points that will help identify modified BM cap examples:

a. **Brim wrapping**: Brim wrapping on BM Other Ranks caps is *wider* than on World War II originals. The wrapping on original caps looks neater (more carefully done). The best way to learn the difference between the two is by actually viewing original Kriegsmarine caps and *un-modified* BM caps.

b. **Bullion Embroidery**: The thread used for the embroidered oak leaves on BM staff officer (Korvettenkapitän through Kapitän) visors is of a slightly thicker diameter than that on pre-1945 caps. The individual threads are more visible.

c. **Cover/top Material**: The white cover cloth has a high percentage of synthetic yarn. The texture is also incorrect – Bunsdesmarine visor cap top material often has very pronounced thick ridges, which makes the texture quite different from original Kriegsmarine white top caps (though newer covers have finer ridges). The cover may also have holes in it from the cockade prongs (old form) or a hole fitted with a grommet (new form used with a screw post cockade).

d. **Front Support**: At the front of the cap is a vertical, stiff, padded support panel covered in white cloth which holds the shape of the cap crown and cap front. On new form caps, there is also a cloth loop attached to the top rear of this support pad. A white painted, wide, flat band (with a chromed adjustment clip) fits through this loop. It is this band that holds the lightweight cover in the correct circular form. Such structures do not exist within wartime Kriegsmarine caps.

e. **Chin Strap**: If left unchanged, the BM chinstrap is an easy giveaway since it does not have the left side snap always found on wartime original chin straps; in fact, there is no snap at all. The strap is also much too thin, made as it is from vinyl rather than the correct (and more expensive) patent leather used in wartime caps.

(EAST GERMAN) NAVY CAPS

The only similarity that East German navy caps share with pre-1945 Kriegsmarine caps is the color, and (on officer caps), the visor embroidery. In other respects these caps differ markedly from Second World War caps. The East Germans did, however, use both blue and white top caps, and either type can be modified.

CONSTRUCTION

At first glance the Soviet-influence is immediately noticeable in the cap top, with cover side panels that angle sharply upward to the crown piping, and a flat, sloping, plate-shaped top. This cap cover form is very typical of late 1940-1950s-pattern Soviet military cloth headgear designs. The cover is not sewn directly to the cap band, but rather to a narrow strip of material, the bottom edge of which is attached to the cap band. The presence of this strip makes the cap cover *look* like it might be removable, but such is not the case.

The exterior cap band is black, but it is not made in the mohair of Kriegsmarine visors. The horizontal ribs on the thin East German cap band are so narrow and have so little height that they are all but invisible from any distance.

Blue top visor cap for a junior grade East German naval officer. Cloth covered visor with vinyl edge wrapping, and gold wire scallop. Black cap band with gilt metal wreath and metal/enamel cockade. The chinstrap is leather, but matte (not shiny) and does not have a working snap. The cap top is round and flat, with sharply angled sides – clear hallmarks of Soviet-influenced design.

Interior color scheme, materials and sweatshield are all classic signs of East German manufacture. Remember:"Black and Blue."

These two photos show an extremely rare, complete and original visor cap carrying box. The box color provides an excellent impression of metal – which it is not. *Private Collection*

East German blue caps were manufactured of a very dark (almost black) Navy blue, ribbed polyester blend material. They diverge from wartime Kriegsmarine originals here: Kriegsmarine caps did not use ribbed fabrics for the blue top, and absolutely no synthetic fibers. The cap interior differs completely from wartime caps as well, and is very easy to identify. Simply remember the phrase, "Black and Blue." Black visor reverse, black vinyl sweatband, blue lining material. There are many other construction differences, but these are enough to identify the cap as East German. The vinyl sweatband alone places the cap in a post-war time frame. If any further proof is needed, the translucent sweat diamond (permanently affixed to the lining through heat bonding) with its embossed NVA (*National Volks Armee* – National People's Army) should remove any doubts!

By using reproduction or original insignia and removing the cap spring to relax the Russian shape, East German Navy headgear can certainly be modified into a reasonable facsimile of a Kriegsmarine blue visor cap, if one overlooks the error in the cap band, for example. Though such a cap may be suitable for use by reenactors, there is no way to hide its true origin from a collector with basic knowledge.

PACKAGING

Most World War II vintage German caps survived as war souvenirs taken either during the fighting, or during the postwar occupation of Germany. Over the many years since the end of the war caps have continued to turn up in attics and cellars in the United States and Europe, or as turnover from existing collections. The main thing all of these caps have in common is that they are without any packaging. When originally purchased, a cap would certainly come in a bag and likely also in a box – perhaps with the store, or manufacturer's name emblazoned on it. In most cases, the original owner had no use for these after removing the cap, and threw them away. Thus, very few examples of original bags or boxes have survived.

Three U-Boot commanders take a short pause. All three are wearing the *blaue Mütze*. Standing is U-201 commander Kapitän-Leutnant Schnee [snow], whose crew wore a snowman badge on the side of their garrison caps. The man in the middle is Fregattenkapitän Rheinhard Suhren, Commander of U-564. *Courtesy of Dale Paul.*

Fine example of a blaue Mütze (blue top cap) for a Kriegsmarine NCO (Petty Officer). The visor is bare of any decoration (scallop or oak leaves). Note the aluminum cockade. *Courtesy of Bill Shea.*

Junior officer's blaue Mütze. *Courtesy of Bill Shea.*

Gilded anchor button used on naval officer and NCO caps.
Courtesy of Bill Shea.

The extravagantly high cost of permanent storage/transport cases (even during peacetime) undoubtedly discouraged many German soldiers from ever purchasing these and, as a result, few of them exist. Occasionally, however, good fortune smiles and brings one of these rare items to light. The pictures on page 110 show a purchaser delivery box for a Kriegsmarine cap, which was meant to serve also as a convenient carrying case. The box is made of cardboard cleverly designed and colored to look like a steel box such as might be found aboard ship. An amazing and extremely rare item!

Chapter 7

Die Offizier Feldmütze alter Art - Officer's Old-Style Field Cap,
Army/Waffen-SS

Among German Army officers the Officer's Old-Style Field Cap was perhaps the most popular of all forms of cloth headgear. Despite the introduction of the highly practical, easily foldable *new type field cap for officers* (M38 garrison cap*), the *old-style field cap* remained a strong favorite (*in English, the terms garrison cap, overseas cap and side cap are interchangeable). In fact, many of the officers fortunate enough to possess one of these caps, such as Kurt "Panzer" Meyer, second commander of the 12th SS Panzer Division, often wore them right up until the end of the war in complete disregard of the official wear out date of April 1,1942. The popularity of this cap was so great that some officers went so far as to modify their service dress visor caps to imitate it. They did this by removing the chincords and buttons (the shaping wire in the crown was probably already long gone), and *breaking* the front of the cap. The stiff Vulkanfiber visor, however, could not be altered.

The amazing popularity of this type of headgear lay perhaps in the fact that it took the broken, front line veteran look that so appealed to German soldiers, far better than any service visor. The machine-woven regulation insignia for this model was also much more distinctive from a distance (though this was perhaps a liability in a combat environment!) compared to the service visor's metal or bullion insignia, and this certainly enhanced its visual appeal all the more. In fact, the old-style field cap seemed almost *designed* to take a perfect broken look no matter what was done to it. A modified *Schirmmütze* (service visor cap) with no cap spring or chin cords on the other hand, was probably no more than a marginal imitation for a real old-style field cap – a testament to the esteem in which German Army officers held this type of headgear.

In the militaria market, a *Schirmmütze* service visor (cap) modified to look like an old-style field cap is a very risky item to purchase – there is simply no way to determine exactly who removed the cords and buttons, or when this was done.

Commonly called a *crusher* or *crush cap* by collectors and dealers alike, these caps are much sought after. The Luftwaffe and Kriegsmarine in general had no comparable official version of this cap, though very limited photographic evidence suggests that a few Luftwaffe officers may have special-ordered an otherwise standard Luftwaffe cap made with a flexible visor. Unlike the Army model, these Luftwaffe caps were still worn with the aluminum officer chin cords in place.

Fine example of an officer's Old-Style Field Cap manufactured by August Schellenberg. Doeskin top with a shorn nap, and thick-style crown piping. Waffenfarbe: *Gendarmerie* (military police). Correct machine-woven aluminum thread insignia. The visor surface is either lacquered, or painted with some type of asphaltum-like surface coating.

A nicely broken Army Artillery *crusher*, as the Old-Style Field Cap is often called by collectors. The visor surface is heavily scored, but there are no major ridges. Correct regulation machine-woven insignia in aluminum thread. Very fine grade doeskin cover (top) with stiffening wire (cap spring) removed.

This cheerful, portly fellow is a Wehrmachtsbeamter im Offiziersrang – an administrative official with officer's rank – evidenced by the officer-style Litzen (collar tabs). Note that the Litzen are not the normal Army officer type but rather the piped border version worn by administrative personnel. This man's Old-Style Field Cap would be piped in forest green with machine woven, aluminum thread insignia. *Courtesy of Eric Mueller.*

Machine-woven national emblem in aluminum thread, applied with a horizontal stitch across the top above the wings while the insignia is positioned in reverse and upside down. The horizontal securing stitch line is covered when the eagle is folded down over it. Note that the oak leave wreath insignia reaches to the top edge of the cap band.

While the majority of old-style field caps still in existence are Army caps, Waffen-SS officers did use them – though to a far lesser degree. This makes for an intriguing contrast between these two branches of service: when studying any given group of period photographs, Waffen-SS officers appear far less frequently wearing this type of headgear than do Army officers. The reason for this is not entirely clear; the *Offizier Kleiderkasse* (officer's clothing bank, see Chapter 15) after all, supplied caps to both Army and Waffen-SS officers and so there should not have been any difference in the degree of cap *availability* to officers of either service. That far fewer SS officers seem to have chosen to purchase an old-style field cap (when these were available) as compared to Army officers, however, raises an interesting question: *Did Waffen-SS officers feel that their own service visor was sharper-looking than a nicely broken old-style field cap?*

INSIGNIA

Regulation Army insignia for this headgear was machine-woven (flat) in aluminum thread for officers and for Army administrative officials holding officer rank. The insignia underlay (base cloth) was a dark bluish-green – though it often tended more toward dark green (i.e. somewhat less blue). For administrative officials without officer rank, the thread was white cotton.

Ideally, the top edge of the national emblem (eagle) was placed flat upside down and face down near the crown of the cap. It was then sewn with a horizontal machine stitch straight across the top of the insignia, just below the upper edge. Once this horizontal stitch was completed, the insignia was next flipped down over the new stitch line. It was now in the correct position, and right side up. The rest of the raw edges

General's old-style field cap. A fine example in high quality Trikot cloth. Note the aluminum thread insignia wit gold general's piping. Manufacturer: *Leparo* (Leonhard Paulig) of Rothenburg-Oder. *Courtesy of Ruben Lopez.*

were tucked under the insignia, and the stitching was completed either by machine or professional hand.

The oak leave wreath and cockade insignia edges always reached the top and bottom of the cap band – that is, it took up the entire vertical width of the band. In the photograph at right, the insignia is not the same width as the cap band.

Generals were not affected by the wear out date set for the old-style field cap. Rank doth have its privileges in some things, and this was one of them – the continued use of this form of headgear by general rank officers was authorized past the wear out date. Thus, the gilt insignia regulation by which all aluminum insignia on general's caps and uniforms was to be replaced with gilt versions, effective January 1, 1943, should have applied to the old-style field cap as well. It does not appear that this change was actually carried out.

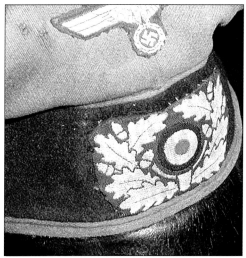

The oak leave wreath insignia has been reapplied. When this was done is impossible to determine. The insignia should fill the cap band vertically (from bottom piping to top piping), but does not. In fact, the stitching is actually visible along the top right edge (as viewed), which is not good. The stitches are uneven, the thread white: all incorrect.

Knight's Cross with Oak Leaves wearer Generaloberst Karl Hollidt poses with his general's grade, old-style field cap (shaping wire still in place) in this autographed photo. Hollidt was awarded his Ritterkreuz (Knights Cross) on September 8, 1941, as Commander of the 50th Infantry Division, and the Oakleaves on August 17, 1943 while Commander of the reconstituted 6th Army (which was later destroyed). *Author's collection.*

Ritterkreuz holder Oberst (colonel) Hitzfeld sports a fine example of an officer's old-style field cap made of doeskin or Eskimo wool. Notice the height of the oak leaf wreath, which fills the cap band. The visor has no apparent surface coating applied. For reasons unknown, the Colonel's cap piping and the Litzen (collar tab) piping do not match one another. Perhaps the cap was a studio prop kept on hand for emergencies?

FLEXIBLE VISOR

The single most noteworthy physical characteristic of the old-style field cap is its flexible visor. This being the case, a hard visor is not usual on such a cap – except in the case of a modified service visor cap which can be easily recognized as such.

Old-style visor cap visors were made from a very thin, very supple (flexible) black leather. The visor reverse is often fluted with parallel ribs and grooves running from the brim to the rear edge, though the reason given for such fluting is not known – perhaps it was simply a means of increasing the visor's flexibility?

Another type of visor material made from compressed paper or cardboard – much thinner than the leather visor – may also have been used, but these do not appear very often. There certainly seems little logic in the use of such a material on a cap that was intended primarily for the rough conditions of regular field use. The wear out date of early 1942 for this model of cap was also ahead of any major shortages in critical material stocks that might later have made such a material truly necessary as a substitute. Nor for that matter is it clear whether or not any cap makers were willing to produce an old-style field cap to order after the official wear out date had passed – i.e. during a time when a substitute visor material might actually have been necessary.

VISOR SURFACE – Coated, or not?

Crush cap visors often have a coating painted on the surface of the visor (and occasionally on the reverse side as well). The purpose of the coating – if in fact, it is a coating rather than simply the lacquered surface of the visor itself – was probably to prevent glare, which was always a very important consideration in a combat environment where an inadvertent flash could easily give away one's position. Some visors, particularly those on general's and Waffen-SS officer crush caps, have only a very thin lacquer coating which tends to develop a fine network of superficial cracks as the leather ages.

This Army signal officer's crusher is a perfect example of an ideal exterior appearance for one of these caps. Correct insignia, a body of fine grade Trikot with a medium gray colored, smooth-textured lining (never had a sweatshield). The lacquered visor surface – or the coating on the visor surface – has been deeply and repeatedly scored and scratched. *Author's early collection.*

If, in fact, the visor surface was coated with something thicker than just lacquer, then it may have been a material that looks quite similar to the soft black asphaltum/tar puddles that children find on the side of the street and play with on a summer day, or that is used to seal tar paper joints on roofs.

This material is quite malleable in hot weather, but not outright sticky provided the surface skin has not been broken. When the air temperature is cool, on the other hand, the asphaltum becomes quite hard and can no longer be easily manipulated, if at all. The visor surface on many old-style field caps looks and feels a great deal like hardened asphaltum. The hardness, however, is perhaps simply the result of aging. The surface may have been much softer once, when new – soft enough to score quite easily; or, it may have become soft in warm temperatures. Whatever the cause, a great many of these coated visors clearly display numerous scars, scrapes and gouges which add a great deal of character to the cap.

ORIGINAL CAPS WITH POSTWAR VISORS

Beware of well-worn caps with a newer-looking visor. Obvious mismatches *are not* a positive sign! Authentic crush caps do occasionally turn up which at one time or another have lost their original visors, and have had a replacement installed. While the attachment work is often quite neat, the visor itself is either a very thin, early post-war Vulkanfiber visor for a police cap (i.e. not flexible enough), or a very lightweight, thin *plastic* post-war police cap visor (not flexible enough, and plastic as a visor material is not acceptable).

The problem with the cap pictured at right is that the original visor apparently came off and was lost at some point, whereupon a creative individual attached a replacement. While the job is very well done, the visor is not original to the cap. In fact, it is doubtful that it is even pre-1945 period.

Though the lighting and clarity of the original jpeg picture are not as good as could be hoped for, the flaws in the visor are still quite clear:

1. WIDTH: It is too wide, from the front edge of the bill, to the bottom edge of the cap;
2. BRIM RIDGE: There is a brim ridge (crush caps do not have ridges), with a sharply defined inner edge;
3. BASE RIDGE: There is a distinctive ridge at the base of the visor, just below the cap cloth edge. While such a ridge is common to Vulkanfiber visors, it is *not* found on the crush cap.

All of these visual clues *are common signs typical of postwar cap visors!* Old-style field cap visors, regardless of type, had none of these characteristics.

Could this then, be a wartime Vulkanfiber visor perhaps, attached to this cap body? The answer is NO. Brim ridges on authentic Vulkanfiber visors (for the service visor cap) were always gently curved (not flat). The inner edge of the ridge curved softly, and is never so sharply defined. The flat ridge at the visor's base provides some support for the chin cords, and in any case is – once again – far more pronounced and defined than on wartime Vulkanfiber visors. The visor on this cap was intended for either a police or military post-war service visor cap – not for use on a pre-1945 old-style field cap, or any Second World War cap, for that matter. It is most likely made of plastic, and plastic, of course, should immediately sound warning signals. *No type of authentic Second World War Wehrmacht military headgear ever used a plastic visor.*

This and the next two photos are examples of a crush cap offered on the Internet. The cap was presented as an authentic piece. While the cover (top) and insignia appear correct, the visor has been reapplied and is not original to the cap. Note the wide, flat brim ridge, and the very vertical ridge at the visor base (below the cap edge).

This angle provides an even better view of the very sharp vertical ridge at the base of the visor, which is not found on old-style field caps, but is typical of post-war examples. An original visor would also be shorter than this one.

The fact that this very thin visor is molded from plastic is perhaps clearer in this picture. The pronounced base ridge is once again quite obvious. Stitching for the visor – although very well done – does not match the rest of the sweatband, and it looks as if the band was purposely split at both ends of the new visor to accommodate the increased width of this replacement. The wear and age evident in the cap lining and sweatband are missing from the visor.

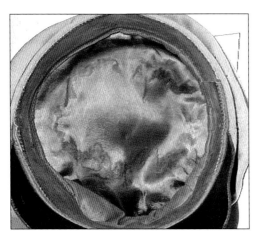

Moderate exterior wear, with heavy sweat staining to the lining of this cap. The overlap for the sweatband ends is at the side in this case, rather than at the rear. Sweatband toning and wear is nice and even. Note also the two-tone cap lining, with one color for the sides (usually orange on such examples) but a slightly brownish shade for the top.

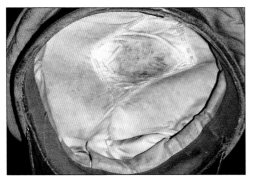

Crush cap interior. Waffengattung (branch): Artillerie. This cap is a perfect example of exactly the type of interior condition one expects (and to some degree, *wants*) to find on such a cap. *Private collection.*

Old-style field cap sweatband. *Stirndruckfrei* [free of forehead pressure] and the D.R.G.M. number, 1 383 756, are clearly embossed into the sweatband leather. The *Gebrauchsmuster* identified by this D.R.G.M. number is for a special cushion insert that is attached directly over the interior visor edge, from end to end (behind the sweatband). The legal owner of this Gebrauchsmuster is the same company that made the cap, *August Schellenberg* of Berlin. (See Appendix C)

INTERIORS

The very nature of the old-style field cap is a good indication of what to expect from the interior of one of these highly desirable pieces: *field wear.* These caps were worn constantly in rough, combat conditions, through hot, damp, or rainy weather. Most therefore, will show a significant amount of wear and tear especially to the interior, which was the weakest link in terms of the materials used in construction. These caps were no strangers to heavy sweat, dirt and dust and they often show the signs of this – and no less should be expected. Interiors in particular tend to turn up in very bad shape, but exteriors also commonly show hard wear. In fact, crush caps often appear quite worn out.

The key factor to consider when viewing this type of headgear – or any cap that shows wear, for that matter: *what sense of age does it give when taken as a whole?* There are certainly methods available for artificially aging any cap, whether it be moth damage, hair grease or oil stains to the sweatband, dirt stains, or a host of other possibilities. Any one of these alone can often be reproduced quite convincingly – but a real cap ages *evenly and equally* throughout its lifetime.

The difficulty for forgers lies in tying all of these individual factors together in a convincing whole. The Artillery crusher at left is a fine example of what a good interior should look like: the sweatshield is no longer present except for small, yellowed fragments; the lining top, sides and forehead area are stained and dirty from wear and age. The 'V'-type sweatband (sewing is hidden) shows very even wear and a dark toning (originally reddish-brown) and two tears. The tear on the left (right temple area) is directly above the visor's corner edge (which is a typical stress point). The cap's bottom edge is quite dirty and well-worn; the wool is also buffed to a glaze in places from rubbing (and perhaps hair oil). All of these many individual factors are extremely hard to fake as a convincing, coherent whole, on this cap however they come together in an excellent and authentic harmony.

SWEATBANDS

Sweatbands for the crush cap were nearly always real leather. The substitute *Ersatzleder* perhaps, could not hold up to the level of wear and abuse these field caps were expected to endure. The leather sweatbands appear in both the perforated and non-perforated versions, depending on the individual manufacturer. The color was usually tan, dark brown, or reddish brown (rarely anything lighter), which is generally well-darkened from sweat, dirt and age. The toning should be even around the band, though perhaps slightly darker at the forehead. Sweatbands were often padded at the front to relieve forehead pressure (Stirndruckfrei) in the same manner as usual with a service visor cap. This again, depended on the maker, however.

The ends of the band generally overlapped and joined at the rear, but overlaps along the sides are also quite common.

LINING COLORS

Old-style field cap linings are usually in gold or pale brown colored rayon, though sometimes a cap may be lined on the sides with a different color (usually orange) to give a unique two-tone effect. Charcoal gray rayon is also a common combination for certain makers such as **Leparo** (**Le**onhard **Pa**ulig of **Ro**thenberg-Oder). Unlike the service visor, however, officer old-style field cap linings did not come in as wide a variety of colors.

Charcoal gray lining for a general's crush cap. This color is typical for caps made by the respected company, *Leparo. Courtesy of Ruben Lopez.*

Army Panzer crusher. The visor on this cap offers a good example of what appears to have been a coating applied to the visor surface, perhaps to reduce glare reflections. This cap has a high quality Eskimo wool cover. The piping is faded in some areas – normal in a cap worn for a long time in harsh field conditions. The lower cap edge is slightly discolored from sweat bleeding through the sweatband, the lining and through the exterior cloth. The rippling of the cap band is from wear and weather damage, not from alterations – and matches well with the overall wear to the rest of the cap. Note the neatly applied bullion cap eagle – not the regulation insignia, but nonetheless original.

SWEATSHIELD

Privately purchased caps (ERELs, for example) often have sweatshields – though more often than not these either cracked and tore out in the field, or were otherwise removed by the owner for one reason or another; only the imprint remains. Many old-style field caps in fact, do not have any indications of having had a sweatshield at all.

WAFFEN-SS OLD-STYLE FIELD CAP

Waffen-SS crush caps are truly a rare breed. Few exist, and with availability so limited this type of cap is a very lucrative target for modified originals, reproductions passed off as authentic caps, or complete forgeries made from unrelated headgear.

Waffen-SS caps appeared in two basic styles:

a. Leather visor (officers)
b. Cloth covered visor (NCOs)

In general, photographic evidence suggests that the crush cap had far less appeal among Waffen-SS officers than it did among their Army cousins. The availability of these caps is sorely limited, and this is especially true of the cloth covered NCO version. The Waffen-SS old-style field cap made by EREL shown here is extremely rare, both by virtue of its being what it is, and by the fact that it has any maker mark at all. In

Waffen-SS crusher, Waffenfarbe: Artillerie. No surface coating on the visor. Average grade regulation Trikot wool and nicely matched Leichtmetall insignia with superb detail. Note the creative touch given to the eye holes of the Totenkopf! Well-padded, overhanging sides give a nice, thick appearance. The crown is still sprung quite straight up (no attempt was made to break it). This cap is an EREL-*Sonderklasse* sold through the Offizier Kleiderkasse. *Courtesy of Bill Shea.*

Waffen-SS crusher interior. SS crush caps are nearly impossible to find – let alone one with the EREL "Offizier Kleiderkasse" logo. Wear and aging are nicely even throughout. The sweatband is damaged, which is to be expected with this type of headgear. The EREL maker's mark is clearly legible and puts this cap near the pinnacle of headgear collectibles not named to a specific historical individual. Only external condition could impact the value of this piece (there is, in fact, some light moth damage). An extremely rare cap by any standard. *Courtesy of Bill Shea.*

this case, not only does it have a mark, but the mark is *EREL-Sonderklasse*. There is also the EREL Offizier Kleiderkasse logo as well, indicating that this cap was purchased by a Waffen-SS officer through the officer clothing sales system. Such caps rarely appear on the militaria market – or at least, not authentic examples. There are simply too few in existence, and those that do exist already occupy prized places in someone's headgear collection.

The Waffen-SS field cap with cloth covered visor is rarely ever seen, and there is some question as to exactly how long it was actually in use. The popularity of this cap seems likewise to have been very low.

Due to the rarity of both types of cap, some collectors have come to believe that the cloth visored cap was simply another version of the field cap for officers, but this is not correct. A question remains as to whether the Waffen-SS officer's old-style field cap was always piped in the wearer's Waffenfarbe (branch color), or whether it was originally piped in officer's universal white like the service visor cap, was then available in branch colors during the six month period that these were authorized. More research is needed on this question.

The wear out date set for these caps was the same as that established for the Army.

INSIGNIA

Several types of insignia were available for the Waffen-SS officer's old style field cap and the NCO version M1938 field cap:

1. Leichtmetall – the same insignia as worn on the service visor cap
2. Machine-woven (BEVo, 'flatwire') insignia, in aluminum thread on a thin black rayon backing cloth.
3. Other Ranks (enlisted) machine woven insignia (aluminum, or matte gray cotton thread), for the NCO cap.

The insignia could be worn with both the eagle and Totenkopf in metal, both pieces machine-woven, or in a mix of the two. The machine woven Totenkopf in this case was usually combined with a metal eagle; a metal Totenkopf with machine woven eagle combination rarely appears. Enlisted grade insignia was most likely used on officer caps only when the correct insignia was unavailable.

Modified Waffen-SS Originals

Modification of an original cap is a risky business for the person doing the work, as well as for the potential buyer. Any cap used for such modification will lose its original value regardless of whether or not the modification works. If successful, the cap may bring a considerable profit.

The most dangerous form of modification – being the simplest – is re-coloring the Waffen-SS officer's white (infantry, or universal) piping on an original cap. The goal: to create an even more valuable Panzer cap, Signals cap, etc. Dyeing or staining the piping in some way requires no physical modification of the rest of the cap. In keeping with the tried and true axiom of 'KISS' (Keep It Simple, Stupid!), the less changes made, the less chance there is of any tell-tale errors that may alert an otherwise unwary collector. A Waffen-SS Panzer branch crush cap will bring in an amazing amount of money over the already high price of the infantry crusher cap itself. The original value of the infantry cap of course, is completely destroyed by the alteration. Use the piping

Close up. The Trikot rib pattern is instantly recognizable; the width of the ribs indicates the wool yarn was an average grade. Many Waffen-SS officers preferred the Totenkopf insignia correctly centered, but set quite high on the cap band close to the piping (on service visor caps as well). The crisp detail on this Leichtmetall insignia is superb, with the well-rounded Totenkopf skull dome typical of originals. *Courtesy of Bill Shea.*

test method on any questionable example, to be sure the base color is the same as the outside. Look also for a darker shade on the cap band upper piping along the sides of the cap under the overhang – little light reached this area, and the piping there should be brighter.

The key point to bear in mind is that a considerable initial expense is required to purchase the original cap to be modified, and such an action is no casual activity. It takes a dedicated individual who truly enjoys cheating customers – and has no problem with damaging an original historical artifact (the infantry cap) in order to do so.

Altering Original Service Dress Visor Caps

The fact that insignia for Waffen-SS crush caps could be either the metal style found on service visors, or machine-woven cloth insignia, makes alterations of a service visor cap an easy thing for unscrupulous people. An altered Other Ranks visor cap will, of course, have a felt cap band rather than the correct black velvet. (Some individuals may attempt to produce more of a 'field' look by using the rougher quality wool common to enlisted caps.)

Removing the chin cords and buttons from a Waffen-SS service visor and claiming it to be an original owner-modified crush cap, well – there is a limit to how many of these could exist, and given the apparent low popularity of crush caps among Waffen-SS officers in general, more than a very rare few such modified examples are unlikely. The main weakness with any claim to being a modified service visor cap however, is that there is simply no way to prove who, in fact, modified it, and when this was done.

Waffen-SS NCO field cap with cloth covered visor. Cap band in black felt. Note the lack of a ridge at the front of the upper visor brim, such as appears on M43 caps. This is correct – these caps did not have a rim on the upper side of the brim. Note the indentation left by metal insignia (removed).

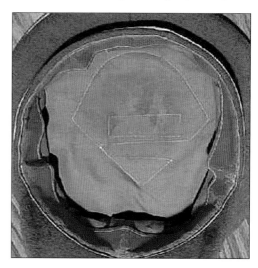

Interior, showing the gray cotton lining. The cap may represent a Government Issue piece, manufactured with a sweatshield but without a maker's mark; the sweatshield is gone, so no indication of authenticity can be gained from that source. The sweatband is the thin, composition type. Worth noting is the indication of uneven stitching where the sweatshield was once located – particularly in the area of the name slot as well as around the border of the square-shaped area (purpose unidentified).

Knights Cross winner Waffen-SS Hauptsturmführer Jacob Lobmeyer (1918-1989), who was awarded his Ritterkreuz while commanding SS-Jagdpanzer Detachment 563 in Russia. A young Lobmeyer wears his SS officer crush cap in this autographed, pre-war photo. The two-piece insignia is machine-woven aluminum thread on black rayon underlay cloth. He appears to have put a significant (and successful) effort into breaking the cap. *Author's Collection.*

Waffen-SS Sturmbannführer Vincenz Kaiser of the Der Führer regiment wearing an SS crush cap with aluminum (Leichtmetall) insignia. He wears the German Cross in Gold and four individual tank destruction badges, among other awards. *Author's Collection.*

CONCLUSION

The crush cap is a true example of the – now perhaps lost – German art of producing superb cap designs. In general, the potential danger to collectors seeking this type of headgear collectible is greatest from reproduction caps. A very high quality reproduction – in any branch color and size desired – can be obtained for as little as $160. A reproduction cap, if a good one, needs no modification or other alteration beyond artificial speed aging, and the possibility for a serious financial mistake by an inexperienced collector reaches a critical threshold here.

The Janke Tailoring firm in Germany, for instance, offers individually made reproduction caps of any model including crushers, that are nearly identical in both quality and appearance to an authentic original costing between $4000 to $7000 – yet are available for no more than a tiny fraction of that price.

These caps are first class copies of original designs, manufactured without any insignia of any kind. See Chapter 14 for more information on reproduction headgear, and pointers on how they differ from originals.

Chapter 8

Part 1: The Standard Field Cap M34 [Continental, Panzer, Tropical]
Part 2: The Officer New-Type Field Cap M38

Part 1:
DIE HEERES EINHEITSMÜTZE M34 –
ARMY STANDARD CAP M34 (Overseas cap)

During the first four years of the war, the M34 field cap was the most common item of German cloth military headgear in use among enlisted men and NCOs (i.e. Other Ranks). The M34 served, in fact, as the standard headgear for all troops on leave from the field. Visor caps rarely accompanied enlisted men into combat areas and front-line troops almost never appear wearing them in period photographs taken after 1940. Visor caps were limited primarily to soldiers assigned to a military garrison quartered in Germany or in one of the occupied territories. The only cap to equal the M34 in sheer numbers (and which may have topped it in popularity) was the M43 standard field cap, which was not introduced until mid-1943.

As a garrison cap (also correctly called an overseas cap or side cap in English), the M34 was easily carried and stowed anywhere on one's person, weighed next to nothing, and had a relatively sharp appearance. The apparent simplicity of design, however, makes this cap a popular target for modifiers.

Such individuals make it their business to take almost any known, contemporary or older model garrison cap if it looks similar in cut and color to an M34 and modify it to appear as original wartime German issue headgear. This includes black French Foreign Legion garrison caps converted to look like the black M34 for Panzer troops. Some of their efforts are downright laughable to anyone with any experience; others are quite a bit more convincing, particularly to novice collectors. With an original M34 valued at over $175 in the 2001 militaria market, selling modified contemporary caps to unknowing buyers is truly a bargain for unscrupulous dealers and "target-of-opportunity" sellers.

Even more astounding, however, is the number of people willing to pay for one of these fakes. *The Germans did not make sloppy quality goods,* even in the final year of the Reich's existence. People who buy obviously bogus caps may, in fact, be new to the field of headgear collecting and don't know what they are doing. Nonetheless, there is simply no excuse for laying out a sizable sum of money on an item without first having at least a minimal idea of what the correct original looks like. Lower-end versions of

Army M34 cap. The trapezoid insignia is unusual for this type of cap. Machine-woven in mouse gray thread on an oxide green backing cloth. The top edge of the skirt scallop is not tacked to the cap body in this example. *Courtesy of Bill Shea.*

Side view. The single air vent grommet is quite clear. The V shaped dip in the cap front is quite noticeable – a characteristic of the M34. *Courtesy of Bill Shea.*

Standard M34 with normal two-piece machine-woven insignia. The cockade is correctly positioned at the center of the skirt front. The wide, relatively flat top panel with an inward V dip at the front is the most notable characteristic of the M34. The upper edge of the skirt scallop is tacked to the cap body in this example. The soutache branch color is for Transportation. *Courtesy of Dale Paul.*

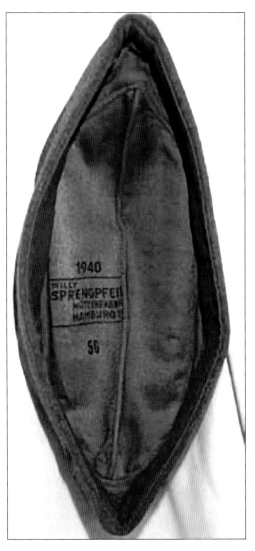

Gray lining on a field gray M34. Note the center ridge at the top of the cap. All the main seams are sewn together with the lining and form thick ridges. Note the width of the top panel – much wider than on the boat-shaped garrison caps used by other services. There are no unit stamps. The makers mark is for Willy Sprengpfeil of Hamburg (this company survived the war). *Courtesy of Bill Shea.*

these modified/converted caps are usually so flawed – their true nature so obvious – that they should be recognizable as fakes to anyone with even a rudimentary knowledge of Third Reich military headgear. A caveat to remember: if it looks like a fake, it probably is. And the corollary: If the price is too good to be true, then nine times out of ten there's a very good reason why.

CONSTRUCTION

A. Continental

The M34 field cap for Other Ranks issued for continental use was usually made of average quality woolen yarn, with a short nap finish. Late war examples were often produced in a characteristically coarse, low-grade wool with a heavily felted finish. The wool surface on such caps tends to look slightly 'hairy.'

This model cap was issued in field gray for regular troops and in a rich, warm black (to match the black tanker uniform) for Panzer troops. The field gray color varied between makers, though within a limited permissible range of acceptable variation. This color remained essentially the same throughout the war, despite the major shift in tint for uniforms introduced with the M-44 field blouse. By that time, most field caps issued were the M43 model – which did show more color variation.

The thread used in the body construction of *early* M34 caps was usually either field gray or some other equally dark color. Some caps, particularly those made after 1941, may use a more noticeable, off-white thread. Though this color is also correct such caps need more than the usual amount of care during any inspection. A cap manufactured with light-colored thread should be judged against a date stamp, if any, that may appear inside the cap. Off-white thread on a cap with a pre-1941 date stamp is questionable.

The M34's side skirts were not attached to the cap body at any point along their upper edge except at the front, where they were sometimes secured to the body with heavy threads to the left and right of the vertical center seam. These threads are not present on all caps, but they do appear quite frequently when the cap has a soutache.

Construction seams appear at the center front and rear of the cap body and on the skirt arm (the scallop). There is also a fore and aft seam along the top center of the cap running from front to back. Not actually a seam, but rather a sewed-in crease, it is formed by tucking the top material downward along the center line, pinching it together on the inside of the cap together with the lining, and then sewing along the length of the ridge thus formed. In comparison with the interior ridge on Luftwaffe and Waffen-SS garrison caps, the M34 ridge is not very pronounced, depending on the type of lining (a thicker lining material resulted in a more pronounced ridge).

Unlike visor caps where stitching is not normally visible on the cap exterior, M34 caps do show external stitch lines. These are sewn parallel to each other on either side of a given seam, and parallel to the top and bottom edges of the side panels. The cap's top center crease shows no external stitching. The material hem on the inside of the side skirt's upper edge seams is left unfinished, since the skirt was never actually meant to be turned down.

The top of the M34 cap tends to spread apart when worn often appearing almost flat. Thus, the 'V'-shape top so characteristic of the boat-shaped Luftwaffe and Waffen-SS garrison caps is less obvious on most M34s.

LINING

Materials and Colors

M34 linings were made in cotton using a smooth weave pattern (a satin weave, for example). Other materials are not unknown or necessarily incorrect, but are not standard and so should be viewed with caution until proven authentic. For both field gray and Panzer versions of the continental M34 cap, the most common lining colors are:

1. Field gray
2. Grayish brown
3. Steel gray

Early Panzer caps in particular nearly always have steel gray interiors.

Beware of plaid linings or anything other than solid colors. The cardinal rule in collecting is always *"guilty until proven innocent."* Anything outside the ordinary – that is, "off spec" – should be considered with this thought in mind!

Private purchase M34 caps had better quality linings of high grade cotton or rayon, in various colors including gold.

Note: *In all cases the M34 lining either covered the interior completely, or reached to within two or three millimeters of the bottom edge of the cap. There should be no more than two or three millimeters of exposed wool along the interior bottom edge.*

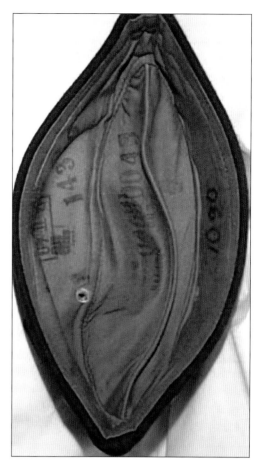

Black Panzer M34 (interior). Gray linings are particularly common on early Panzer M34 caps. The center top ridge is quite noticeable in this example. The interior grommet washer is unpainted, and German unit numbers are stamped in a box on the left side of the cap (the cap front is facing down). The other, large-sized numbers are Russian inventory markings as this cap spent time under Soviet control. *Courtesy of Bill Shea.*

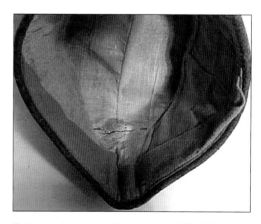

Gold rayon lining on a private purchase M34 (private purchase M34s are relatively rare). Another indication that this M34 was not standard Government Issue is the presence of a partial sweatband. Government Issue caps did not have this feature. The center ridgeline is hardly noticeable on this example. *Courtesy of Bill Shea.*

Black Panzer M34 with machine woven insignia and correctly applied soutache, though no securing thread loops are present at the the apex. Hoheitszeichen (national emblem) in gray cotton thread, with the cockade on dark green backing cloth. *Courtesy of Bill Shea.*

Closeup of the soutache end insert point.

Maker Marks

Any maker marks in an M34 cap usually appear stamped in black ink on the left side of the cap lining, and include the maker's name, city (though this is not always present), the year of manufacture, and the cap size. In addition, there may also be unit markings, again in black ink.

Padding

A thin layer of padding was sometimes inserted between the wool and the lining material at the front of the cap in order to give it more body. Private purchase caps are the most likely to have such padding.

SOUTACHE

The soutache is the colored, inverted 'V' chevron located on the lower front of the cap, which served to indicate the wearer's *Waffengattung* or branch of arms (Artillery, Panzer, Infantry, etc.). The soutache appeared on M34 caps until its use was officially discontinued in the summer of 1942. It continued to appear – albeit unofficially – even after that date, however.

The black M34 for the Panzer troops was not limited to the pink soutache representing that specific branch; the black cap was authorized for wear by anyone assigned to a panzer unit regardless of that individual's actual branch. Thus, a soldier of the *Nachrichtentruppen* (signals branch) assigned to a panzer unit would wear the black panzer cap, but with a lemon yellow signals soutache rather than the pink of the Panzertruppen.

On contemporary caps that have been modified to fool potential buyers, the soutache is the most common item on the cap to have errors. The Germans discontinued the use of the soutache relatively early in the war, and so caps with this feature are more desirable as collectibles than caps without it. The cap modifiers may take an original cap and add a soutache, or if modifying a contemporary cap they invariably want to add a soutache in order to cater to this demand. Yet, they almost always get one thing or another wrong. The single most notable flaw in modified caps is incorrect attachment of the chevron ends.

Key Point:

On authentic M34 caps, the chevron (soutache) ends actually pass through a seam near the base of the cap. The ends of the chevron are thus only visible on the reverse (inside) of the front skirt arm. This was normally done at the factory during manufacture, and all the stitching is therefore done by machine.

Other key points to watch for:

Correct: *Potential Error* (in italics)
- The soutache is made of cotton: *wrong material*
- Securing stitch is in same color thread: *different color thread*
- Securing stitch runs down the center line of the soutache (chevron): *stitching not centered*
- Chevron is attached with a 90 degree angle at the apex of the inverted 'V': *chevron attached at wrong angle*
- Securing thread loop often appears across the chevron's apex.: *securing loop appears in different position*

• Ends of the chevron are inserted into the seam, just above the lower edge of the cap: *ends are sewn to the bottom edge or holes are made in the seam in order to insert the ends.*

Since the soutache is without doubt the weakest link on an M34 cap, it is wise to always give it particular attention.

The most important items to note in the previous close-up example:

1. The top of the soutache chevron forms a 90 degree angle, which provides a seat for the cockade. The cockade's backing cloth in this case is dark green (correct).

Front close-up of a Panzer M34 with lemon yellow soutache for *Nachrichtentruppen* (Signals). Machine-woven insignia. The seam insertion point for the chevron ends is quite clear. This soutache has a single loop of yellow thread at its top, used to secure the apex to the cap. *Courtesy of Bill Shea.*

Excellent shot of another young Panzer private, this time with a later model Panzer wrapper (no collar piping). He also wears the black M34 with soutache. The cap is not positioned correctly on his head in accordance with regulations. *Author's collection.*

A young Panzer man in an early Panzer wrapper – the collar is piped in Waffenfarbe, which practice was later discontinued – wearing a black Panzer M34 with soutache (the field gray M34 cap was not authorized for wear with this uniform). *Author's collection.*

The apex of the soutache reaches the stitch line just below the upper edge of the skirt arm (scallop).

2. The thread used to secure the soutache is a very close match – if not the exact same color – as the soutache itself.

3. The attachment stitch line runs lengthwise along the centerline of the chevron.

The ends are (correctly) sewn *directly into* the lower stitching seam. This is normally done at the factory when the cap is being made; thus, the sewing is perfect. The thread loop which is sometimes used to secure the apex of the soutache to the cap, is not present on this example.

GROMMETS

The M34 cap was ventilated to allow air circulation over the wearer's head, and there should be air hole grommets on the cap. These were arranged with *one grommet per side*, located just above the point of the side skirt turn up. The short grommet shaft went completely through to the cap interior. The exterior rounded portion was normally coated with field gray (or black) baked-on enamel, though this may be worn off on some caps. The vent hole allowed air to enter the cap and cool the wearer's head, as well as to allow moisture to escape. With the cap interior, the grommet shafts passed through a rimmed washer; the protruding shafts were then split and crimped back as small tongues over the washer rim.

Lack of grommets/vent holes is a sign that the cap is incorrect – likewise with 'two too many' air holes.

INSIGNIA

Regulation insignia for this type of headgear consisted of two individual pieces – the *Hoheitszeichen* (national emblem – eagle) and the *Reichskokarde* (national cockade). The insignia was either machine-woven in white cotton thread on a field gray or dark green backing in early examples, or in a mouse gray cotton thread on an oxide green-colored backing on later caps. The gray eagle on this latter combination is difficult to see from a distance.

For the black M34 used by Panzer troops, the machine-woven BEVo-style eagle in white (early pattern) or later a pale gray cotton thread on a black rayon underlay was usual, with aluminum thread for officers. The Reichskokard was machine-woven on a dark green underlay cloth.

Cap insignia was either machine-sewn to the cap at the factory or professionally applied by a skilled individual at the time of manufacture. The insignia edges were tucked under when sewn to the cap – the raw edges of the backing cloth were never left exposed. The stitch work did not normally pierce the cap lining, and should not be visible on the inside.

Once in the field, there was almost always someone in a unit who could do quality semi-professional sewing work. Hand-applied insignia are usually straight-stitched within the insignia border (rarely across it). Ideally, the individual stitches should be relatively close together, and equally spaced; that is, the *distance* between each stitch should not vary greatly. In addition, the *length* of each individual stitch should be the same, again without easily noticeable variations. The key is neatness: Hand-applied insignia with sloppy, irregular, unequal stitches is highly suspect (particularly if the insignia looks very new) and may indicate that it has been post-war applied (or reapplied). Check also for any color difference around the current insignia (darker cloth) which might indicate that some other type or form was originally in place.

B. M34 for Panzertruppen

In all the discussions, books, and vast array of other reference works available on the World War II German Army, the significant point of interest for most people is always the Panzer arm – the tank force. This is quite ironic, since the German Army high command *did not* recognize the real significance of the Panzer arm in terms of independent, high speed mobility and firepower.

All of the famous Panzer commanders understood the capabilities of the tank in conjunction with motorized infantry and thus, the optimal employment of tanks in armored warfare, but the German high command simply failed to do so. The German Army, formidable as it was, could have been a devastating and perhaps unbeatable force had the OKH (and later the OKW) employed German armored forces according to the concepts of mechanized warfare that its armor commanders had always strongly advocated.

The fact is, when people today think of the World War II German Army, their first association is tanks. And so it comes as no surprise either, that collectors show a great interest in Panzer-related items and these caps are always high on the desirability list. The law of supply and demand dictates the result: high, and ever-increasing prices. Black M34 caps and any other headgear specifically associated with Panzer forces will continue to grow more expensive as the years pass and the number of available caps (not already in collections) shrinks to a minimum. The sooner you can add an authentic example to your collection, the better.

Unfortunately, the high demand for panzer headgear makes it a common target of both modifications and reproductions. Modifications are somewhat limited by the color itself: any substitute, whether contemporary or old, must be black. A cap of a different color can always be dyed black, of course, but quite often the depth of the black in re-dyed caps does not look quite right, or the cap lining is dyed black as well (correct Government Issue Panzer M34 caps never have a black lining).

German original black wools are a very rich, deep color, which is entirely consistent throughout the material. There are no uneven, lighter patches, etc. (unless something was obviously spilled on an area, for example, that resulted in fading).

The basic thread used in construction of the black M34 cap was also black cotton, or some other very dark color that blended well with the black. Examples of M34 caps with light-colored or white thread used in their construction are usually fabrications, not originals. Light-colored thread on a Panzer cap defeats the intent of the black cap's design, and consequently would not have been acceptable to most makers, or to military authorities.

C. Tropical Feldmütze

The Army's tropical field cap is a garrison cap very similar in style to the M34, and another very common target for modifications. The correct regulation material for tropical field caps was a cotton twill (the texture is thus ribbed, like Trikot). The color of the tropical material on average is a khaki that tends somewhat toward olive brown; a slightly more mustard tint is also common. Pure olive green is not correct, nor is light brown. With examples of caps that have been bleached to near white by sun fading, this condition should apply to the entire exposed surface.

Bundeswehr lightweight (summer) field cap, modified and offered as authentic by an Internet seller. There was no mention whatsoever in the description that this item is absolutely *not* an original wartime tropical Army cap.

A truly stunning example of a poor modification. No one should be fooled by the reapplied insignia with its incorrect colors, unmatched thread, and total amateur application. The soutache is completely false.

Example of a Bundeswehr Army garrison cap (1960s model) for enlisted troops (Other Ranks). The side skirt can be pulled down in the same manner as the wartime M42 (but without the buttons). Authentic M34 caps, however, did not have this capability. The brownish gray color is also quite incorrect for authentic pre-1945 M34 caps.

Although the tropical field cap was cut to essentially the same pattern as the M34 the cotton cap cloth was much thinner than the M34 wool. As a result, Army tropical garrison caps tend to have a much flimsier appearance than do their continental cousins. In terms of construction, however, seam lines and stitching points were essentially the same.

The side skirts on authentic caps are scalloped like the M34. They were not affixed to the cap anywhere except at the very front, where two thread loops along the top edge of the flap (one on each side of the center seam) were used to secure it to the cap body.

The modified Bundeswehr ([German] Federal Armed Forces) cap shown at the top of page 129 has several flaws that are easy to spot and recognize, and which should immediately alert a potential buyer to the fact that it is not authentic. The most important of these to be aware of are:

1. *Boat shape*: No original Army garrison cap was cut in this shape.
2. *Two grommets* (original caps have *one* per side)
3. *The side skirt is simulated – it is actually part of the cap body* (authentic tropical caps have separate, movable side skirts).
4. *The swastika portion of the national emblem is sewn down over the top edge of the front skirt arm* (scallop). This is totally incorrect and was simply *not done on authentic caps*). The eagle itself is not the correct size (it is too large). In addition it is woven in white thread, while tropical insignia was usually light blue thread on a tan base.

This example is nothing more than a modified Bundeswehr Army Other Ranks cap, and does not even come close in either cut or dimensions to either an original

M34, or to an original tropical cap. The gently curving shape, in fact, is much closer in form to a Luftwaffe overseas cap. The Army pattern eagle would be incorrect in this case. As it stands, the insignia is wholly incorrect in having the wreathed swastika extending down over the top edge of the flap. No matter what form of addition you use, the numbers on this piece simply do not add up.

In the next version of a modified cap, the insignia is again completely incorrect. This is no major surprise since the two pieces are nothing more than poorly made reproductions with the insignia colors – backing cloth included – incorrect even for a continental M34. They have no place at all on a tropical cap. The stitching on the eagle is completely amateurish, and in two different colors of thread. The soutache is no more than a joke: too wide, too high, and made of the wrong material. Original soutache cord is somewhat flat, though thick and with rounded edges; this example looks like a section of sneaker lace, not to mention that it is also incorrectly applied. Original soutaches were never folded over on themselves at the apex of the 'V', as is the case here, nor did they have raw ends hanging below the cap edge.

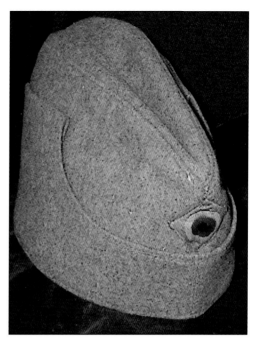

Current model Bundeswehr garrison cap for enlisted troops (Other Ranks) in gray. This color is also totally incorrect for Third Reich era caps.

LINING

Tropical field caps had smooth textured cotton linings in red or in some cases maroon, depending on the maker. Tan linings do occasionally appear; the tan is, in fact, very close to the color of the cap exterior. This color is fairly rare, however, and most examples bear the maker stamp **Berolina**, a company owned by a fellow named Ernst Hoffman. Herr Hoffman also owned the **G.A. Hoffman** company, a respected visor cap manufacturer located in Berlin. Grayish tan linings can also be found from time to time, but again, these are rare and usually bear the mark of Clemens Wagner (Braunschweig).

Lining attachment methods were the same as with the continental M34, and these caps also had the usual center ridge inside the cap. Makers marks also followed the same pattern, sometimes showing only the maker name and cap size, or the name, location, cap size and year of delivery, stamped in black ink (rarely purple) on the lining. Remember to match the surrender date of the Afrika Korps against any date stamp appearing in a cap, if the seller claims it is an Africa Korps piece.

GROMMETS

The tropical cap, like its continental cousin, had one single grommet per side. The exterior grommet ring was coated with a brown, baked-on enamel which, though it can be chipped if rapped hard enough against another hard surface, is very difficult to simply scratch off.

INSIGNIA

Tropical insignia is a complex topic that deserves independent study due to the variety in colors, etc. On Afrika Korps caps, the *usual* Army tropical machine-woven insignia is in matte blue gray cotton thread on thin, tan-colored backing cloth. Wool-backed insignia is *incorrect* on Army tropical headgear. Application methods varied, but the insignia was normally attached at the factory either by machine, or by a professional seamstress. As with the continental version, insignia stitching did not pass through the lining and should not be visible inside the cap.

The most common stitch pattern is a zigzag machine stitch. Ideally, this should appear on both the eagle and the cockade, but in an imperfect world such is not always

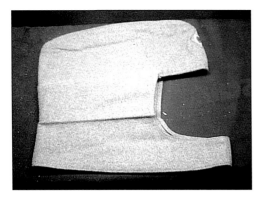

Cold weather version Bundeswehr Other Ranks field cap with the side skirt pulled down. The brownish gray felted wool feels quite thin. Note the female snap (aluminum) on the cap body just to the right of center (in line with the cockade). The matching male snap is located on the right corner of the skirt near the face (just above the crease line).

An assortment of German post-war enlisted rank garrison caps. The first three are [West German] Bundeswehr issue 1) Army winter model; 2) Navy Bordmütze; 3) Army (lightweight) warm weather model (note the grommets). The last (top) two caps are East German NVA issue. Note the cockade positioning.

the case. Straight-stitch machine-sewing was also used by some manufacturers, and again this pattern should ideally appear on both the eagle and the cockade. Hand applications did occur of course, but the work was usually done by someone in the owner's unit who was reasonably proficient in sewing. While cruder efforts do appear, caps with decidedly amateurish-looking stitching should be thoroughly inspected before any purchase is considered.

CONTEMPORARY BUNDESWEHR CAPS

Caps most commonly chosen for M34 modification are Bundeswehr garrison caps, since the material and cut is superficially similar to that of wartime originals. In fact, the field version of the Bundeswehr garrison cap actually has more in common with the wartime M42 cap than with the M34, since the Bundeswehr cap has a side skirt that can be pulled down completely around the face.

Wartime M34 caps were never made in either the coarse brownish gray (early field model), or gray color (dress model) wool used in the contemporary Bundeswehr garrison caps for enlisted men and NCOs.

The Bundeswehr combined certain design aspects of the boat-shaped *Schiffchen* cap used by the Luftwaffe and Waffen-SS, together with other aspects taken from the pre-1945 Army M34. The result is a hybrid cap that is a little of both. There are many telltale points that identify a Bundeswehr Army cap beyond question. The easiest to recognize are:

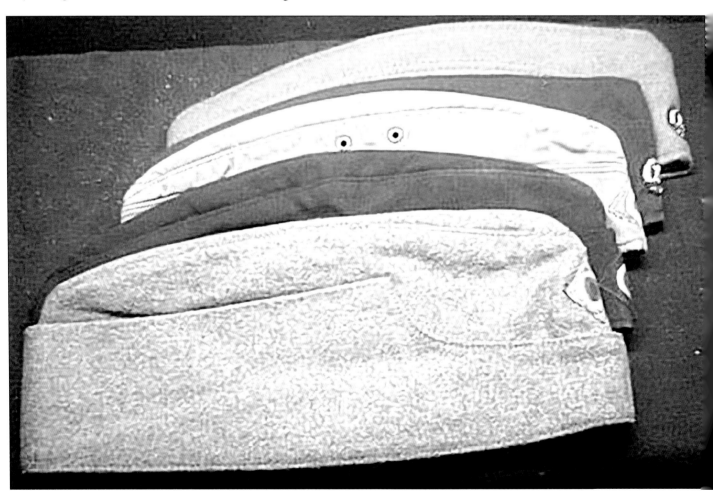

Army officer's M38 cap with *Gebirgsjäger* (mountain troops) branch soutache. The officer BEVo insignia is aluminum thread on a dark green underlay. Hand-embroidered, thick (3D) bullion cockade. *Courtesy of Bill Shea.*

1. **Skirts**: The side skirts extend a longer distance from the front of the scallop to the point of the upturn, and the front arm of the skirt itself is far narrower than on authentic M34 caps.

2. **Vent holes**: Authentic wartime caps have an air vent grommet located just above the skirt turn up on the cap body. Bundeswehr wool version caps *do not have* air vents.

3. **Skirt snaps**: Bundeswehr garrison caps are fitted with small snaps behind the corners of the skirt turn up, which serve to secure the skirts tightly to the cap body. Original M34 caps *do not* have snaps.

4. **No top seam**: The Bundeswehr cap top section is much narrower than the M34. The latter was so wide that it looked almost flat when worn. Bundeswehr cap tops are made from one piece of material and *do not* have any center seam, while authentic M34 caps on the other hand, always have a fore and aft-running center seam.

5. **Insignia**: Any World War II period national emblem (eagle) insignia would be at the front crown of the cap, exactly where the German federal cockade is positioned on a modern Bundeswehr cap. Thus, there may be signs of a federal cockade having been removed, such as differences in cap cloth color and so on, which might indicate that another insignia was on the cap prior to the eagle.

With these differences in mind, there should be no difficulty in spotting or identifying any modified, contemporary Bundeswehr caps.

EAST GERMAN CAPS

East German National People's Army (NVA) caps are all boat shaped, and thus not well suited for modification into M34 look–alikes. They simply don't ... look alike, or for that matter, even look close to an original. On the other hand, some people will, in

Note the double line of parallel stitching just below the scallop piping. The upper of the two stitch lines secured the cloth piping tail to the inside edge of the skirt. The lower stitch line hemmed the edge of the skirt itself, which has been folded over and inward. *Courtesy of Bill Shea.*

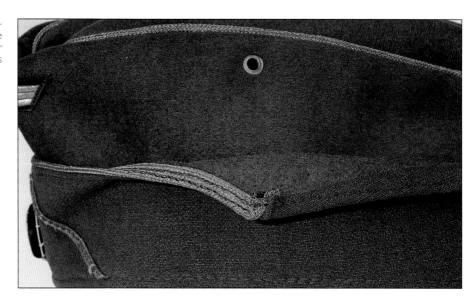

fact, try anything (and for that matter, will buy anything). The picture on page 132 provides some examples of East German and Bundeswehr caps for comparison.

The main points to remember are:

1. All East German caps are boat-shaped, while the M34 was not.
2. East German caps are a lightweight blend of polyester and wool, while the M34 was heavy wool.
3. East German caps have Soviet-influenced high sides and *simulated* side skirts. M34 caps dimensions were not as high, and the side skirts were movable – not part of the cap body.

Part 2
DIE FELDMÜTZE NEUE PROBE FÜR OFFIZIERE - OFFICER'S NEW TYPE FIELD CAP (M38)

The officer's new-type field cap was introduced with the intent of providing officers with a convenient, easy to carry piece of headgear of practical use in a field environment. US Army officers, as a point of comparison, rarely wore caps in the field – but when they did, it was nearly always a garrison cap.

The German Army felt that this type of headgear would be equally useful for its own officers, and far more practical than the officer's old-style field cap (which would be phased out of service). It had the incidental added benefit of requiring less material for production. The goal of completely replacing the old-style field cap with the new type did not prove entirely successful. The great popularity of the older cap model ensured that at least some officers continued to wear it throughout the course of the war (generals, for example) – wear-out date notwithstanding.

According to the industry trade paper *Uniformen-Markt*, Volume 4, February 15, 1939, the primary material for the new-type officer's field cap was field gray doeskin (rather than the regulation Trikot). The doeskin of choice was usually a relatively lightweight, short-napped material with a lightly brushed finish. Regulation Trikot wool

General's M38 cap, front view. A fine example in superb condition, with a pink Panzer branch soutache. The insignia is hand-applied (machine-woven in aluminum thread). The cap material is high grade brushed doeskin. *Courtesy of Bill Shea.*

was an option rather than a requirement on this type of headgear. Caps made with Trikot cloth were usually a fine grade, with thin ribs and a slightly fuzzy surface.

To help maintain the cap's shape, a 2.5 cm wide strip of material was attached inside the cap along the line of the aluminum crown piping. In addition, a 3 cm wide strip of oilcloth or other waterproof material was inserted along the lower edges of the side panels behind the lining (presumably to prevent sweat damage).

Inside the cap, the fore and aft top center ridge is far more visible on the M38 than it is on the M34. This is because the cap top is narrower, and the crease therefore, slightly deeper on the M38.

LININGS

Official requirements called for a lining of field gray satin, or rayon, but this actually varied between makers for private purchase caps. Most common is gray or black rayon; other colors also appear from time to time, however, and include brownish-gray as well as dark green.

SWEATBAND

Though not always present, official regulations called for a 23 cm long partial (forehead area only) sheep's leather sweatband on M38 caps. Government stock M38s thus may also have a partial band. Very few manufacturers offered the option for a full circumference sweatband, even on private purchase grade caps.

M38 with red (artillery) soutache. The cap itself is original, however the insignia has been re-applied. The soutache ends are not through the seams, the chevron apex is too sharp. The angled, cross-border stitches used on the eagle are highly suspect, with completely non-matching olive drab colored thread. Note the suspicious damage to the cap cloth around the cockade.

This picture from a wartime German publication shows Army *Ritterkreuzträger* [Knights Cross holder] Feldwebel [Sergeant] Hans Strippel in a black Panzer M34 with soutache.

Major Friedrich-Wilhelm Breidenbach (Knights Cross awarded on September 30, 1944), CO of Panzer Brigade 101, poses for his award photo wearing an officer's Panzer M38. Despite the late date (1944), his cap still carries a soutache.

INSIGNIA

Regulation insignia for the officer's new-type field cap was a machine woven (BEVo) aluminum thread Hoheitszeichen (eagle) on a dark green backing cloth. The accompanying Reichs cockade was in thick, hand-embroidered bullion. The insignia was normally hand applied (at the factory), rather than machine-sewn.

Although examples of forged M38 caps are not that common, when they do appear the errors that may help you to identify them usually involve the insignia and the application thereof as well as the officer piping (which may be a post-war addition). Pay especial attention for errors, particularly if the cap has a soutache. If you find one inconsistency, then the entire cap becomes suspect, and it is likely there will be more.

Machine woven (BEVo) insignia reproductions have reached impressive levels of accuracy these days, and when applied to a cap, it is no easy task to tell them from originals. The only guide you may have in this situation is the quality (or lack of same) of the attachment sewing. For a photographic example of a machine-woven officer's insignia, see Appendix D.

Chapter 9

Part 1: Waffen-SS M1940 Schiffchen [Continental, Panzer, Tropical]
Part 2: Polizei Schiffchen

Part 1

DIE WAFFEN-SS FELDMÜTZE (M1940/SCHIFFCHEN) –
WAFFEN-SS FIELD CAP (Little boat)

Colloquially labeled *Schiffchen* (little boat) by soldiers, this cap was introduced in 1940 as the standard field cap for wear by Waffen-SS troops on front line duty. Closely akin to the Luftwaffe's Fliegermütze (introduced in the mid 1930s), the M1940 had moveable side flaps (attached to the cap body only at the base).

The Schiffchen served as the Waffen-SS answer to the Army's M34 cap in purpose, if not in form. Schiffchen caps were also used by officers, but were manufactured from finer grade, high quality wools (doeskin or Trikot) than the enlisted version. Officer models were also piped in aluminum mesh or white cord around the top edge of the side skirt. This concept was quite different from that chosen by the Army, which had introduced an entirely new cap model for officers (the M38) independent of the enlisted troops' M34.

The Schiffchen was also produced in black for Panzer troops. Wartime photographic evidence suggests that the black Panzer version of the Army officer's M38 cap (with SS insignia) was quite popular among Waffen-SS Panzer officers in particular. Prior to 1940 this is understandable, since before that year there was no Waffen-SS alternative to the Army M38 in either field gray or black. Even after the introduction of the SS Schiffchen in 1940, however, many Waffen-SS officers – especially tankers – continued to wear the Army cap rather than the Schiffchen. They simply did not convert to the SS model – or, if they purchased one, they didn't wear it. The reason for this is not clear, but perhaps these officers simply didn't want to spend more money while they still had a serviceable M38, or had become too accustomed to the Army headgear. Perhaps they simply preferred the double row of officer piping on the Army cap (around the crown and the edge of the front scallop), which the Schiffchen did not offer (it was piped only around the top edge of the side skirt).

A dashing, *very* young SS-Mann wearing an Other Ranks Schiffchen. Perfect example of two-piece machine-woven insignia in white thread on black backing cloth. The two parallel ridges on the cap top remain close together (pinned, perhaps?). The cap looks quite sharp.

Waffen-SS Schiffchen for Other Ranks (enlisted) soldiers in field gray. Two-piece machine woven insignia in white cotton thread on textured black rayon backing cloth, standard for this type of cap. *Courtesy of Bill Shea.*

A. Continental (field gray) and Panzer (black) Caps

CONSTRUCTION

The design followed essentially the same pattern as that of its Luftwaffe cousin, the Fliegermütze, however, the gradient of the side skirt curve differed slightly. The top front corner of the Schiffchen was slightly rounded when worn, whereas the Fliegermütze stood a bit straighter (and thus appeared almost to lean forward).

It is when worn that the Schiffchen's boat shape was most pronounced, and this characteristic form is what allowed for a good, snug fit that reduced the tendency to slip off in windy conditions, or during jarring movements (common to all garrison caps).

The center of the cap top formed a deep 'V'-shaped crease running fore to aft when viewed from the front. This shape also had the unfortunate side effect of generating one of the more colorful, if cruder, nicknames for this type of headgear (which out of propriety can't be divulged here). Schiffchen interiors show the usual pinched center ridge in the cap top common to all German-made garrison caps.

Although the crease was fairly deep, the top of the cap still tended to spread apart and the parallel 'V' ridges lost height when the cap was pulled down too far on the head. The crease flattened out and became less prominent as the top spread apart (though never as flat with the top of the Army's M34), and the normally crisp appearance of the cap suffered. To prevent this, many soldiers pinched the crease together on the *inside* of the cap and then pinned the material in place with a safety pin. This held the sides of the top crease together from the interior, and prevented the exterior 'V' shape from being lost when the cap was worn.

A nearly mint example of a Waffen-SS officer's Schiffchen with aluminum piping. The wool is only an average grade (not doeskin), with a very short nap. Machine-woven aluminum thread insignia. Unusual is the fact that the eagle is applied with a zigzag stitch. *Courtesy of Bill Shea.*

Safety pin, used to hold the V-crease together. This pin is somewhat rusted, and the material around it has aged together with the pin (which has never been removed from its original position) – a very good indication of authenticity. Normal grayish colored lining material. *Courtesy of Bill Shea.*

Below: Side view. The cut of the Waffen-SS Schiffchen (officer or enlisted) and the Luftwaffe Fliegermütze were essentially the same, though the Schiffchen had the very slightest bit of rounding to the front top corner, and the degree of arc on the side skirts was also different. *Courtesy of Bill Shea.*

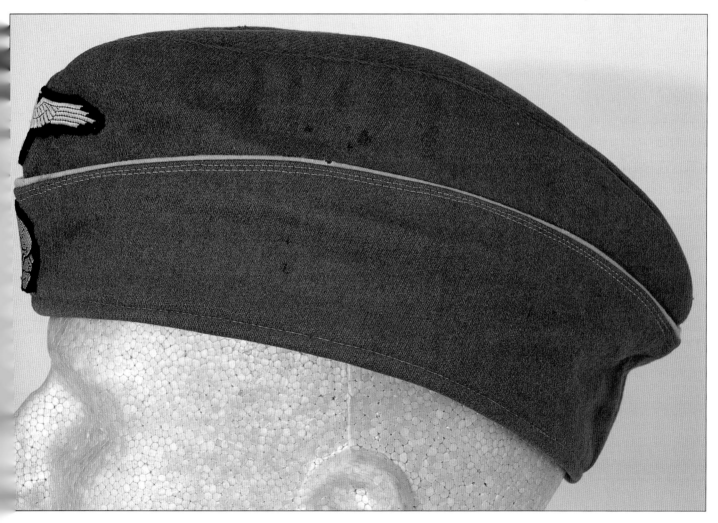

Safety pins found on original caps usually show varying degrees of rust. Make sure that the apparent age of the pin and the surrounding material match: If there is rust on the pin, there should also be indications of rust on the surrounding material where the pin for years has pressed tightly against the cloth.

The material for the Other Ranks Schiffchen was field gray wool (black for Panzer units). The actual quality of the cloth varies on issue caps from a heavy, somewhat coarse, felted grade to a shorn surface with a slightly softer *hand* (feel).

Officer models were made in various doeskin grades (but all with a short nap), and in Trikot.

LINING/SWEATBAND

Linings on continental M1940 Schiffchen caps were normally field gray cotton fabrics for enlisted troops, though other colors – brownish gray in particular – also appear. Green rayon was the standard lining color for average material quality officer Schiffchens (which probably represented government off the shelf stock caps), with other colors also possible, though less common, on higher quality private purchase caps.

INSIGNIA

The two most critical and distinct features to keep in mind when evaluating the insignia on a Waffen-SS cap are:

1. Hand application: 90% of all Waffen-SS field cap insignia, regardless of field cap model, was professionally *hand-applied*. This is quite different from the other military services, where machine-sewing was usual.

2. Padding/Puffiness: Waffen-SS cloth insignia was padded slightly when applied. Ideally, the insignia should not lay flat against the surface of the cloth. Instead, it should appear slightly puffy – a characteristic easily noticeable on close inspection particularly with the Totenkopf.

The standard insignia for the Schiffchen was the Hoheitszeichen (national emblem) and Totenkopf, both machine-woven in off-white cotton thread on black artificial silk (rayon) backing cloth for Other Ranks, and in aluminum thread for officers. Officer insignia stocks were not always available however, and when shortages occurred officers often used enlisted insignia – or Army insignia (as a last resort). If an officer cap with enlisted insignia is encountered, the insignia sewing work should be carefully proofed for evidence of recent application, and the base cloth (underlay) surrounding the insignia should be checked as well for any indication of an earlier application. Insignia stitching *did not* normally penetrate the lining, and ideally should not be visible inside the cap.

If the insignia is machine-applied, use extra caution during any pre-purchase inspection. Though authentic machine-sewn examples do occur, the majority of insignia was hand-sewn and this is the ideal from which to judge all else.

Most of the reproduction insignia currently available (Army or Waffen-SS) is of the machine-woven 'BEVo' variety (simpler to produce than embroidered forms). Poor quality reproductions generally show obvious flaws, for example, the top of the eagle's head is much too high, or it is on an even line with the top of the wings or even below them (i.e. too short). The correct position is just above the horizontal line formed by the wing tops. High quality reproduction BEVo (machine-woven) enlisted and officer in-

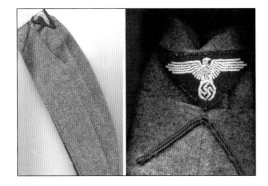

Internet auction offering from 1999. If the cap is in fact a pre-war style Waffen-SS M1934 Schiffchen (not covered in this book), then the correct position of the trapezoidal SS eagle would be on the left side skirt, *not* the front of the cap.

signia, when viewed from the front, is often nearly identical to the originals and is readily available on the militaria market (See Appendix D).

In the previous photo, the cap itself presents a question. At first glance, it appears to be a pre-war, early model Waffen-SS M1934 cap. The top 'V' crease is slightly off-center to the right, which is correct for this model. The peak ridges on each side of the crease appear to be the same height, however, while on an original cap the left ridge (closest to the centerline) should be slightly higher than the outer (right side) ridge – not the case on this example.

On an original M1934 cap, the trapezoid eagle would be located at the top of the left side skirt, beside the scallop. The eagle on this cap is therefore both incorrectly positioned and slightly crooked. In addition, there is no sign at all of the usual front button (pebbled, or with a raised Totenkopf), which was used on the M1934. The ends of the soutache – though quite neat – do not pass through the seam line as they should.

If you suspect possible high-quality reproduction insignia, the easiest place to look for a confirming error is in the negative pattern formed on the insignia reverse by the weaving process. Unfortunately, if the insignia is mounted on a cap this will not be possible, in which case only the newness of the insignia (as compared to the cap itself) may give any useful indication. Visit militaria shows or a trusted dealer, and study as many authentic insignia examples as possible. As of this writing, the most technically accurate – thus, the most dangerous – reproductions of machine-woven insignia are made somewhere in South Korea.

ARMY M38 CAPS

The common use of the Army officer black Panzer M38 by Waffen-SS officers makes it likely that an M38 Panzer cap will be encountered with Waffen-SS machine-woven insignia. In fact, collectors would be just as likely to come across an SS-style M38 than they are the black officer's Schiffchen if not for one fact: the popularity of the M38s meant they were often used until they wore out, leaving only the Schiffchen model behind. Field gray M38s on the other hand, seem to have been less popular than the field gray Schiffchen. Some officers did retain their field gray M38s after the Schiffchen was introduced, but far fewer, it appears, than did so among the tankers. In one interesting period photograph, Hauptsturmführer Jochen Peiper of the 1.SS Panzer Division *Leibstandarte*, who often appears in photographs with the black Panzer M38, wears instead a field gray M38 with an Army eagle and a metal SS Totenkopf – a rare combination.

Army Panzer M38 caps – quite rare in their own right – can be easily modified into an SS piece simply by replacing Army insignia with SS. The thread used may be a giveaway if it is a modern synthetic.

Such a switch is far less likely with the Army black M34 cap, since Waffen-SS enlisted soldiers were all issued the SS Schiffchen after its introduction and thus there was no need for the M34.

Some faked caps are easily recognizable, and the picture at right provides a perfect example. Can you recognize the flaws?:

1. Mixed Insignia *(with the eagle an obvious reproduction)*: It is extremely unlikely to find a machine-woven eagle in combination with a visor cap Totenkopf on an enlisted man's headgear. Enlisted soldiers were not usually allowed to violate insignia regulations in this manner. The machine-woven eagle is a reproduc-

Waffen-SS Knights Cross holder Oberführer Kurt Scholz, in a very unusual combination, wears a field gray Army M38 officer's cap with the Waffen-SS Totenkopf on the front skirt, and a machine-embroidered trapezoidal eagle (intended for the M1934 SS field cap) on the side skirt. *Author's collection.*

An example of an Other Ranks Panzer M34 with SS insignia. This cap has all kinds of problems! Can you identify them?

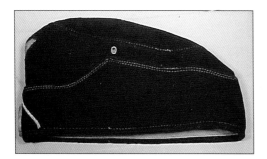

Side view. The unpainted air vent grommet is a definite eyesore.

Officer's Panzer Schiffchen (Internet offering). The weak lighting does not help in making an assessment, but this cap is likely a high quality reproduction. The aluminum officer's piping is just a bit too round (tubular) for comfort. The eagle is unquestionably a reproduction and incorrectly applied, as well – which casts doubt on everything else.

Close up view of the insignia on a Panzer Other Ranks Schiffchen. Note the very even borders and the slight puffiness characteristic of authentic, wartime-applied insignia. Only black thread was used in the manufacture of this cap. *Courtesy of Bill Shea.*

tion in any case: the head is too high and thin, the beak is misshapen; the width of the wing roots varies from right side to left. The eagle's tail feather/leg area is also incorrect.

2. White thread: White thread on any kind of black panzer cap is totally out of character and would simply not be used. In addition, though neatly done, the individual stitches vary slightly in length – all sewing work appears to have been done by hand.

3. Cut: The shape of the cap is subtly off. Where the skirt scallop rounds the cap front, the width is too narrow, which makes the Totenkopf insignia appear much too large for the cap. The skirt arm is wider on an original cap.

On the positive side, the soutache attachment is good, if somewhat uneven on the left side.

The officer's Panzer Schiffchen shown at left is a good example of how insignia flaws can be a major help in identifying an incorrect cap. Superficially, the cap itself looks fairly convincing, but the eagle has an incorrect head and beak, an uneven, flawed chest feather pattern, and an incorrect tail feather/foot area. The wreath is only a single ring. The edges of the underlay cloth are not tucked under the insignia – a major fault.

Inspect the SS eagle and Totenkopf on an M38 cap very closely; look for evidence of a previous Army eagle, or the diamond shape left by a cockade insignia. The likelihood of the cap's authenticity increases if the insignia is padded, but this is still no absolute guarantee. Inspect for any flaws indicating the insignia may be a reproduction, and the cap as well.

A word of caution: *Never* let yourself be railroaded by a dealer into foregoing an inspection if you feel one is necessary. It is *your money*, and it's *your responsibility* to be both thorough and cautious. Any seller who claims to feel insulted by such an inspection, or who is unable to understand your entirely reasonable care and caution given the potential expense, is not someone to do business with. Beware of Internet sellers who conclude an offer with "all sales are final." They should offer at least a three day inspection period on such expensive items. The intent of this phrase in most cases is to leave you, the buyer, with no recourse when you discover you've received a bad item.

OFFICER PIPING

Piping on Army M38 caps used by Waffen-SS officers was no different from the Army version. With the SS officer's Schiffchen, on the other hand, piping was applied around the top edge of the side skirt, but *not* around the cap crown. Officer Schiffchen piping was initially in woven aluminum cord, but this was later changed to white cord with aluminum to be used only by Waffen-SS generals. This policy was not entirely successful. Many officers continued to wear their aluminum-piped caps. Both aluminum and white piping are therefore correct on an officer's Schiffchen and as a result, it is no readily apparent whether a cap piped with aluminum cord belonged to a general, or to a lower ranking officer.

SWEATBAND

Partial leather sweatbands in the forehead area are usual on both government stock (i.e. in average grade Trikot or doeskin material), and private purchase officer caps (high grade Trikot or doeskin). Artificial leather or other leather substitutes were rarely used

Rear view of an officer's Schiffchen with the side skirt turned inside out to show the white cloth piping tail. Note the double (parallel) line of stitching used to secure the tail to the skirt and to hem the raw skirt edge. *Courtesy of Bill Shea.*

Army type Panzer M38 worn by Knights Cross winner Untersturmführer (Lieutenant) Karl-Heinz Worthmann. The cap has Army insignia, with the jawless cavalry skull used on early SS caps rather than the regulation later version Totenkopf. Awards: Knights Cross, Iron Cross 1st Class, Infantry Assault Badge, Wound Badge in silver. *Author's collection.*

Above left: Waffen-SS Obersturmführer Alois Kalls, another tanker who, interestingly enough, is also wearing an M38 with SS eagle and jawless skull. Awards: Knights Cross, numbered Tank Assault Badge, Wound Badge (silver). *Author's collection. Right:* Waffen-SS Untersturmführer Karl Kloskowski in an M38 with machine-woven SS eagle and Totenkopf. Kloskowski (died in 1945) won his Knights Cross while still an NCO (Oberscharführer). *Author's collection.*

Right: For a change, a rare example of an SS officer wearing the SS Panzer Schiffchen. Obersturmführer Karl-Heinz Himme's Schiffchen has machine-woven insignia. Awards: Knights Cross (not shown), German Cross (gold), Iron Cross 1st Class, Panzer Assault Badge (silver), Wound Badge (silver or gold). *Author's collection.*

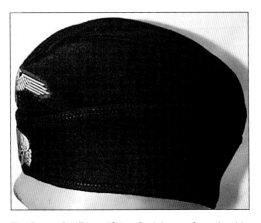

The Panzer Schiffchen (Other Ranks) seen from the side. The curve that gave it its name is clearly visible. Notice the very straight top edge of the eagle – an example of perfect application. *Courtesy of Bill Shea.*

for Schiffchen sweatbands. Full circumference sweatbands hardly ever appear on German garrison caps, though they are standard fare on all *post-war Bundeswehr* models – wise to keep this point in mind. *Private purchase* Other Ranks caps (high quality material) may also have a partial sweatband in reddish tan or gray leather, but such caps are seldom encountered.

B. Waffen-SS Tropical Schiffchen

The tropical cap was constructed in the same manner and using the same design as the field gray and black models, but of cotton twill instead of wool. Other lightweight materials including captured stocks were also used, but for caps made with other than the normal materials only experience will help the collector to discern real from fake.

The standard model tropical SS Schiffchen had the same red lining as other German tropical caps, though once again, tan linings also appear. Makers marks, when present, are usually in black ink stamped on the lining. One maker in particular, a company named Schmid & Menner, seems to have manufactured Waffen-SS garrison caps exclusively – at any rate this name has not been seen (to date) on any other model of headgear, nor on caps from any other armed service.

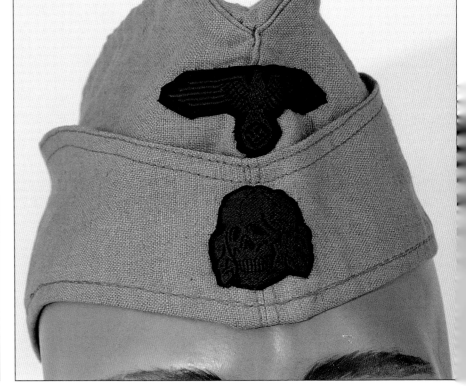

The Schmid & Menner maker stamp on a Waffen-SS tropical cap interior. *Courtesy of Bill Shea.*

Waffen-SS tropical Schiffchen in cotton twill. Two-piece tropical insignia is matte mustard brown cotton thread on black rayon (artificial silk) backing cloth. Note the color of the thread used for the hem stitching. *Courtesy of Bill Shea.*

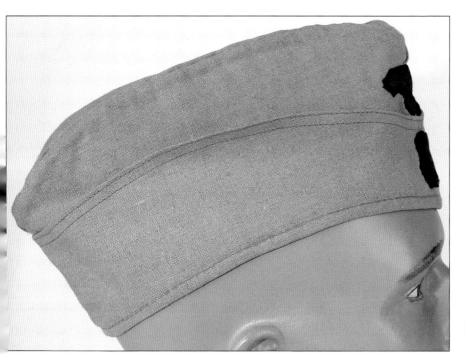

Side view. Pay particular attention to the curve gradient of the side skirt – note the curve's highest point, and compare it with the Luftwaffe example below. *Courtesy of Bill Shea.*

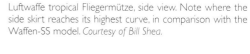

Luftwaffe tropical Fliegermütze, side view. Note where the side skirt reaches its highest curve, in comparison with the Waffen-SS model. *Courtesy of Bill Shea.*

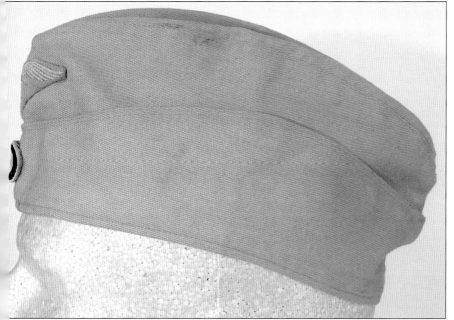

AST GERMAN CAPS (MODIFIED)

The East German NVA (*National Volks Armee* – National People's Army) garrison
ap follows the general lines of the wartime Schiffchen with its gently curving top and
de skirt, but any similarity stops there. Although the relationship to the Schiffchen is
ear, there is a definite Russian influence in the characteristically high body sides – a
sign tendency common to all Soviet military garrison caps. East German cap linings
e always gray, with NVA stampings in white ink. The East German state cockade is
wn directly to the cap front, centered above the lower edge without any backing
oth. The insignia is applied so tightly that there is certain to be evidence of it remain-
g if removed.

East German garrison cap for Other Ranks. The front seam is very distinct; the cockade is sewn directly to the cap without any backing cloth.

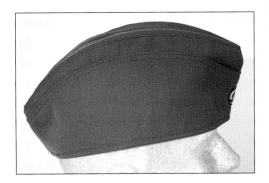

Side view. The sides of the cap are very high, and the distance between the material ridge that forms the top of the [simulated] side skirt and the cap top itself is very narrow. Such high sides are typical hallmarks of Soviet cap design.

Note the following differences from the wartime Schiffchen on this East German (summer weight) cap:

1. *Material:* The cap is made of a lightweight Trikot cloth, visibly thin due to the fact that it is a 55% polyester blend with only 45% wool.

2. *Shape:* While it resembles the Schiffchen superficially, the East German *Einheitsmütze* (standard cap) is cut with noticeably higher sides than those on an authentic Schiffchen. This increase in side height is directly attributable to Soviet military design influence.

3. *Color:* East German garrison caps are a mouse gray color (standard for East German uniforms). This color has absolutely nothing in common with the wartime field gray used by the pre-1945 German Army and Waffen-SS.

4. *Skirt:* The side skirts are not real – they are an integral part of the cap for design only. Wartime Waffen-SS garrison caps – whether continental or tropical – had movable skirts.

5. *Interior markings:* Most East German caps are clearly stamped NVA in white ink on the cap lining.

Part 2
DIE POLIZEI FELDMÜTZE (SCHIFFCHEN) - POLICE FIELD CAP

The Schiffchen used by the combat police was produced to the same construction specifications as its Waffen-SS cousin, both internally and externally. The only significant differences between the two were the color of the wool and the insignia: All poli

Standard issue police garrison cap. The most noticeable characteristics of this type of headgear are the color, and the police insignia. *Courtesy of Bill Shea.*

Side view. Note that the cut of the cap is exactly the same as the Waffen-SS model Schiffchen. Hem stitching is done with off-white thread. *Courtesy of Bill Shea.*

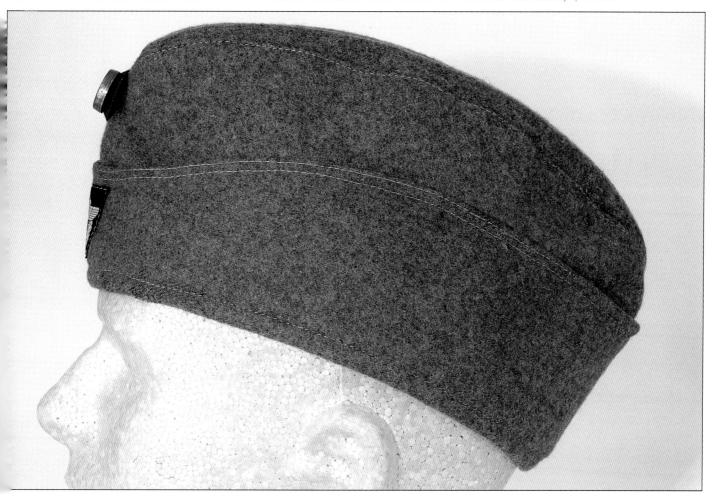

Side view of the field gray Waffen-SS Schiffchen for color comparison. *Courtesy of Bill Shea.*

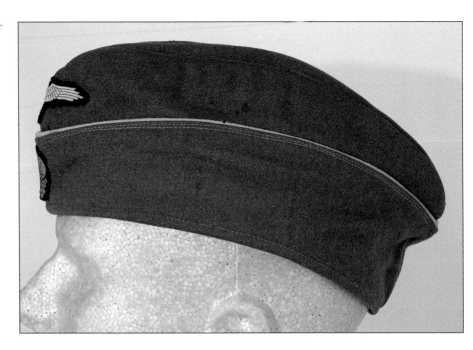

Police NCO/enlisted man's insignia on dark green backing.

cloth field headgear was a bluish tint quite distinctive from the Waffen-SS or Army field gray, or the police green color used for police visor caps. Why this particular shade was selected remains something of a mystery, but so it is.

Linings were cotton (satin weave, thus smooth textured) for enlisted caps, rayon or silk for officers. Occasionally, an Other Ranks cap will have waffenfarbe piping on the skirt edge, but these are rare.

INSIGNIA

The usual Police insignia for the Other Ranks Schiffchen was two-piece: The machine-woven police eagle and wreath, and an aluminum cockade. The eagle and wreath were woven in off-white thread on either a very dark green, or a light greenish-blue rayon underlay cloth.

The eagle and wreath was centered on the front part of the skirt, with the cockade at the cap crown just below the lowest point of the 'V' crease. This is the same position taken by the federal cockade on some modern Bundeswehr caps.

Though a machine-woven cockade was possible, the usual configuration was an aluminum version affixed with prongs. For officers, enlisted insignia could be used, or a hand-woven, bullion eagle/wreath with a bullion or stamped aluminum cockade. It is possible to encounter the combined (one-piece) eagle, wreath and cockade insignia on a pale bluish background, but this insignia was intended primarily for the M43 cap and was not standard for the police Schiffchen. Inspect any such cap very closely for other details that do not fit, and which may be cause for rejection.

OFFICER PIPING

Police officer caps were piped in the same fashion as the Waffen-SS officer Schiffchen, in aluminum mesh cord along the top edge of the side skirt. Officer model caps are extremely rare.

MODIFICATIONS/REPRODUCTIONS

The peculiar blue of authentic police caps makes modifications to contemporary caps from Germany's military services, from other wartime Wehrmacht services, or from other countries a very difficult proposition. Any collector experienced with this particular shade of blue (quite different from the grayer shade used by the Luftwaffe), will not be fooled by a cap in any other color. A much greater danger, therefore, lies in reproduction caps. High quality reproductions will be in the correct form, will be of similar quality, and may be manufactured in material with a color that comes quite close to matching the original police shade. A brand new-looking, "mint" cap deserves special attention. It may be correct, but be sure to check for any small errors that taken together might add up to a reproduction.

Chapter 10

Die Luftwaffe Fliegermütze - The Air Force 'Flyer's Cap'

The Luftwaffe Fliegermütze, introduced in the mid-1930s (in its final form by 1936), was the first service-wide use of a boat-shaped garrison cap. Actually, the Kriegsmarine's field gray garrison cap issued to fortress (i.e. shore) troops used the same form, and in fact, pre-dated the Fliegermütze. The Navy cap, however, was used by only a limited number of personnel, while the Fliegermütze was issued *throughout* the Luftwaffe. Its characteristic curving, boat-shaped top, deep fore to aft center crease (front to back), and gently curving side panels were later applied to the Waffen-SS M1940 Schiffchen as well. The front top corner of the Fliegermütze, however, was just a smidgen straighter than the SS (or the Navy field gray) cap and as a result, tended to angle ever so slightly forward on the wearer's head. The Fliegermütze was worn by all Luftwaffe ranks, with minor modifications to indicate officer grades.

With aluminum piping and officer quality insignia added, the Luftwaffe created its officer model without any need for a new design. The Fliegermütze is without doubt a

Luftwaffe Officer's Fliegermütze in high-grade Trikot. Aluminum cord piping around the skirt. This cap has an Other Ranks (enlisted) national emblem insignia in an unusual, slightly brownish tint, and a standard Other Ranks machine-embroidered cockade. *Courtesy of Bill Shea.*

Side view. The side skirt curve gradient on the Luftwaffe cap reaches its highest point toward the rear, just past the center of the cap (compare with the Waffen-SS Schiffchen). *Courtesy of Bill Shea.*

sharp-looking cap, and has become a popular type of headgear among collectors. With demand brisk, modified caps and reproductions are also on the market, of course. The modifications are primarily authentic enlisted (Other Ranks) caps which have been upgraded to officer. Modifications to actual post-war German military caps (e.g. Bundeswehr), on the other hand, are *less likely* for a number of reasons which will be clarified later. Reproductions are also a concern, though more so in the case of officer caps.

Of particular interest when studying period photographs of Luftwaffe frontline units is that officers appear wearing the Fliegermütze nearly as much as they do wearing the service visor cap. More so than officers of any other arm of the Wehrmacht except perhaps, the Kriegsmarine, Luftwaffe officers wore the garrison cap in the field – that is, at least until the M43 model cap was introduced. Only Panzer officers (from any arm of service) equaled this usage level – quite understandable, considering how much more convenient than a visor cap the garrison cap is for wear on a tank. Another interesting aspect of Luftwaffe usage is that the majority of Luftwaffe *officer* caps were made of fine grade Trikot rather than doeskin (though this does not apply to Luftwaffe general's caps). Among Luftwaffe officers, Trikot material seems to have been much preferred as the cloth of choice for the Fliegermütze – and this is what most of the officer caps you will encounter as headgear collectibles will be made of.

This Luftwaffe enlisted man provides a perfect example of the standard Fliegermütze for Other Ranks personnel. Wool backed, machine-embroidered two piece insignia. *Author's collection.*

CONSTRUCTION

A. Continental/Panzer

Construction of the Fliegermütze was essentially the same as previously described for the Waffen-SS Schiffchen and there are no major differences with the exception of the color and material used, and the gradient of the side skirt's (flap) curve. The Fliegermütze was the first cap model to really deserve the nickname "Schiffchen" due to its boat shape, and this carried over to the Waffen-SS and Kriegsmarine caps as well once these were introduced. This model of Luftwaffe cap was produced in Fliegerblau (gray blue) wool, or in Trikot. The wool used for enlisted caps was a slightly rough, medium weight material often having a felted finish (i.e. appears slightly hairy). Caps produced with average grade Trikot (for enlisted men's caps and uniforms) had a percentage of rayon fibers blended with the wool yarn and were often used for warm weather periods, since this material was thinner and lighter.

The main seam stitching threads on these caps – as always – *should not be visible*. Thread used for the pair of hem stitch lines parallel to the top edge of the side skirt and to each side of the vertical seams was usually in a color matching the Fliegerblau base cloth, however much more noticeable off-white thread was also quite common on Luftwaffe caps and is completely correct.

OFFICER QUALITY/OFFICER PIPING

The officer version of the Fliegermütze was made in either very good quality, thin-napped blue-gray doeskin wool, or in a high grade (fine ribs) blue-gray Trikot. Trikot was of course, the regulation material for these caps; Luftwaffe officers seem to have preferred Trikot over doeskin with the Fliegermütze. This is an interesting contrast to Army officers and the M38, where the reverse was more often the case.

Luftwaffe Other Ranks caps were not piped – with the possible exception of a very short period in the mid-1930s during which branch piping along the upper skirt edge may have been tested and some samples produced. Authentic examples of these caps are not likely to be encountered by most collectors.

The top edge of the side skirt on officer caps was always piped with either a smooth surfaced aluminum mesh cord (gold wire mesh or celleon cord for generals), or twisted aluminum wire cable cord like that used on officer tunic collar edges.

The aluminum mesh cord is normally a bit flatter than the cord commonly used for visor cap piping: It is slightly oblong – that is, vertical (flat) on the sides, rather than perfectly round.

The piping tail strip was usually a lightly textured white cloth that was machine-stitched to the inside of the skirt with two parallel lines of stitches. *None of this sewing work was done by hand.*

One of the two parallel stitch lines actually served to hem the top edge of the skirt itself (which was turned inward). The thick, rounded skirt edge thus provided a seat for the aluminum cord to rest against. The ends of the piping cord met at the rear of the cap. The entire piping arrangement including the cloth tail is visible if the side skirt is pulled down. The raw material edges of the skirt's rear seam are also visible.

Panzer Version

The black Panzer version of the Fliegermütze was only issued to Luftwaffe personnel serving with the Luftwaffe's Hermann Göring field division. The limited number of these caps means that they are very seldom encountered and very much in demand, and thus have a very high monetary value in the militaria market (particularly the officer grade cap). The potential for significant gains has resulted in many fakes, most of which have their origin in reproduction caps.

The design of the Luftwaffe Panzer model was exactly the same as the fliegerblau version, but in a warm, deep black. The Panzer cap was made of smooth-napped wool, *not* Trikot. For officers, the wool often had a doeskin finish, but in other respects differed little from its blue-gray cousin. The rarity of these caps places their market value at a high level. Fortunately, the only other black vintage garrison cap *in Schiffchen form* was the Waffen-SS M1940 – itself a very valuable cap and thus not likely to be altered.

Luftwaffe Knight's Cross winner Arnold Dörning (NCO) and friends strike a pose in front of the flight line. All wear the standard Fliegermütze for Other Ranks. *Courtesy of Dale Paul.*

LININGS

Regulation lining material for Other Ranks (including NCO grade) caps was blue gray cotton in a straight twill weave (that is, a straight, flat ribbed weave, although herringbone pattern cloth was occasionally used as well). Smooth textured cotton linings also appear from time to time.

Twill linings are also found on officer caps, though these probably originated from government stocks. While a cut above the quality level of the standard issue Other Ranks cap and likely available at a very moderate price, government stock caps for officers were no match for a top grade, private purchase piece. On the other hand, a twill-lined officer's Fliegermütze made from low to average Trikot, might actually represent an example of a field upgrade from an enlisted cap once the owner received an officer's commission. The aluminum mesh or cable cord officer piping would thus be added to an enlisted man's quality cap, but would be applied by an individual with at least semi-professional skill.

Fine rayon, or satin linings are the provenance of private purchase caps in which the overall material quality – both inside and out – is noticeably better than on government models. Linings are again predominantly in the regulation blue gray, though other (solid) colors could be used. *Caps with out of the ordinary linings in more than one color, or material with patterns (plaids, etc.), should be avoided.*

Linings on issue caps were sometimes stamped with the name and city of the maker in black ink, as well as with the year of manufacture. Such stamps are absolutely no guarantee of authenticity, however, as they are faked with great ease (see Chapter 14).

An unidentified Luftwaffe enlisted man in a studio shot, most likely for the family, wearing the standard Fliegermütze for Other Ranks with machine-embroidered insignia. *Courtesy of Eric Mueller.*

SWEATBAND

The Fliegermütze *issued* to Luftwaffe personnel during the Second World War did not have a leather sweatband. Private purchase officer caps, and perhaps officer caps from *government stocks* as well, were often fitted with a *partial* sweatband at the forehead. Such bands should always taper downward from the center toward each end until the point where the band joins the cap's bottom edge. They are usually made of real leather (Ersatz [substitute] leather is not common). Complete sweatbands are practically never encountered.

INSIGNIA

Regulation insignia for this type of headgear was the two-piece machine embroidered type (national eagle and the cockade) on a blue-gray wool base (or black wool for the panzer cap). The eagle was in off-white cotton thread for Other Ranks, or in hand-embroidered bullion wire for officers (gold eagle for generals, with the usual cockade). A great many small variations occur with this insignia (particularly with the bullion) owing to the number of manufacturers, but some points they all have in common:

1. **Detail**: Quite good. Distinct feathers, distinct talons.
2. **Eagle's neck**: The eagle's neck is long on authentic insignia. It is a separate piece, not a continuation of the right wing root (viewer's left). Very short necks are indicative of reproductions.
3. **Swastika arms**: With machine-embroidered insignia, the vertical main bar of the swastika crosses *over* the horizontal bar in nine out of ten examples. This is actually a tendency rather than a rule of thumb, and a reversed arrangement does not necessarily disqualify a cap.

The cockade was usually applied to the center of the skirt front without any material (i.e. diamond-shaped) border – that is, the underlay was tucked out of view beneath the cockade.

Machine-woven (BEVo) forms of the national emblem are so rare on the Fliegermütze that any example encountered must be given a very detailed inspection/examination before approval.

Luftwaffe officers were authorized to wear bullion insignia, but seem to have preferred the enlisted eagle version together with a bullion cockade, or both eagle and cockade in the enlisted style (off-white cotton thread) – and so officer caps with Other Ranks insignia are nothing unusual. The only cause for alarm would be visible signs of insignia tampering – or a very unusual insignia type.

BUNDESWEHR MODIFICATIONS

Though the boat shape of the modern Bundeswehr Luftwaffe garrison cap is similar enough to that of its wartime ancestor to allow modifications (particularly of officer models), there are several very noticeable and instantly recognizable differences that may alert a potential buyer of such a cap's true nature:

1. *Color*: The most obvious difference in a Bundeswehr cap (and the most difficult to overcome), is the very intense, deep blue color which, if not altered, is far too dark to ever be mistaken for the much lighter, wartime fliegerblau by any collector with even a modicum of experience.
2. *Side Skirt Curve*: The curve gradient is gentler than that on wartime caps: That is, the Bundeswehr cap's side skirts do not arc as high at their center point as they do on wartime originals.
3. *Branch piping:* Bundeswehr enlisted rank (Other Ranks) garrison caps are piped along the top edge of the side skirt in the wearer's branch of service color (e.g canary yellow for flight or paratrooper personnel). The piping is very thin and quite narrow, made of a flat strip of cotton cloth in Waffenfarbe (branch color) World War II vintage Luftwaffe caps for enlisted personnel *were not normally piped around the skirt*. Only the *soutache* (chevron) on the front of an authentic cap served to identify the wearer's branch, and this practice was discontinued by all the military services in 1942. Bundeswehr officer caps do have aluminum cord piping very similar to that used on pre-1945 caps, but it is slightly thicker.

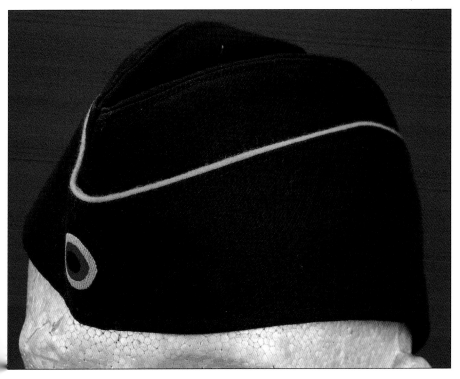

Interior showing the maker's label *sewn* to the right side of the cap lining (indications of which will remain if removed), as well as the very thin top ridge. Linings for these modern caps are in a very dark, bluish gray material (for all grades, including officers).

Bundeswehr Luftwaffe cap for Other Ranks manufactured in 1989 by *Bamburger Mützenindustrie* (a common post-war maker). Very dark blue color, with the federal cockade centered on the front skirt arm, and branch piping along the skirt edge.

Interior. There are no significant interior differences between Other Ranks and officer grade caps; the lining materials are the same, as is the color. The maker tag also appears on the right side — standard for these caps regardless of the wearer's grade. This tag is dated 1992. The maker is *Albert Kempf* (Teunz is the town of manufacture) — a former wartime producer.

Bundeswehr officer's cap. The front section of the skirt is higher than on original Third Reich era headgear, since there is no need to provide space for an eagle on the top section. The aluminum piping is also available in a thicker, (twisted) cable form. The 3D bullion wire federal cockade is attached to the cap with metal prongs. The enlisted-style machine-woven cockade on a diamond-shaped dark blue underlay cloth (as in the Other Ranks cap shown above) is also common (machine-applied).

4. ***Material***: Bundeswehr caps, whether issue or officer model, are usually made of a very light weight polyester/wool blend Trikot cloth. They have a smooth textured, dark bluish gray liner. The cap cloth feels very thin, whereas pre-1945 period Government-Issue Luftwaffe caps were made of a distinctly heavier, felted or lightly brushed wool. (Luftwaffe work caps on the other hand – very rare items, these – were produced in either black or crème cotton herringbone twill material for which there is no comparable Bundeswehr cap.)

5. ***Internal Center Ridge***: All Bundeswehr garrison caps, like their Wehrmacht forbears, are made with the protruding interior center ridge which runs from front to back at the inside top of the cap. Since the Bundeswehr cap material is quite thin, the ridge is also very, very thin (unlike wartime caps). In addition, the Bundeswehr cap ridge is about 3 cm deep – much deeper than ridges found on any model of authentic headgear.

6. ***Insignia***: Bundeswehr insignia is limited in most instances to nothing more than the federal cockade, the colors of which are gold/red/black (black forming the central dot). It is normally machine-woven on a thin backing cloth in the same dark blue as the cap, and applied to the center of the skirt front (the backing is shaped into a diamond). The insignia is machine-sewn to the cap and signs of its removal (diamond outline, for example) may still be visible on any modified cap. Officer caps are available with either the machine-woven cockade or a three dimensional bullion wire cockade (applied with prongs).

EAST GERMAN MODIFICATIONS

For the same reasons as explained earlier in the Chapter on the Waffen-SS Schiffchen, East German Luftwaffe caps – which are no different from the East German Army model – are easily recognized if modified with World War II Luftwaffe insignia.

REPRODUCTIONS

Of far greater concern to collectors are reproductions, which are available from an ever-increasing number of sources. Fortunately, most of these are intended for wear by reenactors, or as a filler for an open space in a collection until the original comes along. They can just as easily be offered as originals, however – as indeed, they often are. The highest quality reproductions differ hardly at all from authentic, wartime caps. Materials are often nearly the same, the wool finish, the stitching – all accurately recreate the original piece down to the RB numbers and maker mark stamps on the lining. Now throw in a little field wear from a reenactor and the danger increases dramatically. The only defense against such caps is the sense of age/time that most collectors develop and possible flaws in the insignia.

Reproduction caps usually use reproduction insignia, and although the quality of these in general has improved considerably in recent years, it is still usually possible to recognize them through some flaw or other – no matter how insignificant – or by incorrect application (stitching). Reproduction caps with *authentic* insignia (which is generally, if not readily, available), are a more difficult proposition and here again it is the thread, as well as the application, that might help to identify such replacement work. Threads made of 100% cotton or a cotton/rayon blend are becoming increasingly difficult to locate, most companies having shifted long ago to producing synthetic thread with a higher tensile strength. The color selection among the cotton threads still available is likewise very limited. Thus, the thread used to attach insignia (unless original period thread), may provide a clue.

The age sense on the other hand, is a very subtle, subjective thing that depends on many factors added together to form an overall impression of the authenticity of an item. This sense usually comes only with experience in seeing and handling actual pieces, both authentic and fake.

Visual, tactile sensations, the sense of smell; all contribute to building a subjective – *not* analytical – assessment of the true age of a given piece. Physical factors that can be identified (e.g. wrong shade, faulty sewing, etc.) may then confirm the initial assessment. For example, although absolutely mint condition unissued headgear does appear on the market from time to time, it usually smells old, in some sense; you can *feel* that the dulling of colors, if any, is attributable to real age, not to artificial aging procedures. With caps that have been used by reenactors on the other hand – *which show much of the same kinds of wear as the originals* – you may still catch a scent of sweat. After fifty plus years, however, no sweat scent should still remain on an authentic cap.

A combination of many impressions taken together form an overall *subjective* impression of a given piece. With a good age sense, you either don't feel good about an item, or you do.

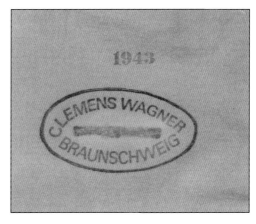

A grayish tan lining on a tropical cap for Other Ranks manufactured by Clemens Wagner. The oval stamp form was typical of this company – but *any* maker stamp can be easily faked. *Courtesy of Dale Paul.*

B. Tropical

The tropical version of the Fliegermütze was cut to the same design as the continental cap, but was manufactured from khaki brown colored cotton Trikot (i.e. a straight-ribbed material). Take a moment here to remember that Trikot is a twill weave fabric with parallel, diagonally-running ribs. Though diagonally arranged, the ribs themselves are straight and unbroken. Trikot cloth *does not* have the same offset, repeating *V* pattern characteristic of the cotton *herringbone twill (or drill)* used for the Luftwaffe black and tan work caps – which material the Germans referred to as *Drillich*. (work caps are not covered in this book).

Luftwaffe personnel also used a blue gray colored warm weather tropical cap in the exact same cotton Trikot fabric. Examples of these caps appear very seldom in the militaria market, and their value is thus quite high.

The shade used for tropical headgear varied slightly between makers, from a khaki brown to a slightly more mustard color, and on authentic caps various degrees of sun-fading must also be taken into consideration. Evaluating the correctness of the material color quickly becomes a somewhat problematical exercise for a novice collector. With this type of headgear, experience is definitely required. Initial purchases of tropical items should be from a trustworthy, experienced dealer (while at the same time, do not forget to improve your knowledge by asking for detailed explanations).

Side skirts on wartime tropical caps are independent and moveable – an important point to remember when confronted with a cap having only *simulated* side skirts. The smooth textured cotton linings found in these caps were normally a standard red color used by all the military services for tropical headgear, with RB numbers and/or maker names (sometimes also the city) and year of manufacture stamped on the lining. Tan or grayish tan linings that appear from time to time are also correct, particularly when attributable to a specific maker such as the **Berolina** company (Berlin) for the tan color and **Clemens Wagner** (Braunschweig) for grayish tan.

INSIGNIA

Standard insignia was the machine-embroidered national emblem in off-white cotton thread and a separate black/white/red thread cockade, both on a khaki background. This insignia was applied in the same fashion as insignia for the continental cap and

Light khaki color cap, with a machine-embroidered national emblem in a triangular shape, but with its top edge cut to match the curve of the wings. *Courtesy of Bill Shea.*

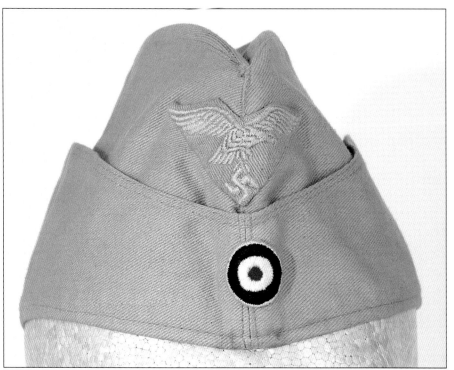

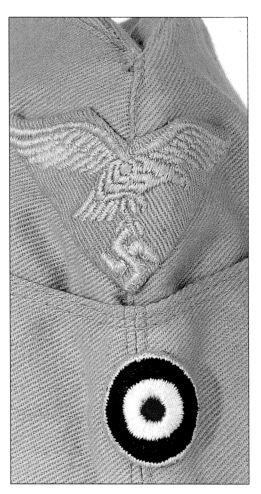

Insignia close up. National emblem (only) applied with the backing cloth trimmed and tucked to form a straight-edged triangular shape (similar to the shape of the combined eagle/cockade insignia later used on the M43). *Courtesy of Bill Shea.*

was normally done at the factory after production. The eagle could be applied with the backing cloth cut to shape (i.e. a narrow, even border all around the eagle), or with the top cut to shape and the bottom angled down to a point (i.e. somewhat trapezoidal). It could also be applied with the backing cloth trimmed and/or tucked to form a complete triangle with the base side up, and collectors may come across one or each of these forms on the militaria market.

The cockade, available in either a padded (puffy-looking) or flat form, could also be sewn directly to the cap with no background cloth showing – and indeed this seems to be the most common variety. It could also be applied with the background cloth edges cut or tucked to form a vertical diamond shape (the cockade was usually flat – *not* padded – on this style).

Officers usually used the standard enlisted/NCO (Other Ranks) machine-embroidered insignia, with a definite tendency toward the padded-style cockade (without any visible underlay). As mentioned earlier, hand-embroidered bullion insignia on a khaki underlay was authorized for wear, but rarely used due to the difficulty in cleaning that it presented. There is no reference in German publications thus far researched – or in other reference works for that matter – to a flatwire (machine-woven) or a machine-embroidered eagle (in aluminum thread on a khaki base) ever having been used on Luftwaffe officer caps. This is not to say that such insignia was *never* used; "never" in connection with German militaria is a risky claim since exceptions if anything, are the rule. In this sort of situation, however, a lack of evidence *for something* is, in effect actually evidence *against it.* The following example shows a superb Luftwaffe officer's tropical garrison cap that is *almost* perfect – but in all probability is an excellent reproduction designed to fool the unwary. The material is first rate, the weight is good, the cut is perfect. The lining is also dead on, both the color and the attachment (with the central ridge). It even comes complete with the ink stamped makers mark Otto Schlient (a valid maker), though makers marks are no guarantee of authenticity.

The twisted aluminum cable-style piping (as on the edge of the tunic collar) is excellent, and extremely well applied (by machine). In fact, the overall quality of the cap's construction is above reproach. The insignia is nicely (and professionally) machine-applied with a fine, even, tight zigzag stitch (in matching color thread). The cockade is straight machine-stitched (thus not an exact match in style between eagle and cockade). While impressive, the national emblem insignia itself, however, raises questions. It appears to be an example of a machine-embroidered piece in aluminum wire on a khaki background. The shape of the eagle is correct; feather detail is lacking, however, and the surface is somewhat scratchy (rough). The insignia also feels much too stiff.

It is worth noting here that much of the recent reproduction insignia originating out of Pakistan, for example, also tends to be far stiffer than the originals – or the backing at least, is stiffer. Whatever the case may be – the original insignia was both lighter and more flexible.

Interior – as it should be. The lining is pristine – too good, in fact!

Below: Front view. Note the zigzag machine stitching around the entire circumference of the insignia, and the officer aluminum wire strand cable piping (perfectly attached). A fine looking cap – unfortunate, that appearances in this case, may be deceiving.

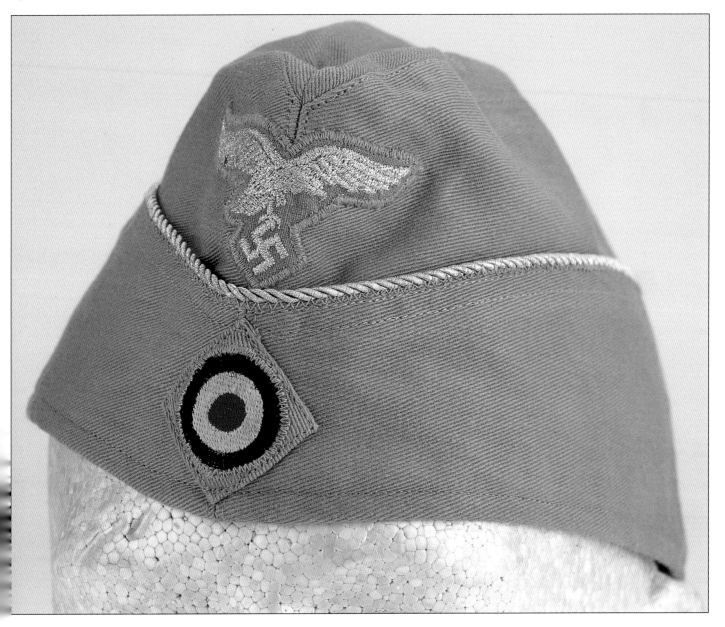

The cloth underlay for this aluminum thread eagle is an exact match with the cap material color. The general shape of the eagle is excellent, and the insignia is very nicely applied, note however the lack of any feather detail, and that the thread looks quite stiff.

On the other hand, this eagle *could* represent a form of heretofore unknown insignia; but the likelihood of this is so remote that the only reasonable conclusion remaining is that the cap is *not original*. One final detail of note, is that the khaki underlay cloth for both cockade and eagle is a perfect match with the cap material. While certainly possible on a sun-bleached cap (the sun after all, is a great equalizer), it seems very unlikely on a new cap with insignia and cap normally coming from two separate sources (this cap looks absolutely unissued – right off the production line). While German production *was* good, this would indeed be stretching things a bit much.

The undocumented insignia and the peculiarities associated with it – be they color or texture, etc., raise too many questions for comfort. The cap's absolutely nagelneu (as the Germans would say – brand new) appearance merely adds nail in the coffin of doubt.

Remember also that authentic examples of the Luftwaffe tropical officer's cap are extremely rare by any standard, and thus very expensive. With this in mind, the uncertainties raised by a cap such as this one are too great to risk any large sum of money on it.

As a reproduction, however – offered at the price of a reproduction – it is certainly a beauty.

Chapter 11

Die Kriegsmarine Feldmütze, und Die Bordmütze M/1939 - The Navy Field Cap (Field Gray) and ('Shipboard') Cap Model M/1939

The Kriegsmarine first issued a field gray, boat shaped garrison cap to its shore personnel (those working primarily at fortifications), as early as 1935. This cap was a contemporary of the initial form of the Luftwaffe Fliegermütze. Which military service therefore had the distinction of introducing the first garrison cap in Schiffchen form is not clear. The field gray cap and the dark navy blue Bordmütze (which entered service officially in late 1938) differed little from each other, despite the three-year difference between them. The Bordmütze was officially designated the M1939 due to modifications made after its initial introduction, which were not completed until 1938. The Bordmütze had a higher front than the Luftwaffe or Waffen-SS models, which was reinforced (prior to the 1939 modifications) to help it keep its shape (as it was intended for use with headphones and other devices worn across the middle of the cap).

The high front of the Bordmütze, arcing backward to become the cap's top ridge, genuinely looked like the prow of an overturned boat hull. This cap – if any – was truly deserving of the nickname Schiffchen, since it looked more like a boat hull than any other model despite the fact that the top of the Bordmütze actually formed a straighter line (even when the cap was being worn) similar to the Army's M34. Luftwaffe and Waffen–SS caps on the other hand, appeared to curve just a bit more when actually atop a wearer's head.

The Bordmütze's side skirts were gently curved, with the highest point of the arc lying just to the rear of the cap's centerline (similar to the Luftwaffe Fliegermütze).

The officer field gray cap and Bordmütze were the same as the enlisted grade version with the exception of better material quality and gold mesh cord piping along the top edge of the side skirts. As with any Kriegsmarine item, these caps (including the khaki and white versions, which are not covered in this book) are much in demand with the supply very limited. At any time this short supply and high demand situation exists, modified caps or complete reproductions are guaranteed to appear on the market as people out to make a dishonest dollar seek to cash in on high price values by cheating unwary collectors. As always, be very cautious of prices offered for a rare item that are far lower than the usual market value for such a piece.

When confronting a cap in the militaria market that you suspect may be modified from a postwar cap, keep in mind that the likeliest candidates for this are either the Bundeswehr blue Bordmütze, or the East German navy equivalent. On the other hand, reproduction caps are also very likely, and will be more difficult to spot in some cases.

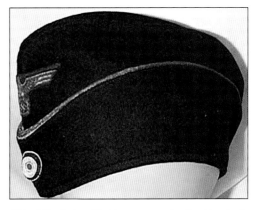

Officer's Bordmütze. Machine-woven national emblem in gilt thread. Note the high arc to the side skirt curve. The very high front corner, with empty space between eagle and cap top, is characteristic of all Kriegsmarine garrison caps. *Courtesy of Bill Shea.*

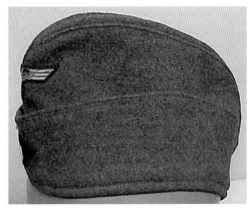

Field gray Kriegsmarine Feldmütze for Other Ranks. This cap is without any soutache, the use of which was discontinued after 1942. *Courtesy of Bill Shea.*

Officer's Bordmütze, insignia and piping close-up. Machine woven national emblem in gilt thread (dark blue rayon underlay), bullion cockade: note the individual bullion ring encircling the red bullseye. The mesh cord piping is slightly flat on the sides, rather than tubular, and its off-white cloth tail strip is visible where the blue cap material has pulled free of the hem stitching. *Courtesy of Bill Shea.*

CONSTRUCTION

The garrison cap for coastal units was made of field gray wool (average grade), though the Kriegsmarine version of field gray leans decidedly toward the green.

The wool was fairly short-napped, which made the cap appear thinner (lighter weight), than other models. As always, there is the standard protruding interior ridge, with the lining stitched together. The navy blue Other Ranks Bordmütze was also produced in a very short-napped wool with a fairly smooth hand (feel) – even though the wool was not doeskin. The most significant overall impression of Kriegsmarine garrison caps is that they are lightweight.

In keeping with standard German practice, the main seam stitching on both the navy blue and the field gray cap *is not visible*. Thread for the parallel hem stitch lines on the top edges of the side skirts and straddling the vertical seams was usually in a color matching the base cloth (e.g. field gray, or navy blue), however off-white thread found on field gray caps is also correct.

OFFICER CAPS

Officer caps were made in high quality, shorn (i.e. short-napped) field gray or navy blue doeskin wool.

The top edge of the side skirt on officer caps was piped with a smooth, gilt mesh cord (gold wire or celleon cord). This can cause some confusion with Army caps for general officers. The M38 Army officer cap, however, was piped around the cap top *and* sometimes the front edge of the side skirt scallop in gold cord for Generals, while Navy officer caps were piped *only around the top edge of the side skirt*. The mesh piping is normally a bit flat on the sides, since it did not have a heavy round cord at its center. The piping's tail strip was machine-stitched to the inside of the skirt using two lines of parallel stitches, one of which simultaneously hemmed the top edge of the skirt (the raw edge was folded over to the inside). The entire piping arrangement is visible if the side skirt is turned down, as well as the raw material edges of the skirt's rear seam.

LININGS

Regulation lining material for Other Ranks (including NCO grade) field gray caps was a gray twill cloth. For the Bordmütze, the standard lining was black moiré (moiré is a smooth-textured material with a wavy pattern visible from different lighting angles). Officer caps have linings in the same basic colors, but of higher quality materials including rayon (top quality rayon is very similar to real silk in both sheen and feel hence the nickname "artificial silk"). The linings on Kriegsmarine caps are in one solid color, again, the standard practice on all German headgear. *Avoid any cap with a multicolored, or plaid lining.*

Right: Officer's Bordmütze, *not* original. The front piece of the side skirt is much too high, while the curve of the skirt is not quite as high as it should be, otherwise a very good copy. The machine-woven (BEVo) Other Ranks version insignia is correctly positioned, with the eagle's stylized wreath just above the edge of the piping.

Far right: Same cap from a slightly different angle. Compare the arc of the skirt curve with that of the original cap at the beginning of this chapter. The tradition badge (always on the left side) is for the U-404. Tradition badges in general, are very susceptible to counterfeiting; many were hand-made onboard ship, and the "homemade" look can be convincingly copied with relative ease.

SWEATBAND

Other ranks caps issued by the government from government stock were without sweatbands. It was possible for an individual to secure a private purchase cap with this feature, but in this case the sweatband was nearly always a partial leather band covering the forehead area only.

INSIGNIA

For the field gray and navy blue cap, two-piece machine woven insignia on a field gray (also dark green) or dark blue underlay (respectively), was regulation. The underlay material was usually rayon. The eagle was woven in golden-yellow thread for Other Ranks insignia. Officers were authorized to wear a gilt wire bullion (or celleon) hand-embroidered national emblem; some officers, however, wore the enlisted version machine-woven eagle instead.

The cockade was usually applied with a straight machine stitch to the center of the skirt front with the underlay cloth folded into a diamond-shape. Hand applications certainly occurred, but in general, such examples should be very neatly applied with relatively equal stitch lengths. The officer cockade was usually a high relief, bullion type (hand-applied, with no stitching visible). No threads should be visible inside the cap (i.e. they should not come through the lining).

BUNDESWEHR MODIFICATIONS

Modifications of Bundesmarine caps are certainly possible. The boat shape remains the same today, and the color is still navy blue (Navy traditions do die hard). The Bundeswehr has gone through at least two cap models since the Federal Armed Forces was created in the 1950s, but both models differ from the World War II Bordmütze in material. The Bundesmarine caps are not made purely from wool, but rather from a polyester/wool blend. As with the modern Luftwaffe Fliegermütze, the Navy caps have a much thinner, lighter feel. The federal cockade is centered on the skirt front above the cap edge. For enlisted (Other Ranks) caps, the cockade is the usual machine-woven type of gold/red/black on dark blue underlay cloth, shaped into a diamond and machine-sewn to the cap. Removal may leave signs, but if the Kriegsmarine cockade is applied over the exact same spot, any indications of previous insignia may not be visible.

Neither the early or current model Bundesmarine Bordmütze has the black cotton moiré lining that was standard on wartime caps (though modern linings can be replaced, of course). Linings on Bundesmarine caps are a solid color (without any water pattern), with a large, white manufacturer's label sewn (as usual) to the lining on the right side of the cap. As of this writing, it is does not appear that the Bundesmarine has ever fielded a cap specifically for shore-based personnel (i.e. an equivalent to the field gray Kriegsmarine cap).

The Bundesmarine officer's Bordmütze, like its pre-1945 cousin, is piped in a thin gold cord along the top edge of the side skirt.

EAST GERMAN MODIFICATIONS

Modified East German Navy caps are among the easiest fakes to identify. This is due to the Soviet influence evident in the very high sides of these caps – something that simply can not be disguised. Other factors are equally damning:

Kriegsmarine Knights Cross bearer Leutnant Albrecht Achilles wears an officer's Bordmütze with tradition badge. The gilt officer piping on the skirt is easily visible. *Courtesy of Dale Paul.*

East German [People's] Navy officer Bordmütze in its original form (unmodified). Note the high cap sides. Thin gold cord piping is applied to the cap crown, but not the skirt edge – quite incorrect for a real Kriegsmarine cap (piping is the reverse: skirt edge, *not* the crown). The side skirts are secured to the cap along their entire length and can not be moved.

1. **High Side Skirt**: The side skirt on East German caps is very high, leaving only an extremely narrow strip of material between the top edge of the skirt and the top edge of the cap.

2. **Simulated Skirt**: The skirt on East German caps *is simulated only*. This feature in and of itself is sufficient to destroy any illusion that such a cap might be an original.

3. **Material**: East German caps are made of a lightweight 65% polyester blend (i.e synthetic) Trikot fabric. (Bundeswehr caps also have a ratio of 65% polyester).

East German Navy officer caps show no material quality differences from Other Ranks caps. In fact, the only difference at all is thin gold piping around the crown of the cap, which is completely incorrect for a World War II Kriegsmarine Bordmütze (original caps *were not* piped around the crown).

East German caps are so easy to spot in fact, that only a person who has made no effort at all to familiarize himself with the collectibles he is planning to invest in, could be fooled by such a modification. For those of you who wish to know (easily) your enemy, study the photos!

REPRODUCTIONS

Reproductions of the Kriegsmarine field gray caps and the Bordmütze can be a serious concern depending on the amount of effort made by the manufacturer to be faithful to the original – and of, course, access to original quality raw materials. A very good reproduction passed off as an original can be a challenge to spot, though 'field wear' should be of less concern (not much call for Navy reenactors, after all). With this in mind, consider again the 'age' sense. Does the cap look old? Does it look worn? Does it *smell* old? Age does have a smell, after all – one that is hard to fake: that slightly musty scent that comes from years locked up in an attic, or a closet. A brand new cap

Interior of the same cap. Black ribbed lining. The white tag at the top of the cap actually forms a loop (purpose uncertain). Note the white inked NVA mark stamped on the lower edge of the lining (National People's Army – used for all services)

unless you are convinced otherwise beyond a shadow of a doubt by any number of other confirming factors – is probably just what it looks like: a brand new, *reproduction* cap. Check the width of the side skirt, and the height of the skirt curve. Study the sewing – look for inconsistencies; inspect the insignia for the same (and the quality of sewing work used to attach it). Poorer quality reproductions are bound to show this fact through one flaw or another.

Chapter 12

Die Feldmütze M42 - The M42 Field Cap
Field Gray Version/Panzer Version

FIELD GRAY Version

The M42 field cap was originally intended as a winter replacement for the M34, but quickly became an interim model between the M34 and the M43. It was produced in only moderate numbers and was only issued for a limited time before being replaced by the M43 standard (general issue) field cap. The M42 in fact, was essentially an M43 without a visor. It had a wide, doubled-over side skirt that could be pulled down around the face which was also characteristic of the later M43. To make repositioning of the

M42 cap with correct machine woven insignia (eagle in matte gray) on an oxide green, T-shaped backing cloth. *Courtesy of Bill Shea.*

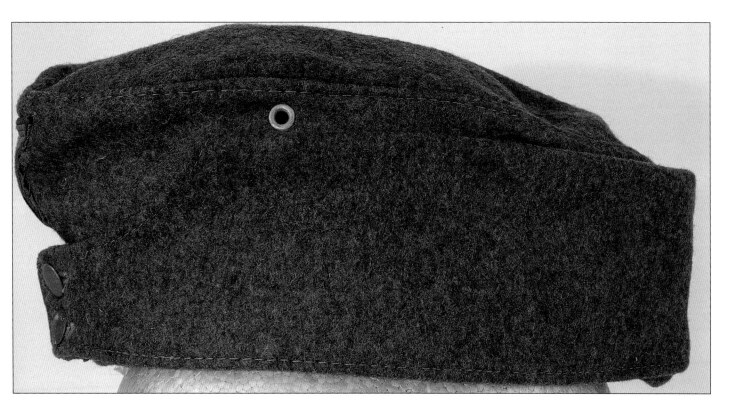

Side view. *Courtesy of Bill Shea.*

side skirt (down around the face) somewhat easier, the two narrow left and right (wearer's view) front arms of the skirt – which together formed the front scallop section – were *not* sewn into a complete, unbroken piece (as was the case with the M34).

Instead, the length of the skirt arms was extended slightly on each side to permit an overlap, with the end of the left arm overlapping the end of the right. The flaps are secured by two vertically aligned buttons sewn to the end of the right arm. The buttons were painted field gray and were made of pebbled aluminum. Black M42 caps issued to Panzer troops used black painted, pebbled aluminum buttons. Other button types were not standard on this cap and thus may not be original.

There were only two significant differences between the M42 and the M43 cap. The first was a pair of field gray (or black) enameled air vent grommets, which were positioned one on each side of the M42 body above the scallop turn up point. The other difference was the continued presence of a center seam in the cap top, which was eliminated on the M43.

Intermittent discussion continues among collectors regarding whether the Germans re-issued M42 caps with visors added – essentially converting them into M43 caps. There is no definitive evidence documenting such a practice, however, or how it might have been conducted. It is more likely that existing stocks of the M42 were issued until exhausted. Any re-issue would have entailed a recall of all caps still on hand at military depots to whichever manufacturer(s) had been assigned the task, or else the work would have to have been done on the spot; in either case, no easy thing to accomplish from a cost-effectiveness/quality standpoint. Though it is certainly within the realm of possibility that some caps may have been converted, the likelihood of the average collector actually encountering an authentic example of such a variant is extremely low.

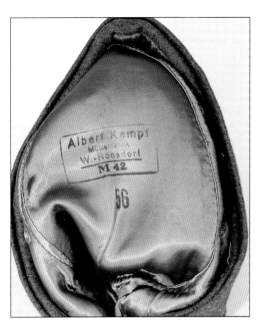

M42 interior. Regulation rayon lining in a twill weave. The silky texture of the rayon produces very fine, faint ribs. *Courtesy of Bill Shea.*

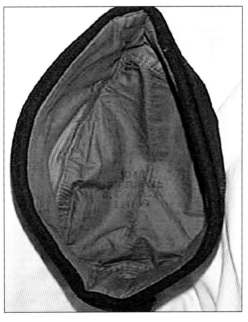

Interior of a Panzer M42. In this case, the lining fabric is gray cotton in a smooth (linen) weave. With use, the gray often takes on a slightly brownish tint. *Courtesy of Bill Shea.*

CONSTRUCTION

Exterior

The base cloth for M42 caps was produced in an average grade wool/rayon blend colored field gray, or a deep, warm black for Panzer troops. As with the M34, the black wool used for the Panzer version generally had a somewhat softer nap than the field gray cap, though why this was the case is not clear. The fabric was thick, with a felted finish common on field gray-colored material. As the name suggests, this finishing technique gave the final fabric a slightly felt-like texture when the nap was shorn, or a slightly hairy texture on versions with the nap left long. Nap lengths varied depending on the textile manufacturer that supplied the cloth, and the selection made by the cap manufacturer.

The M42 cap was only issued to Army troops, and this relatively limited distribution also meant that there little to no opportunity for variations in the width of the skirt arms to occur, such as was later the case with the M43. Thus, M42 caps do not appear in a one button, narrow flap style.

OFFICER VERSION

Nearly all M42 caps were produced primarily on government contracts for issue to Other Ranks troops. *No official version for officers was ever introduced.* It is clear from photographic evidence, however, that some officers (including generals) did, in fact, wear the M42 cap. In such cases, the insignia was normally the standard enlisted version machine woven type (dull, mouse gray on green oxide underlay). On very rare occasions an M42 cap appears with aluminum officer piping around the cap crown. Such pieces almost certainly represent a field upgrade of a cap worn by an NCO who had been promoted to Leutnant (Lieutenant) – that is, to officer rank.

LININGS

Cotton or rayon linings were usual for the M42, in a gray or brownish gray-colored twill weave (fine, straight rib lines). Caps produced toward the end of the M42's production life (early 1943) might also have gray rayon linings in a herringbone pattern. Produced as they were for government issue stocks, M42 caps do not have any form of leather sweatband.

PANZER Version

The black M42 cap for Panzer troops was the same as the field gray version except for the color of the base cloth, the black enamel on the air vent grommets, and black painted buttons. Production of the black M42 cap for Panzer was even more limited than that of the field gray version, and thus these caps – though many collectors do not find any particular attraction to the M42 design – are very rare. With rarity comes high prices and the Panzer M42 price range, in year 2000 dollars, is between approximately $1000 to $1300.

INSIGNIA

Regulation insignia for the field gray M42 was a combined national emblem (eagle and cockade machine-woven on a triangular (trapezoidal shaped) piece of underlay cloth. The backing was woven slightly thicker around the eagle and cockade, forming a distinctive *T* shape. When applied to the cap, the excess material was trimmed and folded under the insignia. Thus, the standard, correct form of insignia that you should expect to find on the M42 is the machine-woven *T* shape (standard insignia for the Bergmütze as well). The thread was usually dull mouse gray on a green oxide-color underlay. Visually, the eagle was quite difficult to at a distance. Standard insignia on the M42 Panzer cap was again in *T* form, machine-woven in pale gray thread on a black (rayon) underlay cloth.

Less widely issued (especially the Panzer version) and in service for a much shorter time than other cap models, the M42 provided correspondingly less opportunity for non-regulation forms of insignia to occur. Any examples of the M42 encountered with other than the *T*-style insignia should be viewed with caution until inspected and proven authentic. No metal insignia of any type was ever authorized for wear on this cap.

In the instances when officers made use of the M42, they appear to have remained with the enlisted insignia – which was most likely already in place on the cap (machine-sewn, or neatly hand-sewn) when received.

SOUTACHE
The M42 entered service at nearly the same time that use of the soutache was discontinued. This cap model was never intended to take a soutache, and indeed, there is no place to conveniently mount one.

MODIFIED BUNDESWEHR CAPS
Modifications of Bundeswehr winter model field caps are possible, but are also fundamentally flawed by the fact that they have no separate skirt ends. That is, the side skirt forms an unbroken whole as on the M34, without any separate ends that can be buttoned together. Authentic or reproduction buttons can, of course, be added for appearances' sake; the front scallop skirt has no button holes, however, and remains one complete piece – a fact that simply can not be hidden.

REPRODUCTIONS
Reproductions are the most significant danger with the M42. The cap is not very difficult to reproduce in terms of design, and the main consideration in a reproduction lies therefore, in the quality and weight of the wool used in the cap's construction. Fairly accurate lining fabrics are somewhat easier to come by than the wool, and do not represent much difficulty for reproduction manufacturers. The principal weakness with such caps remains the body wool, which is usually of a somewhat lighter weight than original fabrics on authentic caps. The color may also be very close, but looks too new or too green. When these factors are present take a close look at the insignia as well, keeping an eye open for signs of a reproduction. If at first glance the insignia is incorrect (i.e. not in *T* form), then the entire cap becomes suspect and requires careful scrutiny.

Though it may seem like such an obvious error no one could possibly make this mistake, some reproductions have, in fact, appeared with the buttons mounted on the left side skirt arm (i.e. the flap) rather than on the right. A cap with this arrangement is definitely wrong. *The left skirt arm should always overlap the end of the right arm.* The buttons are always mounted on the right, the button holes are always on the end of the left skirt arm.

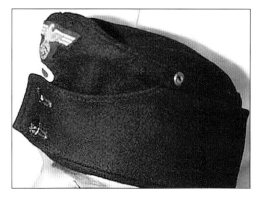

M42 for Panzer troops. Regulation *T*-pattern insignia, black pebbled aluminum buttons. This example has unpainted grommets. *Courtesy of Bill Shea.*

The *T*-pattern combined eagle/cockade insignia is very neatly hand-applied on this cap. Note the black paint on the buttons has been buffed off in areas. The soft, felt-like finish of the black wool used for the Panzer cap is clearly seen in this photograph. *Courtesy of Bill Shea.*

Chapter 13

Die Einheitsfeldmütze M43 - Standard Field Cap M43
Die Bergmütze

INTRODUCTION

The M43 model standard or General Issue field cap is one of the most popular forms of headgear ever issued by the German military. This is as true for collectors of military headgear today as it was for German soldiers during the Second World War. This cap was universally adopted by all four military services: Heer, Waffen-SS, Luftwaffe and the land-based forces of the Kriegsmarine. The basic design remained fairly consistent between them with the exception of the front skirt flaps (i.e. the skirt arms, which buttoned at the front).

With this relative uniformity in mind, M43 models are grouped together within this one chapter, but are presented separately for each military service. Cap materials and construction etc., were essentially the same for all services and are therefore fully described only in the first section (Army M43) in order to avoid repetition, unless there is a significant difference that needs to be noted.

Tropical visored field caps were not variations of the M43, nor were they known by this designation; most, if not all, having been fielded much earlier. Since the design of these caps was based on the Bergmütze and the appearance is similar to the M43, they are also included with the M43 section under the appropriate service. The Bergmütze itself – though actually the predecessor on a timeline to both M43 and tropical caps – is the last item covered.

A. I. DAS HEER – THE ARMY
Part 1: Continental
Part 2: Panzer
Part 3: Tropical

Part 1:
CONTINENTAL

Introduced in June of 1943, the M43 cap is so representative of the German soldier that it is a regular in war movies. Sgt. Rolf Steiner in the film *Cross of Iron* wears a

Authentic general's M43 cap. Trapezoidal machine-woven Other Ranks matte gray insignia (not unusual). Gilt mesh cord piping (tarnished), gilt pebbled buttons. The side skirt's top edges have only one line of stitching rather than the usual two. Note the relatively square visor (rounded corners, but essentially square). Very high-grade doeskin wool, in a somewhat rare bluish tint. *Courtesy of Bill Shea.*

M43 cap throughout the entire movie. A similar-style cap appears in the motion picture *Enemy at the Gates* (released in 2001), but the events portrayed in this film were set in Stalingrad in late 1942 – much too early for the M43. Use of the model M43 cap in this film would be incorrect. The two-button style Bergmütze, however, was nearly indistinguishable from the M43 at a distance – and there were certainly some individuals at Stalingrad (and on the Eastern Front in general) who had previously served with a mountain unit and no doubt continued to wear their Bergmütze caps in their new regular units. This would account for the occasional odd photograph dated to late 1942 in which one or two soldiers *appear* to be wearing the M43.

The M43 was issued first to all field troops, then to soldiers serving in rear areas as available stocks permitted. Initial supplies of the new cap appear to have been sufficient, since the cap appeared in relatively short order among the troops on all fronts, most of whom found it both reasonably sharp in appearance and, more importantly, very comfortable to wear. It quickly became a favored form of headgear. This was no

In contrast, a modern Bundeswehr Army general's *Bergmütze* – as the old M43 design is now known. General's rank is indicated by the golden sabers and gilt buttons; gilt crown piping is an option, but is not used on this example. The color is a medium gray, though it appears darker in the picture. The visor looks much thinner than wartime originals, though still fairly square.

Modified postwar cap in an M43-style cut, offered through the Internet. The origin of this cap is not certain, but it is definitely not a pre-1945 piece and the mismatched insignia is completely incorrect. The cap color is wrong, the visor length much too short (more like a Bergmütze).

Rear of an officer's M43 showing the seam used to join the two skirt halves together. A very small portion of the rear seam for the cap *body* is visible directly above the rear skirt seam (the same on all M43 caps, regardless of service). Note the overlap method used to sew in the ends of the officer piping. *Courtesy of Bill Shea.*

real surprise: the basic design for such a cap had, in fact, long been available in the form of the *Bergmütze* (originally developed in Austria) and the original civilian version of the mountain cap had always been popular with the German public.

For the Army model M43, the visor was extended slightly from the Bergmütze design, and the skirt arms were widened to hold two-buttons. The skirt arms were usually squared off at the ends (though some makers preferred beveled corners). The M43 crown was lowered slightly from that of the Bermütze, and unnecessary interior padding (i.e. horsehair support in the front) was dispensed with. In other respects, the M43 remained essentially unchanged from its mountain cap predecessor.

The Army was far more consistent in its M43 design than its sister services, and this is evident in the relatively minimal variation in skirt arm width, or in the number and type of buttons found on Army Government Issue caps (i.e. manufactured on contract).

CONSTRUCTION

Exterior

The base cloth for Army M43 caps was made of average grade woolen yarn in a flat weave (i.e. not Trikot). The fabric was thick, and fairly heavy in comparison with the cloth used on officer versions. Naps on these caps varied, ranging from short and relatively smooth naps on earlier caps to coarser, heavily felted naps on caps manufactured later in the war. Lightweight summer versions also appear (not common) in a non-wool Trikot material.

The M43 side skirt is fully serviceable as a pull down face protector. The skirt is unlined (bare wool), since the cap was designed to be worn over a *Kopfschützer* (stretch headcovering), in extreme cold weather conditions. Buttons are affixed to the end of the *right* skirt flap, with the left flap overlapping the right. This arrangement is exactly the same as that found on the earlier M42 model cap. Period photographs occasionally show examples with the flap overlap reversed (right flap over the left), but this was unusual.

M43 cap for Other Ranks in lightweight material for wear during warm weather. Standard insignia, hand-applied. Note the beveled ends of the skirt arms. *Courtesy of Bill Shea.*

Narrower, single-button style arm flaps (some with rounded ends) are far less common on Army M43 caps issued by the government, than is the case with Luftwaffe or Waffen-SS versions of this headgear. This is particularly true for caps issued during the initial production period (June 1943 through mid-1944). The skirt was not constructed from one complete piece, but rather from two halves: one attached to the base on the left side of the cap, the other to the base of the right side, with the seam that joined them together being centered at the rear.

M43 caps issued to enlisted troops were *not* ventilated, and *should not* have any air vent grommets. The cap top had no central seam. No stitching for major construction seams should be visible on the outside of the cap, though external reinforcement (e.g. hem stitching lines) do appear on each side of, and parallel to, the main vertical seams (just as on the M34 or M42). On Army caps, this stitching was normally done in field gray or some other, dark-colored, matching thread.

No branch piping of any kind was used on wool M43 caps, nor was there a soutache.

During the spring and summer, officers often wore a private purchase M43 in cotton or a very fine wool Trikot, since this material was generally lighter and cooler than the normal wool fabrics (including doeskin). Lightweight caps of this type are rarely found for enlisted personnel.

VISOR

The M43 visor was a cloth-covered pasteboard (or heavy cardboard) stiffener. Toward the end of the war, even pressed paper was used. An additional strip of material was affixed to the top front edge of the visor prior to the cloth covering being added. Once in place, the strip formed a thin, fairly flat ridge along the top leading edge of the visor on authentic, Government Issue caps. The strip was intended to give added support to the front of the brim and thus reduce any tendency for it to bend and crease when held by the edge. At the same time, a thickened leading edge served to strengthen the entire visor. This ridge (or lack thereof) is easy to feel between thumb and forefinger.

The number and spacing of sewing lines on the cap visor (bill) varied depending on the manufacturer. The closest there is to a standard is two parallel stitch lines on the *underside* (reverse) of the bill only, running close to the front edge.

No sewing should appear on the upper side of the visor – with the inevitable exception found only on some private purchase officer caps, and these examples always seem to be made of Trikot material (whether wool, or cotton) rather than doeskin.

On very late war caps, the bottom of the visor was sometimes left uncovered in order to save on material. In this case, the cloth visor covering wraps around the front edge of the brim and is sewn to the visor bottom just behind the front edge, leaving most of the tan, or creme-colored cardboard or pressed paper stiffener exposed.

OFFICER CAPS

M43 officer caps were manufactured to the same design specifications as Government Issue caps for enlisted personnel, but in better quality materials with officer insignia and pebbled aluminum buttons. Officer buttons frequently show obvious traces of field gray or gray paint, and this is not unusual. While field gray painted buttons were regulation on enlisted rank (Other Ranks) caps, it was also a common field practice of many officers to paint over their aluminum buttons in order to reduce visibility under combat conditions, where a tell-tale flash of silver might very well draw enemy fire.

Officer caps also included aluminum mesh cord piping sewn into the circular top seam of the M43 during manufacture, with gilt buttons and gold mesh cord (gilt or celleon) for generals. Some officers also had aluminum piping (using the flatter piping cord found on the M38 side skirts) attached to the upper edge of the front scallop. This addition, presumably, was strictly a private purchase option and would not have been available on caps bought from government officer cap stocks (since it was non-regulation). Though not yet verified, it is likely that an option existed in the Offizier Kleiderkasse catalog for the additional front scallop piping; a Kleiderkasse catalog from 1943 or 1944 has yet to be located for study, however.

Other options were also available to officers on private purchase models that never appear on issue caps, such as a *partial* (forehead area only) leather sweatband. In general, full sweatbands are only found on wartime caps as *exceptions to the rule*. If and when they *do* turn up, the color is usually tan, or reddish brown. On the other hand, full sweatbands are typical of modern Bundeswehr headgear. As mentioned earlier, authentic caps for Other Ranks troops were not usually vented, and no grommets should be present. Private purchase officer caps on the other hand, may have one (more rarely two) grommets per side.

LININGS

Other Ranks cap linings were usually cotton, in a solid gray or brownish gray-colored smooth-textured (satin) weave, or gray herringbone pattern. While other colors occasionally appear, they are always solids. A blue and white plaid pattern in an M43 cap is a perfect example of a very out of place, unacceptable lining. The simple truth is that

Interior of a private purchase officer's M43 lined in dark brownish gray rayon. *Courtesy of Bill Shea.*

the M43/Bergmütze cap design was so popular that it continued in use after the war and is still used today.

Linings for private purchase officer caps were usually in smooth-textured cotton or rayon, in a solid color, including black (rare). The herringbone pattern is much less common on officer caps than it is on those for Other Ranks. If the cap has a sweatshield (only on privately purchased caps and quite rare), the shape is generally a diamond and may or may not have a makers mark. Caps made by the well-known Wuppertal-Ronsdorf firm of Peter Küpper, for example, frequently have a diamond-shaped sweatshield over a smaller black cloth diamond on which appears the *Peküro* trademark logo in silver. Sweatshields in private purchase caps are problematical: some dealers prefer to see only a maker stamp in ink, since this was the standard practice. They feel that sweatshields are a hallmark of all postwar caps made in this style (true), and thus the potential always exists for the cap to be no more than a modified piece.

The sweatshield in Bundeswehr caps, however, is made of a clear, supple vinyl with no stiffness, or else a thin, flexible acetate, both of which are relatively easy to spot. In addition, if you are familiar with the postwar manufacturers, the maker name can also give a clue. The only original Second World War period makers who later manufactured caps for the Bundeswehr are: Alkero (Albert Kempf, Teunz – formerly of Wuppertal-Ronsdorf), Peküro (Peter Küpper, of Wuppertal-Ronsdorf, now closed) and Ludwig Vögele of Karlsruhe (still active). *Bamberger Mützenindustrie* is a postwar maker; the Carl Isken label represents another postwar maker (though in existence since 1900, no pre-1945 *military* caps of any kind with an Isken makers mark have yet been encountered).

Unusual lining examples do turn up. Though not necessarily incorrect (as long as it is a solid color), such divergences from the usual pattern need to be given more than the normal scrutiny.

Government M43 caps for officers also had rayon or cotton linings, but with an exterior cap cloth in a medium quality Trikot material (lightweight), or an average quality doeskin. Private purchase versions on the other hand, were available in very high grade doeskin finishes, or high-grade Trikot (fine, thin ribs).

Interior of an authentic, private purchase Officer M43 with a very unusual "off spec" green quilted lining (such quilting is usually associated with a private purchase Bergmütze). It also has a full sweatband – an extremely rare option which would only be available on a private purchase cap. The visor is quite rounded on this M43. *Courtesy of Bill Shea.*

Part 2:
PANZER

The Army black panzer M43 was no different in design than the field gray version. As usual, however, the black wool used for the average, issue grade Panzer cap had a shorter nap with a smoother hand (feel) than the wool used for the usual issue field gray cap. Just as with the M34 and M38 Panzer caps, so the thread used in the construction of the Panzer M43 was invariably black, and light-colored threads on such a cap are highly suspect. Linings are the same as for field gray caps, though most common is a solid, medium gray cotton without any pattern (though herringbone does appear).

The desirability of black M43 caps among collectors has led to a nearly overwhelming number of modified versions of these caps being offered in the militaria market, at shows as well as the Internet. These, together with reproductions passed off as authentic pieces, make for a serious challenge to collectors who wish only an authentic example. Although modified caps are, in general, poorly done and can be recognized without much difficulty, reproductions present more of a problem. The only

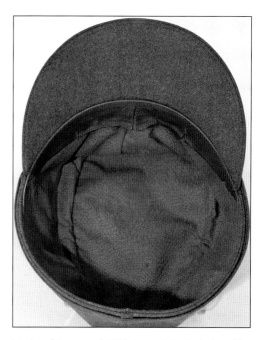

Interior of the general's M43 shown at the beginning of the chapter. Yellowish gray cotton or linen, smooth texture lining, with partial leather sweatband (the sweatband is embossed at the center 'Qualitäts Ware'). *Courtesy of Bill Shea.*

defense against these is to view as many authentic originals as possible, from trusted sources. The following guidelines may also help.

Modified Caps:

1. **Design**: Ensure that the cap design itself matches the correct specifications for the M43 (taking into account minor variations). Something as simple as this is often the first indication that a cap is not original; design inaccuracies are very common, since the modified cap was, of course, *not originally an M43*. Pay particular attention to the length of the visor.

2. **Buttons**: On Army M43 caps – including the Panzer version – any buttons other than the pebbled aluminum type are much more unusual than on Waffen-SS or Luftwaffe (Herman Göring) caps. An Army M43 with another type of button should invite caution, and a very close inspection.

3. **Lining**: Linings are not easy to replace without leaving signs of damage. Leaving an original lining in place involves the least effort by the person doing the modifications. It is unlikely, however, that the cap's original lining will match authentic specifications. Such an error is easy to spot.

4. **Wool quality**: The quality of the wool is a much more subjective thing to measure. You need experience with authentic caps to best develop this sense, but at least initially, consider any wool with a very (low grade) coarse, rough texture to be a cause for suspicion.

5. **Insignia**: Quite often the insignia is another obvious indication that something is not right. The insignia on modified (and reproduction) caps will usually also be reproduction. Poorer quality versions are quite easy to spot. With machine-woven (BEVo insignia) check for poor detail on the eagle and an uneven weave surface. Check wing lengths (many are too short), the body shape (plump bodies are no good), wreath leaves (a common location for errors), and a proportionally too small swastika. A single ring of oak leaves is incorrect.

Reproductions:

These caps are designed to look like the real thing, and are generally made close to the original specifications. Makers marks on the lining are no guaranty whatsoever of a cap's authenticity, and should not be depended on to convince you of the same.

1. **Weight**: Although these caps are essentially the same in cut as the originals, the wool is often thinner. As a result, reproductions will feel lighter than an original. This is usually most noticeable on the side skirt, which as a result may not hold firm top edge (where it folds over) due to the thinness of the wool.

2. **Visors**: Reproduction visors are also usually thinner than originals, and may not have the thick brim ridge often (but not always) found on authentic caps.

3. **Linings**: Linings may, or may not, match those on original caps. Look for newness; signs of wear are desirable.

Knights Cross wearer Feldwebel Rudolf Berger wears an M43 with two-piece insignia. Note that there is no backing to the cockade – this is the correct application method. *Courtesy of Dale Paul.*

INSIGNIA

Other Ranks

Correct regulation insignia for the Army M43 Other Ranks cap was machine-woven in either off-white or matte gray cotton thread on an oxide green rayon backing, usually

Other Ranks M43 with standard machine-woven matte mouse gray eagle on a field gray trapezoid base. Note the oak leaf wreath is formed from *two* rings. *Courtesy of Bill Shea.*

Left: Machine-*embroidered* insignia on a trapezoidal field gray base (more gray than green) for an Other Ranks M43 cap. The eagle is in off-white cotton thread. *Courtesy of Bill Shea.*

Two-button flap *Bergmütze* cap with *T* form insignia and a Jäger badge made of Leichtmetall mounted on the left side skirt. The two-button Bergmütze and M43 are easily mistaken for each other. This cap is worn by Feldwebel Georg Klein [15th Infantry Div., 204th Jäger Regiment]. Klein was awarded the Ritterkreuz on August 8, 1943. *Author's collection.*

he shape of a trapezoid. The mouse gray eagle was quite difficult to see from a distance. Early versions used the same *T*-type insignia as used for the Bergmütze.

Though Army M43 caps show little variation in the basic cap design, the insignia is another issue altogether and is often the source of confusion. Though the standard form was as described above, machine-*embroidered* versions of the trapezoid were also used. These are uncommon on authentic caps, but very common on reproduction and modified headgear.

As mentioned, M43 insignia was borrowed from the Bergmütze pattern. It had a thicker background area around the eagle and cockade which formed a clear *T* shape when the insignia was properly trimmed and applied. NCOs, however, sometimes wore the same two-piece insignia found on the M34 cap (dark green underlay), in which case the cockade may be applied with no underlay cloth visible.

What's wrong with this picture? Easily recognizable, reproduction Other Ranks insignia. Eagle, much too large; Body, too wide; Wings, proportionally too short – and if that weren't enough, there is no cockade. Wreath: incorrect leave pattern (common flaw); Swastika: Tiny! The white attachment stitching is a "no-no" – and the sewer must have been drunk. The skirt arm buttons (not visible), are not made of aluminum.

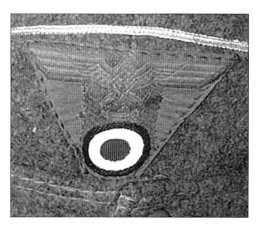

Other Ranks machine-woven insignia used on an officer's cap. The insignia in fact, is in T-form, but the backing is applied as a trapezoid – not trimmed to the T-shape. Very neatly handapplied. Note the buttons are painted field gray. *Courtesy of Bill Shea.*

Standard machine-woven trapezoid insignia (Other Ranks) on this very high quality private purchase cap for an officer in a Jäger unit. *Courtesy of Bill Shea.*

Separate two-piece eagle and cockade insignia (M34 type) or the *T*-form insignia is sometimes found on Panzer caps, but eventually these were supplanted by the trapezoid form, machine-woven (BEVo) on black rayon in off-white or pale gray cotton thread. Trapezoids *without* a cockade are nearly always reproductions.

Officer Insignia

For officers, two-piece insignia in aluminum thread machine-woven on a dark green rayon base cloth was regulation and is the most common variety encountered. German officers could and did use other types of insignia, however, including enlisted grade forms. Uncommon, but sometimes encountered is a hand-embroidered bullion eagle and either a bullion cockade (such as on the M38), or a machine-embroidered version. The eagle in this case should be the smaller size cap eagle as worn on the visor cap – not the larger breast eagle (for a tunic).

Leichtmetall (aluminum) insignia (i.e. the eagle from a service visor cap), was *not* authorized for wear on the M43 with the only exception being tradition (esprit) badges attached to the left side skirt (Mountain Troop Edelweiss, Jäger badge, etc.). That this regulation was occasionally ignored however, is evident in the following photograph of Panzer soldier and Knights Cross bearer Feldwebel Hans Strippel. Perhaps holding the Ritterkreuz exempted him from any complaints about the very non-regulation metal eagle he wears on his Panzer M43?

Even Feldwebel Strippel's cap has a machine-woven cockade, however. A metal cockade (from a service visor cap) mounted on an M43 is not acceptable. Examples like the one in the following photograph are common, and will nearly always prove to be modifications or poor quality reproductions.

Officer's cap with two-piece hand-embroidered bullion insignia. Most likely an enlisted upgrade or a cap from government stocks, late 1944 time frame; the wool has a heavily felted finish that gives it a rough appearance. Note the aluminum buttons are painted . *Courtesy of Bill Shea.*

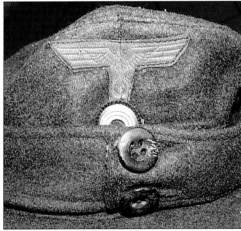

Feldwebel Hans Strippel of Panzer Regiment 1 received the Knights Cross on January 26, 1943, for destroying six of fifteen attacking Russian tanks in a single engagement. After 70 victories and distinguished action at Cherkassy, he became the 485th soldier to receive the Oakleaves to the Knights Cross on June 6, 1944. *Courtesy of Dale Paul.*

The metal cockade does not belong here and the buttons are highly suspect, as well as the peculiar box-shaped stitching line surrounding them. The color of the cap material is far too green. The eagle itself, in any case, is not original (the legs are incorrect, wreath and swastika are way out of proportion to the eagle's size).

MODIFIED BUNDESWEHR CAPS

The Bundeswehr knew when to hold on to a good thing, and so, when selecting field cap designs for the new Federal Army it re-instituted the M43 cap design (with a few changes, of course), which it officially designated as a *Bergmütze* (mountain cap).

There are quite a number of notable points that make modified Bundeswehr caps easy to identify:

Bundeswehr Army officer's Bergmütze (M43 style), in the light gray, dress version. Aluminum buttons and crossed sabers (in silver) under the cockade indicate officer rank. Note the stiff cap front – this is due to an internal support strip that is positioned vertically behind the front seam (under the lining).

Interior of a modern Bergmütze. In keeping with standard Bundeswehr practice, everything is gray – both the lining, and the full sweatband. Note the sweatshield and makers mark (Albert Kempf, a former wartime manufacturer in Wuppertal-Ronsdorf).

1. **Color**: The Bundeswehr cap color differs markedly from the wartime field gray. Of the several versions of the cap once used or currently in use by the Federal Army, the dress version is a very light gray, the field version a somewhat yellowish brown.

2. **Material**: The gray caps are made with a very low wool percentage, thus the material is very thin and without a distinct nap. The field version uses a wool/polyester blend with a short nap. Visors look very thin in comparison to those on original M43 caps, particularly on the dress version. The best description for these caps is *lightweight*.

3. **Visor**: Visors do not have a reinforcement strip along the front edge, and so will not have the thick brim ridge characteristic of many Third Reich era M43s.

4. **Shaping ring**: M43-style caps made for the Bundeswehr sometimes have a narrow diameter plastic rod running around the bottom edge of the cap, within the lining. The ring is intended to help the cap base hold a round shape. Such rods are never found on authentic wartime caps.

5. **Sweatband**: Bundeswehr versions, including issue grade caps for enlisted soldiers, have full circumference, gray-colored sweatbands. This is not the case on original M43 caps, however. The full sweatband gives the Bundeswehr cap interior a very crowded look.

6. **Sweatshields**: All caps (Other Ranks caps included) have thin, flexible postwar sweatshields, usually with a maker mark such as Albert Kempf or Bamberger Mützenindustrie (the latter a postwar manufacturer).

7. **Insignia**: Insignia for the modern caps is nearly always metal, attached by prongs mounted on the rear. If a modified cap originally had a federal cockade attached, the holes in the material caused by the prongs should still be evident, possibly also two more (if the cap had officer's sabers as well).

8. **Crown Support**: Inside the front of the Bundeswehr cap, positioned vertically behind the front seam (beneath the lining), is a thin, flat strip of metal or plastic used to help stiffen the front of the cap. The strip is flexible and can be bent if pushed. Its presence can be felt by pressing on the front of the cap near the top. There should be slight resistance to pressure, but the cap will actually bend and remain in the new position if enough pressure is applied.

9. **Piping**: Early German Bundesgrenzschutz (Federal Border Guards) uniforms also include an M43-style, Bergmütze cap. The BGS caps, however, as well as being of a green base color, also had dark green piping around the crown. Wartime caps , however, were never piped in any type of branch color.

Modified Bundeswehr or BGS caps present very little real threat to any collector armed with the knowledge of these characteristics.

EAST GERMAN CAPS

East German caps differ radically in design and construction from World War II caps – so much so, in fact, that it is nearly impossible to modify them to look like an authentic M43. Postwar Soviet Army influence on East German cap design was significant and is responsible for the difference. East German visored field caps look very much like their Soviet Army counterparts. The most notable point is a side skirt arrangement with ends that overlap *across the top of the cap* (perhaps borrowed from Italian cap designs), and secured by *Velcro*. Velcro, of course, did not exist during World War II.

There is simply no getting around this very major difference, and it successfully prevents any chance at convincing modifications. The cap material is also incorrect (very thin) and quite lightweight.

Front view of an East German visored field cap. The immediately noticeable difference is the side skirts, the ends of which are secured across the top of the cap (this is a common characteristic of Soviet design). *Courtesy of Kay Stephan.*

Part 3: TROPICAL

The tropical standard visored field cap, like its garrison cap cousin, is also a very popular target for modified caps. Fortunately, there are few modern equivalents available that share a design similar enough to pass even a casual inspection. In most cases, reproductions are the more significant danger for this type of tropical headgear.

The major flaw with either modified caps or even reproductions is usually incorrect color, though authentic caps certainly showed a degree of color variation that can cause confusion. It pays to study actual caps owned by an experienced collector or dealer in order to gain an idea of which shades fall within the correct color range.

CONSTRUCTION

Exterior

The tropical field cap, introduced well before the M43 in 1941, was also based on the Bergmütze design – but with a slightly longer visor and a much lower crown. Many of these caps have a thick ridge along the base of the visor and bottom of the cap; this was a maker-specific characteristic and is not found on all caps.

The base color for the M41 varied greatly. The most common shades ranged from olive khaki to olive brown. Green, or pure olive drab, however, are not among the acceptable shades. These caps were manufactured from cotton in a twill weave and therefore have a very fine, ribbed texture. The cotton material produced a much lighter cap, which also dried far more quickly than wool after becoming wet. The top of the tropical cap was one piece, but pinched into a ventral center seam similar to the M42. Caps without this feature are incorrect.

Other Ranks tropical cap, offered Online. Possibly a reproduction; a physical inspection would confirm this. Superficially at least, the design is correct, though the visor rounds out to the sides somewhat more than average. The color is acceptable for an *unissued* (mint) cap. The insignia colors are also correct, but the eagle shape is quirky (wings), the swastika slightly small – likely a reproduction. The rounded ridge along the visor base is a valid characteristic of some makers. The brand new condition, however, is a cause for concern.

Tropical M41 field cap without any side skirts. No official documentation for this type (i.e. "skirtless") has yet been seen. The cap color, material and the insignia are correct, as is the lining. *Courtesy of Bill Shea.*

Below: Side view. Two brown enameled grommets, no side skirts. *Courtesy of Bill Shea.*

In a departure from previous German cap design standards, the side skirts on the tropical cap were *simulated* only. Some makers achieved the appearance of skirts by pinching the cap material into a thin ridge that followed the normal skirt outline, then adding the usual single or double stitch lines below this ridge to complete the impression of an actual skirt. Other makers used real skirts (which formed the lower body of the cap), to which an entire cap top section was then attached. Whichever method was used, the skirts were never movable, and there were no buttons. Authentic caps do occasionally appear which do not have even a simulated skirt line. This may represent a final development in the evolution of this cap, but there is nothing to corroborate this supposition. The few examples of such caps so far studied have every indication of authenticity.

As with the tropical garrison cap (indeed, as with all German field caps), external reinforcement stitching lines do appear on each side of, and parallel to, the main center front and rear vertical seams. Stitching was normally done in light brown or tan-colored thread, though sometimes paler colors were used (but not darker colors). Soutaches were used until officially discontinued in 1942, but no other indication of the wearer's branch appeared on this type of cap.

Air Vents

The tropical standard field cap was ventilated, and so there should be *two* horizontally aligned air vent grommets just below the top seam on each side of these caps, with the first grommet positioned just above the point where the simulated side flaps turn upward. The grommet exterior was covered with an olive brown, baked on enamel; the interior washer ring and grommet shank (inside the cap) were not painted.

Soutache

The soutache on the front of tropical billed caps needs special mention, since it was not attached in the same manner as used on the tropical garrison cap. Though the soutache was stitched down its center line, the lack of independent side flaps on the tropical field cap prevented the chevron ends from being inserted into the side flap seam – there was simply no live seam to insert them into. Thus, the ends were laid against the cap in line with the lower edge and stitched straight across with a hem stitch (the same hem line which secured the bottom of the lining to the inside of the cap). The ends of the soutache below the stitch line on the tropical cap are, therefore, fully exposed and reach to the base of the visor.

Visor

Reinforcement stitch lines on the cap visor (bill) vary depending on the manufacturer. Tropical cap brims are always fully covered in cloth, with no cardboard exposed.

Lining

As with the tropical garrison cap, linings were bright red, rarely a slightly darker shade close to maroon. Some makers again preferred a tan lining material not much different from the exterior color, but it is not certain whether these appeared on issue caps (as opposed to private purchase caps). Linings are always one color only and of a smooth, cotton fabric. At the base of the lining, between lining and cap material, some tropical caps have a very thin, flat rubbery strip which seems to have been intended to keep sweat from bleaching through to stain the exterior cap cloth. There is no official record of any requirement for this insert, and given tropical temperatures it was not likely a very effective device.

Insignia

Regulation insignia was cloth for both Other Ranks and officer caps; metal insignia was not authorized for wear on the tropical visored field cap. Insignia was machine-woven in pale blue cotton thread on copper or tan backing cloth, and applied in similar fashion to that used for the tropical garrison cap (by machine, or by hand). Both the two-piece (eagle and separate cockade) insignia, or the trapezoid eagle and cockade combination are correct for this model cap.

Officers were authorized to use their usual insignia, an aluminum (or gold for generals) machine-woven eagle on dark green, but most seem to have preferred to wear the Other Ranks (light blue) style insignia. Officer caps with Other Ranks insignia therefore are not necessarily incorrect.

Insignia stitching did not pass through the inner lining and should not be visible inside the cap.

OFFICER CAPS: PIPING/SWEATBAND

Tropical field caps were not piped for Other Ranks, nor did issue caps have a leather sweatband. Officer caps had silver cord piping around the crown seam, with the ends overlapping at the rear. Another strip of thin piping was attached along the edge of the simulated scallop on the cap front on some caps and as with the M43, may have represented a private purchase cap option. The piping should always appear professionally applied. Be wary of any cap with hand-applied scallop piping.

Officer tropical visored caps of any kind are rare in the militaria market. There is no estimate of how many may have survived the war, and most of those that did likely did so as war souvenirs brought back by Allied soldiers. Afrika Korps troops taken prisoner after the surrender of that force were allowed to wear their caps in captivity after removing the insignia, and it is unlikely that many of these survived the additional usage. The rarity of officer tropical headgear means that the majority of such caps on the market are simply not authentic. Any cap you encounter should be very carefully inspected before purchase, unless offered by a trusted dealer.

MODIFICATIONS

As mentioned earlier, there are very few modern caps available which lend themselves conveniently for easy modification into an authentic-looking tropical cap; either the design is incorrect, the color is wrong – or the materials are of synthetic origin. Modified caps will most likely give themselves away by being the wrong color, even if the exterior design is essentially correct.

REPRODUCTIONS

Reproductions are again the most dangerous enemy. Fortunately, few companies or individuals seem able (so far) to reproduce this form of cap with the accuracy needed to fool experienced collectors, though no doubt some exist. In general, newness is one of the most critical factors. The majority of authentic caps in current collections have seen use. Mint original caps should be much rarer than worn examples. The frequent appearance of "mint" examples – particularly from the same dealer – is certainly a good reason for suspicion. Insignia is also a key to identification, particularly with regard to

Postwar tropical-style field cap offered over the Internet. The green color is unquestionably wrong, the thin cap material is not a Trikot weave (not ribbed) – probably made of polyester or a polyester/cotton or nylon blend. The grommets are a peculiar green shade, not the correct brown. Though the interior is not visible, it most likely continues with more of the same: wrong color, wrong material.

correct colors. In general, as with all reproductions and modifications, it is not necessarily one single thing that may tip you off, but rather the accumulation of several small inconsistencies which, taken together, do not add up to an authentic cap.

A.II. DIE WAFFEN-SS und POLIZEI

A. Waffen-SS
Part 1: Continental
Part 2. Panzer
Part 3. Tropical

B. (Combat) Police

Part 1:
CONTINENTAL

In this famous photograph, Knight's Cross holder Michael Wittmann wears an M43 made of fine doeskin, with wide flaps and pebbled aluminum buttons. The insignia is most likely two-piece (with the eagle on the left side skirt). *Author's collection.*

Waffen-SS troops were equally quick to assimilate the new M43 headgear. Though the Schiffchen continued in use, the M43 soon appeared everywhere.

Unlike Army production, Waffen-SS M43 caps were manufactured with far less of a standard in terms of design. The many variations found on this type of headgear create serious problems with identification, and the question of, "why were there so many styles?" is one that remains unanswered. It is possible that the shift of some field cap production (for "issue" grade caps) by the Waffen-SS to its own manufacturing centers (based in concentration camps) led to a lesser degree of standardization, but this has not been proven.

CONSTRUCTION

Exterior

The base cloth for the Waffen-SS Other Ranks M43 was the same wool as that for Army caps, though there are often textural differences in the wool, or in the finish. The field gray color used in these caps also tended to vary over a wider range than the Army version, with some late-war shades more akin to the M1944 field blouse color.

Only the skirt arms significantly varied from the standard Army cap, with the narrower, single-button style flaps being quite common. All of the following variations are correct:

1. Narrow arms (flaps), rounded ends, single horn or thick (early) plastic button [field gray and black Panzer caps]
2. Narrow arms, rounded ends, single stone-nut button (Steinnuß) [field gray caps].
3. Wide arms, square or beveled ends, horn or thick plastic buttons [field gray and black Panzer caps].
4. Wide arms, square or beveled ends, pebbled aluminum buttons [field gray caps, black panzer caps, issue or private purchase].

Waffen-SS Other Ranks M43 cap. Single-button style with narrow skirt arms. Standard two-piece machine-woven insignia in white cotton thread on black rayon underlay. Average grade wool with a felted finish. *Courtesy of Bill Shea.*

5. Caps with narrow arms and a buckle closure – most caps of this type were modified from non-military caps, such as those from the Hitlerjugend, etc.

Waffen-SS officers often wore private purchase M43s in a cotton twill material that was lighter than wool Trikot. These caps tend to be gray-colored rather than the standard field gray.

VISOR

The Government Issue Waffen-SS M43 visor had the standard thickened brim not always found on private purchase models. Some visors may also be covered with two-pieces of cloth: an upper and a lower section, with the seam that joined the two running (visible) along the leading edge of the brim. Many late-war period SS M43 caps also have the cost and material saving style of an exposed visor reverse (underside).

Sewing lines on the cap visor varied depending on the manufacturer. While no stitching should appear on the visor obverse (upper side) of *Government Issue* (i.e. applies to most Other Ranks examples) caps, private purchase models for officers once again did not always follow this pattern. This is particularly true of caps made from lighter weight materials such as cotton twill, or wool Trikot.

OFFICER CAPS

Waffen-SS M43 caps for officers were made of fine grade, high quality materials, with officer insignia and pebbled aluminum buttons.

The cap crown was piped in the Army fashion, however Waffen-SS generals also wore silver piping, rather than gold. No caps should appear with gilt buttons, or gold piping, since this was not a mark of general grade officers in this branch of service. Front scallop piping seems to have had little appeal among Waffen-SS officers and rarely ever appears on authentic caps.

Other options, such as a partial sweatband (forehead area) were available on private purchase models, and these will almost always be either a reddish tan or greenish gray color. Some private purchase caps were also equipped with air vent grommets.

LININGS

Usually a solid field gray, gray, or brownish gray-colored cotton in a smooth-textured (satin) weave for Other Ranks. Herringbone pattern linings (gray base cloth) are not very common on Waffen-SS M43 caps, and should be considered just outside the nor-

Waffen-SS officer's cap. Private purchase cap in a lightweight material (not wool), probably cotton or a cotton/rayon blend. Double-button skirt flaps. Two-piece officer insignia (machine-woven in aluminum thread on a black rayon base), with the eagle on the left skirt. Two air vent grommets per side. *Courtesy of Bill Shea.*

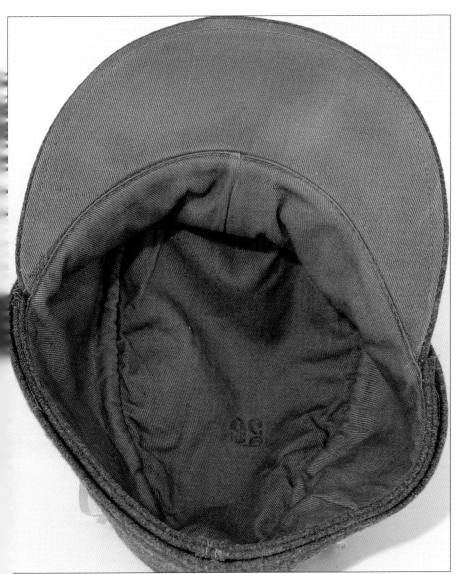

M43 officer's cap in lightweight material for warm weather, worn by a cheerful young Waffen-SS Obersturmführer. The cap has two-piece insignia with the eagle mounted on the left side skirt (out of view). *Author's Collection.*

Interior of a Waffen-SS Other Ranks field gray M43 lined in brownish gray cotton. Unusual on this cap is the visor reverse, which is not covered in the usual fashion with the same cap wool used for the rest of the cap; a different, smooth-surfaced material has been used instead. *Courtesy of Bill Shea.*

Two-piece insignia. Narrow, single-button skirt arms (flaps) with a mustard brown button. The wool has a slight yellowish tint. Hand-sewn, two-piece insignia.

Trapezoidal insignia. Wide, two-button front skirt arms with heavy plastic buttons. Good quality field gray wool. The insignia is machine-sewn to the cap – normal method for the trapezoid.

Another example of the trapezoid, on a late-war cap made in coarse wool with a heavily felted surface (typical of the period). Value around $1500 in 2001. *Courtesy of Bill Shea.*

mal range. Other colors occasionally appear on private purchase pieces, but are always solid (one color only). Linings for private purchase officer caps were normally in gray, field gray or brownish gray smooth-textured cotton, or, more often, in rayon. Solid colors were once again the rule.

INSIGNIA

Given the great variety of reproduction SS insignia, not to mention the variety among originals, it is a wise move to also study a reference work devoted exclusively to this topic, such as Bender's *Cloth Insignia of the SS*. The majority of Waffen-SS M43 insignia is of the machine-woven (BEVo) type – also referred to as "flat wire." Machine-embroidered versions on a wool base were also used, however. Thread color varied; correct and most common are: off-white, matte gray, or silver gray. The only real standard to speak of was that *two-piece* insignia used on wool Waffen-SS M43 caps was always on a black base cloth. Insignia combinations found on field gray caps are:

1. Two-piece, with a large Totenkopf positioned on the cap crown and a trimmed eagle on the left side skirt (that is, not a trapezoid). This pattern is usually found on caps of early manufacture, though some individuals preferred it and wore it throughout the war.
2. Two-piece, with eagle and Totenkopf both positioned on the crown (upper front of the cap). This arrangement replaced the previous pattern and became more or less the standard until use of the trapezoid form became widespread. This was the only acceptable style, if an Edelweiss emblem was worn on the left skirt.
3. Oxide green (or black for Panzer) trapezoid with combined eagle and cockade, machine-woven in mouse gray thread on an oxide green rayon backing (off white thread on black for Panzer) and positioned on the crown. The trapezoid was meant to replace the two-piece version. This insignia type was *machine-sewn* to the cap.
4. One-piece, Totenkopf only (on the crown).

Wear of Leichtmetall insignia (from the service visor cap) on the M43 was forbidden, but some officers occasionally ignored this regulation. If used, either the eagle, or the Totenkopf were in metal (though rarely were both worn at the same time).

Part 2:
WAFFEN-SS PANZER M43

The only real difference in the black M43 for Waffen-SS Panzer personnel was in its being made, as always, from a better quality black wool (better than the standard field gray wool). All visible stitching should be in black or dark-colored thread, never in off-white or light color. Lining colors were the same as for the field gray cap; however, black, or dark gray, is also correct.

INSIGNIA

Insignia for the black cap was the same as for the field gray model, available in either two-piece form, or combined together on a trapezoid base. The trapezoid insignia was machine-woven in off-white thread on black rayon. Reproduction versions of this style insignia are legion, and 80% of them are recognizably flawed. When studying a cap to determine its authenticity, the insignia is always a good place to start.

Waffen-SS Obersturmbannführer Wolfgang Joerchel proves the exception to the rule. His cap sports not only the Leichtmetall SS cap eagle but also a metal Totenkopf in this picture. The Totenkopf may have been doctored by the German photo lab in order to improve visibility. *Author's collection.*

Fine example of a Waffen-SS officer's Panzer M43. Note the highlight on the top edge of the Totenkopf – this is due to padding beneath the insignia which makes the skull dome curve outward – exactly right for original insignia, which in this case is the Other Ranks version. The cap thread is a slightly lighter black than the cap material. Heavy plastic (or Bakelite) button on a single-button flap. The visor on this example is very rounded. This cap was valued at about $2500 in the year 2000. *Courtesy of Bill Shea.*

The other side of the coin: Fine example of a *sloppy* reproduction cap (Other Ranks) with very noticeable reproduction machine-embroidered insignia. The flap buttons, in addition, are much too large. Internet offering.

Waffen-SS officer's Panzer M43 with a lining in black rayon. Most likely a private purchase piece. *Courtesy of Bill Shea.*

Two-piece machine-woven insignia in off-white thread; this is a perfect example of Other Ranks insignia on an officer's cap. The dome of the Totenkopf curves slightly, indicating that the insignia is padded. *Courtesy of Bill Shea.*

Waffen-SS Hauptscharführer Thurner wears a single-button skirt, black M43 with two-piece insignia. His awards include a tank destruction strip, the German Cross in gold, a close combat badge, Iron Cross 1st class, general assault badge, panzer assault badge, an Iron Cross second class ribbon and an eastern front medal ribbon. *Author's collection.*

Ideally, the two-piece eagle and Totenkopf *should each be hand-applied* in such a fashion that the stitches are hidden – or at least, barely visible. The backing cloth was trimmed to shape and the remainder was then tucked under the insignia borders. The Totenkopf Insignia should (ideally) be slightly padded (puffy), with the surface raised above the surrounding cap cloth.

Normal insignia for officer caps is machine-woven two-piece insignia in aluminum thread. There was no officer (aluminum thread) version of the trapezoid – though it is available in reproductions. Authentic examples of officer grade trapezoidal insignia (in aluminum thread) have not been identified. Once again, the most accurate (design) reproductions currently originate in South Korea. Since this form of insignia was apparently never produced, any example should be considered a fake.

The use of Other Ranks insignia (particularly the trapezoid type), on officer caps, however, was quite common.

MODIFIED AND REPRODUCTION CAPS (Continental and Panzer)

Modifications

Some German postwar organizations used a black Bergmütze at one time or another, though it is not clear which. These caps are similar enough to wartime caps that they can be modified and passed off as such. Strangely enough, despite the ready availability of very high quality reproduction insignia, the majority of these fakes instead use very shoddy examples, and this is usually the first means in correctly identifying them.

The machine-embroidered trapezoid shown earlier on the Panzer cap is a prime example; at right is another (machine-woven [BEVo] two-piece):

If you suspect a possible modified cap (field gray or black), check for:

1. Reproduction insignia
2. Design differences
3. Wrong color, or uneven coloring
4. Wrong lining color/material
5. Other factors outside the norm (i.e. buttons on the wrong side), which, combined with one of the above points, make for a questionable piece.

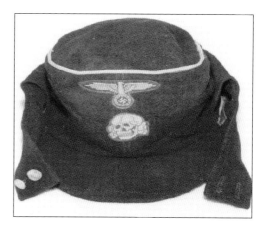

Using what you've learned so far, you should be able to note a number of significant flaws with the insignia on this cap: the underlay is not correctly folded under and the insignia is completely flat (no padding) – it is reproduction. Although the detail is fair, the eagle's body is far too thin – and the backing color much too gray.

Reproductions

RZM Tags

Field gray M43 and black Panzer cap reproductions can be quite convincing when well made. Most are not, but some companies, such as Janke Tailoring (a German tailoring company outlined in the next chapter), produce first-rate caps in materials that are little different than those used in originals.

A common flaw on both modified original caps (wartime vintage, or postwar) and even on the most convincing reproductions, is the attempt by shady dealers to make a cap more convincing by adding a cloth RZM tag. As with Waffen-SS visor caps, however, *no* Waffen-SS field caps ever had RZM tags. The compulsion to add one seems almost impossible for such dealers to resist, if they have a tag on hand. If you inspect a cap and find an RZM tag on it – no matter how convincing the cap may otherwise be – leave it alone. *A Waffen-SS field cap with an RZM tag has been modified, at the least.* Something not original to the cap has been added, which – even if the cap itself is original – decreases the value of an otherwise *clean* (unaltered/untouched) piece.

Pricing below market value

The type of cap most often used to deceive unwary collectors is reproduction Panzer caps in general, and Waffen-SS officer Panzer caps in particular, that are offered as originals. Unscrupulous dealers will offer these at lower than the normal market value in order to attract attention. Enthusiasts will think they are getting a deal at that price: though still quite expensive, it is less than the usual value of a such a cap. The grim fact is, an honest dealer who knows his business will never let a piece go at a major discount: It simply isn't done. A cap offered at a price significantly lower than the market value should automatically give you pause. Honest dealers are simply not in the business to be Santa Claus, as much as we might wish this were so.

Obviously, officer caps are a frequent target for trickery. In this example, the cap has been worn in order to make it look more convincing. It was offered on the Internet at well under the price an authentic officer's cap would command (though the seller made no claim to be Santa Claus). The RZM tag on the inside of the rear side skirt (next the seam) is completely incorrect in any case – RZM tags are never found on authentic Waffen-SS field caps and the contract date on the tag is [19]36 – a bit too early for an M43 cap! Front scallop piping is also quite rare on SS caps, and raises the doubt factor even without the tag.

Waffen-SS tropical visored cap. Two-piece insignia; the eagle is actually a sleeve eagle. A close look will show that both eagle and Totenkopf are lightly padded. Though the black border around the eagle is fairly wide in this example, the raw edges are folded under with no stitching readily visible. *Courtesy of Bill Shea.*

Previous cap, viewed from the left. This item belonged to a soldier in an SS Gebirgsjäger unit, as evidenced by the SS version of the field cap Edelweiss attached to the left skirt *Courtesy of Bill Shea.*

Insignia close-up. Hand-applied and correctly positioned, though some stitches are barely showing under the eagle's right wing (viewer's right). The insignia on this cap is the type for the SS Schiffchen and M43, not the normal tropical insignia. Note that though the skirt flaps are independent of the cap body, the buttons appear to be sewn to (and through) them – there are no visible buttonholes. *Courtesy of Bill Shea.*

In other respects, the pointers already given for Army M43 caps apply here as well.

Part 3:
TROPICAL CAP

A quite rare item on the militaria market, the SS tropical visored field cap was not officially introduced until 1943. It was little different from the earlier Army tropical M41 visored field cap in terms of construction and exterior appearance, with the exception of having no seam on the top panel of the cap, and the issue version not having even simulated side flaps or buttons. It *did* have the two external air vent grommets (brown enamel) and the usual red lining. Examples with moveable skirts exist, with or without air vent grommets, but these are primarily examples of *private purchase* pieces (for NCOs or officers). No soutache should appear on one of these caps, since they were introduced long after use of the soutache had been discontinued.

The same pointers given earlier for spotting modified contemporary or original (Army) tropical caps apply here, as well. The SS tropical visored cap actually conforms much more to a production standard than was ever the case with the regular field gray SS M43, and this makes it somewhat easier to tell real from false.

INSIGNIA
Again, SS insignia requires a separate study. For our purposes here, suffice it to say that the standard insignia for tropical caps was a tan machine-woven eagle and Totenkopf two-piece insignia on textured black rayon underlay cloth, however authentic examples also exist with the off-white on black insignia that was normally used on the wool M43. Officers used both the tan Other Ranks type, as well as the officer aluminum thread, machine-woven version. Insignia was applied in the same fashion as for continental caps.

MODIFICATIONS AND REPRODUCTIONS
Basically the same check points for Army caps apply, however when viewing a cap labeled as a tropical SS M43, the fact that these *usually* have no side skirts should be kept in mind. Check color, lining material and in particular, the insignia very closely (both the insignia itself and the attachment). These caps do not normally appear with trapezoid insignia, and any cap that does is highly suspect as a fake. In general, authentic caps are very rare birds and do not appear on the market very often. Keep this in mind when you encounter one, since that very rarity means the largest percentage of tropical caps you may come across on a collecting mission will be fakes of one sort or another, or reproductions.

B. COMBAT POLICE CAPS

Police M43 headgear differs only in color and insignia from Army and Waffen-SS caps. The skirt arms on police caps tend to be of the wider, two-button type (buttons on the left flap), with single-button arms much rarer. The cloth color leans to a particular blue tint that is peculiar to police caps – no other armed service used this color (the

Fliegerblau of Luftwaffe caps is decidedly darker and is impossible to mistake for the much lighter Polizei blue).

LININGS

The usual lining material for Other Ranks police caps was a slightly purplish-blue cotton color in a flat weave that looks a bit like linen. This is particularly true of later war caps like the one shown below, which is dated 1944. The hometown of this maker, Breslau, was destined to become a part of Poland in the political reconfiguring of East German and Polish national borders following World War II. Under "new management" since 1946 or 1947, the Polish half of the city goes by the name of Wroclaw (Breslau was divided in half).

INSIGNIA

The insignia for police M43 caps is perhaps one of the more unusually-shaped forms found on German military headgear. A machine-woven, wreathed police eagle in light gray thread with the cockade above it were combined on a vertical rectangle with outward-curving sides. The base cloth was again a distinctive shade of light blue.

OFFICER CAPS

Police officer caps bore aluminum crown piping in the same fashion as on Army or Waffen-SS caps. The insignia was the same as used on Other Ranks caps. Linings for officer caps are usually in a green or greenish blue rayon, but it is uncertain whether this held true for both government stock caps and privately purchased pieces.

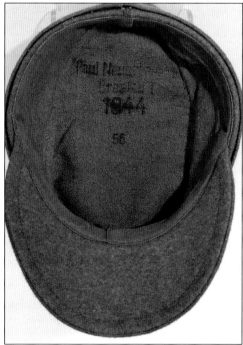

Police cap for Other Ranks. This shade of blue is a distinct characteristic of Polizei M43 caps, and is not found on any contemporary German models, meaning that convincing modification is not possible without some degree of dyeing. Below: Lining. *Courtesy of Bill Shea.*

Note the high degree of detail on the machine-woven police eagle and wreath. *Courtesy of Bill Shea.*

MODIFICATIONS AND REPRODUCTIONS

Examples of modified police caps are relatively unusual. There simply seems to be little interest on the part of those unscrupulous dealers for carrying out modification designed to pass as police caps – though why this is so is not entirely clear. Nor is reproduction police insignia easy to come by, and what does exist is usually of a poor enough quality that it can be spotted with little effort.

A.III. THE LUFTWAFE EINHEITSFLIEGERMÜTZE (Standard Flyer's Cap) M43

Part 1: Continental
Part 2. Panzer
Part 3. Tropical

Part 1: CONTINENTAL

Luftwaffe Other Ranks M43 cap, two-button skirt arm style. Two-piece insignia eagle in off-white cotton thread machine-embroidered on a gray-blue wool base. Unusual in this case is the higher than normal placement of the cockade to where it is immediately below the swastika. Professionally hand-applied, as were 90% of all two-piece insignia examples. *Courtesy of Bill Shea.*

As with the Waffen-SS version, the Luftwaffe M43 Einheitsfliegermütze also showed variations in both skirt arm width and insignia – though perhaps not to the same degree as the Waffen-SS. When initially issued, the Luftwaffe cap had narrow skirt arms (flaps) with rounded ends and a single button – in other words, it was very similar to the original Bergmütze design. This soon changed to the double-button, square-end arm style already in use on Army caps. Buttons on this form are usually blue painted, pebbled-aluminum.

Why any change was made at all, except possibly to allow more standardization (thus simplifying production) within the industry, remains an unanswered question. Suffice it to say that the M43 cap – regardless of its particular form – was as popular among Luftwaffe personnel as it was among Army or Waffen-SS soldiers and particularly for troops on the flight line; perhaps these caps were less likely than the Fliegermütze to sail off one's head if hit by an unexpected prop wash?

3/4 view. Note the rounded, thickened edge of the brim. The side skirt turn-up point begins about 2 cm to the rear of the end of the visor. *Courtesy of Bill Shea.*

CONSTRUCTION

Beyond the issues of narrower skirt arm width (early caps) and a different fabric color (fliegerblau), there was little difference in design on later model Luftwaffe M43s. No piping of any kind was ever used on Other Ranks caps. Note that on the Luftwaffe M43, off-white thread (rather than blue or some other dark color) was commonly used for reinforcement or hem stitching, and the threads therefore, are easily visible in many cases.

Skirt arms and buttons on authentic caps are correct in the following configurations:

1. Narrow single-button (usually metal) arms (on early Bergemütze-type caps)
2. Wide, two-button (blue painted aluminum buttons) arms

VISORS

Visors on issue caps usually include the brim ridge. Early Luftwaffe M43s and the Luftwaffe Bergmütze are nearly identical except for the visor length, which is longer on the M43 (with a higher crown). Visor reinforcement sewing appears only on the reverse (underside), in one or two (parallel) rows located approximately 1.5 cm from the edge of the brim.

Brim sewing on privately purchased caps (Other Ranks or officer) may appear on the visor obverse as well as the reverse – particularly in the case of lightweight base fabrics (such as wool or cotton Trikot). Lighter weight caps do not always have the thickened brim edge.

Heavily damaged, purplish gray rayon lining. The lining damage allows a look beneath at the cap's top panel (cover). This cap is most likely a private purchase piece since it comes with a partial sweatband – issue caps did not have this feature. The rayon lining tends to confirm this supposition. Note the line of reinforcement stitching just short of the visor brim. *Courtesy of Bill Shea.*

Machine-embroidered trapezoid insignia. Note that all the hem and reinforcement stitching is done in a fairly visible, off-white thread. Two-button style skirt flaps. *Courtesy of Bill Shea.*

OFFICER CAPS

Luftwaffe officer M43 caps have the usual aluminum crown piping used by the Army and Waffen-SS. Luftwaffe M43 caps for generals were piped in or gold celleon mesh cord. One, and in some cases two, air vent grommets per side coated with a baked on bluish gray enamel may appear on private purchase caps.

In keeping with the usual German system, material quality of officer model caps is noticeably better than that of Other Ranks caps, though this difference degraded somewhat toward the end of the war. Caps made from fine-grade Tikot or a lightly brushed, high quality doeskin were common through late 1944. Slightly coarser grades also appear, which were probably examples of officer caps from government stocks – slightly higher in quality than standard Other Ranks issue caps, but certainly not as good as that available in privately purchased headgear. For field use, however, these no doubt served just fine. Lower quality wool might also indicate a field upgrade of an NCO cap, once promoted to officer rank.

LININGS

Other Ranks cap linings were normally in a gray, or purplish blue cotton. Grayish blue herringbone twill pattern linings are common on mid-war caps, with patternless solid color linings also common on late-war production. Officer cap linings are usually a grayish blue color, however other colors, including black (rare), do appear. The fabric used was usually rayon, or a very fine grade of cotton.

INSIGNIA

Insignia variation is perhaps the single biggest cause of confusion among novice collectors, followed by skirt arm (flap) width. Initially, Luftwaffe M43 cap insignia for Other Ranks was in two-pieces: machine-*embroidered* in off-white or pale gray thread on a bluish-gray wool backing, cut to shape, and (usually) hand-applied. The cockade could be either of the machine-embroidered type with no visible backing, or a machine-woven (BEVo) type on backing cloth shaped to a diamond. In this latter case, it was common for the upper diamond point to be folded under, slightly overlapping the bottom arm of the swastika when applied.

The most significant difference between machine-embroidered and machine-woven insignia is *texture*. Machine-embroidered insignia has a raised surface in places where the thread is applied thicker – there is an actual three dimensional texture, which can be seen as well as felt.

Visually, machine-woven (BEVo) insignia has a very standardized appearance and differences are minimal (virtually non-existent between examples from the same production run), and practically no texture. The surface is flat – basically two-dimensional. Authentic examples of *machine-woven* Luftwaffe eagles on a dark blue rayon underlay are extremely rare.

Officers were authorized to use the usual, hand-embroidered bullion officer insignia on the M43, such as was used on the visor cap (or a bullion eagle with machine embroidered cockade), however many officers seem to have preferred the Other Rank insignia, instead. Either form is correct on an officer's cap, together with aluminum piping cord around the crown seam.

Left: Close-up of a private purchase Luftwaffe officer's M43 cap in fine quality Trikot. Hand-embroidered nation emblem and cockade. Both are hand-applied; the stitches on the eagle cross over the edge of the backing material. Stitches used to apply the cockade are not visible (which is the correct application method). This cap has a somewhat unusual *all black* lining. *Courtesy of Bill Shea.*

Part 2:
PANZER

There were no major changes in design between the black cap and the blue-gray version except for the usual difference in wool quality (better on the black cap). Period photographic evidence indicates that most of the black Panzer M43 caps used by Luftwaffe armored personnel (Hermann Göring Div.) were in the two-button flap style.

Insignia for Other Ranks caps was the usual Luftwaffe eagle machine-embroidered on a black wool backing, and a machine-embroidered cockade.

Officers commonly used Other Ranks insignia (off-white on black), but were authorized to wear hand-embroidered bullion insignia if they wished. The number of black M43 caps was limited to individuals serving in one single division (and perhaps, the divisional training unit), together with whatever depot stocks were kept for re-supply purposes. This limited usage means that the availability of authentic examples of this type of Luftwaffe headgear today is practically zero (particularly for officer caps). Any black Panzer M43 caps with Luftwaffe insignia that you encounter will, therefore, usually be either a modified cap of some sort, or a reproduction.

Part 3:
TROPICAL

Luftwaffe tropical visored field caps were essentially the same as Army models, complete with simulated side skirts, and air vent grommets coated with a brown, baked-on enamel. A soutache rarely ever appears on these caps, though if one does, the application method follows the Army pattern.

Linings were the standard red color (for both enlisted personnel and officer grades). Privately purchased caps, of course, could vary, though most manufacturers tended to remain with the regulation red.

INSIGNIA

Correct insignia for the tropical cap was two-piece for Other Ranks, in off-white thread machine-embroidered on a khaki (or tan) colored base cloth. The eagle was trimmed to shape and applied by hand, or by machine when available. Machine-embroidered cockades were applied with no backing cloth visible, or in the standard diamond shape for the machine-woven version. Hand applications should be neat, and semi-professional in appearance. Be wary of sloppy, unequal stitching and applications using odd color thread. Trapezoid insignia on these caps is rare (that is, combined eagle and cockade trapezoid).

Occasionally, authentic tropical caps turn up with machine-embroidered insignia on blue or black (Panzer) backing cloth. These are unusual and, as a result, require very close scrutiny. That such exceptions do appear is illustrated by the authentic private purchase cap shown at right. The insignia, in this case, is nearly always two-piece and trimmed (that is, no trapezoid-style backing cloth).

MODIFICATIONS AND REPRODUCTIONS

Modified contemporary caps consist mainly of the lightweight version of the Bundeswehr Bergmütze (not made of 100% wool), which compares well with the weight of original

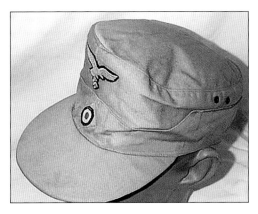

An example of exceptions to the rule: this Luftwaffe tropical visored cap is a private purchase piece. Note the simulated side skirts and the black air vent grommets (normally brown on issue caps). *Courtesy of Bill Shea.*

Unusual to find a *machine-embroidered* eagle on black backing cloth on a tropical cap; this one appears to be either a Panzer or black work [garrison] cap insignia. The cockade, on the other hand, is machine-woven on correct tan backing cloth on the usual diamond. *Courtesy of Bill Shea.*

tropical caps. The color of the Bundeswehr cap is incorrect however, and can't be changed without significant effort. The other indicators of Bundeswehr design given in the Army section also apply here. Modified contemporary German caps, therefore, do not present a major concern to headgear collectors. East German caps, which are not commonly offered on the U.S. militaria market, are completely unsuitable since they differ too greatly in design.

Reproductions on the other hand, present a much greater threat, though at this point in time they are not very common for this type of cap.

A.IV. KRIEGSMARINE FELDMÜTZE 42
(M43-style field cap)

Part 1: Continental
Part 2. Tropical

Part 1:
CONTINENTAL

The Kriegsmarine "M43" was officially known as the *M42* field cap. In design, it was the same as the Army's M43, with standard wide, two-button skirt arms (see the Army section). The wool color was classed as field gray, but leans heavily toward a very distinctive green color, which is noticeably different from the Army's field gray.

INSIGNIA
Correct insignia for this cap was a trapezoidal machine-woven, golden yellow Navy eagle on a field gray or dark green rayon backing. This insignia was not a combined eagle and cockade, however; the cockade was applied separately, with the backing cloth shaped in a diamond. In order to keep the cockade from being hidden behind the flap scallop, it was placed slightly higher on the body with the bottom of the swastika and wreath often overlapping the cockade's top edge – characteristic of authentic Kriegsmarine M42 caps. Officers commonly used the same Other Ranks insignia, but with gilt or celleon mesh cord piping around the cap crown. From a distance, such a cap looked somewhat similar to an Army general's M43 cap, and can easily be mistaken for the same by novice collectors.

MODIFICATIONS AND REPRODUCTIONS
The number of land based Kriegsmarine units that wore these field caps was very limited, and thus authentic examples are quite rare and turn up only infrequently on the militaria market. The fact is that most collectors new to the field are not very familiar with this particular type of cap; there are simply not enough available for novice collectors to become familiar with. The demand for the Kriegsmarine M42 field cap is therefore limited primarily to advanced collectors who are looking to complete an M43 collection and know what they are looking for. The result of this relatively low demand is that very few modified caps, or reproductions are normally offered on the market – though this situation is certainly subject to change.

Part 2:
TROPICAL

Like the Army and Waffen-SS, the Kriegsmarine also had its own version of a tropical visored cap, though this was introduced much later than its Army cousin. Distribution was very limited, and an authentic Kriegsmarine tropical field cap is indeed a rare animal. These caps were produced primarily in cotton twill, but there are also examples made of a lightweight canvas material.

CONSTRUCTION

Similar to the Army cap design, the *simulated* side skirts were in the usual shape, with a front scallop and no buttons. The top of the Navy cap had the same central seam common to most tropical cap versions, but was in fact, one single piece. In the center of the cap top from front to back, the material was simply pinched together inside the cap and sewn along the length of the line so as to create an exterior ventral seam. These caps usually have two air vent grommets per side, though examples with only one have also been observed. The cloth covering for the visor on these caps was in two-pieces, with reinforcement stitching lines appearing only on the visor reverse (underside) on Government Issue examples.

LINING

The most significant characteristic of Kriegsmarine tropical caps is the color of the cotton lining (in a smooth satin weave), which, instead of the red common to caps for the other services, is light green in Navy examples. Lining markings on Government Issue Navy tropical caps are usually RB Numbers, though individual named maker stamps occasionally appear.

OFFICER CAPS

Authentic examples of officer caps are extremely rare, and any cap encountered that the seller claims to be officer grade, should be checked very carefully. Very few original examples, in fact, are known to exist. Officer caps differed only in the addition of golden yellow piping along the top edge of the simulated front scallop (0.4 cm, regulation width).

INSIGNIA

Correct insignia for both officer and Other Ranks caps was a machine-woven (BEVo) eagle in golden yellow thread on a tan backing, applied either trimmed (to the eagle shape) or as a trapezoid. The machine-woven cockade on a tan underlay was applied with the backing formed to the standard diamond shape. The insignia was either machine-sewn or hand-sewn. The stitch used for machine-sewing was usually a straight stitch, though a zigzag stitch was also common. Hand-sewn zigzag stitching is usually very tight, with each individual loop being very short and close together.

MODIFIED AND REPRODUCTION CAPS

Again, these caps are extremely rare (and expensive) and most collectors therefore, have little familiarity or experience with them, or for that matter, even an opportunity to gain experience. Low demand does not encourage modifications of modern caps – which are not really suitable in any case – and reproductions as well, are still hardly ever seen. In over a year of Internet searching through both auction sites and fixed websites, not a single example – authentic or false – ever appeared. Any Kriegsmarine M42 field cap example you may come across, particularly an officer cap, should be checked very closely before making any decision to purchase.

B. DIE BERGMÜTZE – THE MOUNTAIN CAP

The Bergmütze served as the basis for the later M43 design, and as such, it shares a considerable similarity to that cap. Introduced in the early 1930s and modified several times before reaching its final form in 1938, this cap was later issued with two-button, narrow skirt flaps. Mountain caps with pebbled aluminum buttons are frequently mistaken for the M43 – and vice versa. The means to tell them apart lies solely in the Bergmütze's noticeably shorter visor, and higher crown. In some cases, visibly longer skirt arms may also be an indicator – these are found mainly on private purchase caps. A *T*-type combined insignia serves as an added confirmation that the cap is indeed a Bergmütze – though early M43 caps used this insignia as well. A Leichtmetall Edelweiss badge mounted on the left skirt flap may help in identification, but these are available as reproductions and are no guarantee that a cap is authentic.

Army and Waffen-SS Bergmütze's differ hardly at all in terms of construction.

Other (generally earlier) versions of the Bergmütze with narrow, single-button flaps are also quite common, however these are often private purchase caps and are probably based on civilian or Austrian models. The Bergmütze, in fact, originated in Austria.

CONSTRUCTION – ARMY and WAFFEN-SS
The same as for the M43 with the exception of a noticeably shorter visor and a higher, well-padded front. Correct buttons commonly found on the skirt flaps include:

1. Smooth surface aluminum (silver [officers], gold [generals] or painted field gray [Other Ranks])
2. Pebbled surface aluminum (silver [officers], gold [generals] or painted field gray [Other Ranks])
3. Steinnuß (corozo nut)

In 1938, a single air vent grommet was added to each side of the cap body. Bergmützes therefore, appear both with and without grommets.

Officer caps were piped in aluminum cord around the crown, generals in gold cord (gilt wire or celleon mesh). Waffen-SS generals' caps can not be distinguished from those of lower ranking officers since piping for SS generals was also aluminum.

LININGS
A cotton muslin fabric served as the lining on issue caps. Private purchase versions often have quilted linings, frequently in an orange-red color. Most Bergmütze caps (issue caps included) will have either no maker mark or an ink stamped maker mark in the top of the lining; late war caps, on the other hand, may have RB numbers in place of a maker name. These caps often have a thin, crème or dark tan-colored leather partial sweatband, and this includes standard Government Issue caps.

INSIGNIA

HEER (Army)

The correct insignia for the Army Bergmütze is the eagle and cockade combined in the machine-woven (BEVo) *T*-form. On caps with the first version of this insignia, the eagle is in white thread; later versions used gray thread. The Bergmütze was issued to Army troops both well before and after the (combined) trapezoid insignia was introduced. While trapezoid insignia does appear on Bergmützes from time to time, it does not fall within the bounds of regulation insignia for this kind of cap and should therefore be considered suspect.

WAFFEN-SS

Correct insignia for the Waffen-SS Bergmütze is the standard two-piece off-white machine-woven eagle and Totenkopf on black underlay for Other Ranks troops, in aluminum for officers (including generals). Trapezoidal insignia was not used on these caps. If the cap has an Edelweiss badge on the left side skirt, it should be the SS machine-embroidered version. Although some individuals used the Army metal Edelweiss, this insignia was not authorized for wear and is not regulation for an SS cap.

LUFTWAFFE

The Air Force's Bergmütze was also fielded well before its M43 model, and remained in service throughout the war. The Luftwaffe form of this cap differed from the Army and Waffen-SS in being primarily a narrow flap, single-button version, though later models also appeared with two-button flaps (fliegerblau painted aluminum buttons). The side skirt arms on the single-button type, in fact, tend to run a little longer than those on M43s used by other services – that it, the turn up for the side skirt is located slightly more to the rear of the cap. These long arms generally had rounded ends, and were secured by a small metal button (usually painted fliegerblau), or a slightly larger, heavy plastic button (dark blue).

Officer caps used a single large, aluminum button with a smooth or pebbled surface, and the normal aluminum cord crown piping for officers. The wider skirt arm version had two pebbled or smooth aluminum buttons. For generals, the cord was in gilt or celleon mesh cord, and the button(s) gilt. Insignia: Two-piece, machine-embroidered in off-white thread on a blue-gray wool backing is the standard for this cap. The cockade was applied, as usual, with no backing cloth showing on most caps. Trapezoidal insignia is not normally found on the Luftwaffe Bergmütze and should be considered a significant liability when evaluating a cap.

MODIFICATIONS AND REPRODUCTIONS (ALL MODELS)

Modifications are common with Bergmütze caps. The popularity of this design has not diminished among the civilian populace in Germany, and the Bundeswehr calls its own postwar, M43-style field cap a Bergmütze, as well. The Austrian military – original creator of the Bergmütze – also still uses caps in this form. This means that a considerable selection of contemporary caps is available for a modifier to choose from, each with varying degrees of suitability. Be wary of any Bergmütze with:

Army General Dietl (commander of mountain troops in Norway) photographed wearing the two-button style Bergmütze. *Courtesy of Dale Paul.*

Waffen-SS Obergruppenführer and Knights Cross bearer Lothar Debes is shown here wearing a fine example of a Waffen-SS single-button officer's Bergmütze with the SS-style *embroidered* Edelweiss affixed to the left side skirt. Note the position and the angle of the Edelweiss insignia. *Author's collection.*

This Waffen-SS Gebirgsjäger 'Bergmütze' is far outside normal specifications. Offered for Internet auction as an original, it is not. The eagle normally appears only on the front of an authentic cap that has an Edelweiss on the skirt. The placement of both is therefore incorrect – the eagle should not be on the side, and the insignia is very crowded. The cap's condition is terrible (perhaps to confuse the issue?) and it positively screams 'modification'. The angle of the Edelweiss (which is a metal Army version, not the correct cloth SS type) is also wrong – too low, too horizontal, and too much to the rear! The seller never responded to e-mail enquiries about these points. Though the cap is phony, it drew a few bids for a sizable sum. *Author's collection.*

Right: Army Bergmütze for Other Ranks. This cap was offered as a private purchase example, but there are a number of inconsistencies which make it objectionable. The most obvious of these is incorrect insignia: trapezoids are not the correct form for a Bergmütze. It is also hand-sewn – particularly undesirable with a non-standard insignia. The cap's pull-down side skirts are lined on the inside with black felt, whcih was not regulation. The front skirt flaps are secured with a smooth-surfaced Steinnüß button.

1. The longer visor usually associated with an M43 cap.

2. A trapezoid insignia rather than the correct *T*-form.

3. Unusual insignia placement (not in the normally correct position).

4. Incorrect form of the Gebirgsjäger Edelweiss badge (primarily on SS caps)

5. A soft sweatshield with a maker's mark on a private purchase cap – particularly if the mark is in gold ink (common to Bundeswehr caps).

6. Thin, lightweight material (not wool) with a thin visor (Bundeswehr caps are not 100% wool, but rather a synthetic/wool blend).

7. Ear loops on the side skirts.

8. Unusual color linings (e.g. yellow or other off-spec color).

9. Any unusual features that don't fit the norm, especially when several such discrepancies are found together on the same cap.

Interior view showing a quilted orange lining with the ink stamped makers mark "Peter Küpper Wu-Ro" (Wuppertal-Ronsdorf). The makers mark, however, is absolutely no guarantee of authenticity.

Chapter 14

Sammleranfertigungen - Reproductions

REENACTING

Prior to the 1980s there were only two reasons why anyone or any company made reproduction caps:

1. To satisfy the occasional need of some collectors for a temporary filler – that is, a copy of a rare cap to fill a hole until an original can be found. This demand, of course, was not enough to be profitable for the company making the reproductions. (Some in the headgear collecting community disagree with the practice of using fillers.)
2. To sell through dealers without any care as to how the dealers were presenting the item: In many cases, the dealers were of the shady variety that sold these caps as originals.

In the 1980s, there had been scattered reenacting events taking place around the United States, but it wasn't until the decade of the 1990s that an amazing upswing in the popularity of this rather expensive hobby occurred – interest for reenacting in general, and for World War II reenacting in particular. The result was more events being scheduled and conducted than ever before, with increasingly greater public turnouts. The attention and interest these events received in turn drew even more people into the hobby. Concurrent with the increase in reenactor groups has, of course, been a rise in demand for reproduction items of all kinds to supply the needs of group members.

Quick to recognize a highly desirable change in market demand, companies that produce reproduction militaria have raced to meet the need. Increased demand brings higher sales figures for these companies, and the additional funds have allowed some of the wiser firms to invest money in improvements to their product quality. This means that some of the reproductions now available are quite impressive in their attention to detail and accuracy – or frightening due to their ability to fool inexperienced militaria collectors – depending on how one views the subject.

For the most part, the current situation is a satisfactory, if uneasy, state of affairs. Reenactors want realistic copies of Second World War gear and uniforms, and the companies that make these goods market them primarily to the reenacting community. They are clearly identified as what they are: reproductions, and they serve a valid need.

This Waffen-SS officer reenactor wears a reproduction crush cap (officer's old-style field cap) during a road march at a battle reenactment held in the year 2000 at the U.S. Army Armor Center at Ft. Knox, KY (home of the Armor School). A fine quality reproduction, it is likely a cap made by the Janke Tailoring firm. Embroidered bullion eagle, aluminum Totenkopf. *Courtesy of photographer Robert Stevenson and Armor Magazine.*

As long as this remains the case, all is well, but there always have been, and always will be individuals who have no qualms about selling reproduction items with the intent to defraud and cheat collectors. Such individuals are usually exposed sooner or later, but by the time this happens they can do a lot of damage to the hobby. A painful financial loss caused by an unscrupulous dealer can easily and surely derail interest in the hobby for a novice collector – often beyond recall. Part of the purpose of this book is to help combat such dealers. Some people, however, prefer reproductions for one reason or another, and this chapter provides information on this topic.

QUALITY ISSUES

All true reproductions share one thing in common: they are caps that are made from scratch, *not* modifications to an existing piece. Reproductions vary greatly in quality, of course, depending on the maker. Robert Lubstein, the manufacturer of the renowned *EREL-Sonderklasse* caps, would perhaps be proud to know that more than half of the reproduction visor caps on the market today are ERELs. The aura surrounding this name is so great that every reproduction cap manufacturer wants to market its products with an EREL mark – complete with the paper tag used to advertise the unique (optional) EREL ventilation system.

Lubstein might not have been nearly as pleased with the *quality* of the reproductions bearing his *EREL-Sonderklasse* logo. Although some of them are fairly convincing at first glance, hardly any *will actually have* the vent system which the paper tag advertises: reproducing such a system is simply too expensive. In addition, mass produced EREL-lookalikes are usually made without the internal padding of authentic originals; they feel very lightweight, and though the wool used in construction is definitely doeskin, the material itself is very thin and much too soft. The identity of the manufacturer of the most common EREL copies, priced at around $160 in the year 2000, remains unknown to most collectors.

Passable from a distance, this mass-production reproduction Waffen-SS Panzer officer's visor cap (purchased from Reddick Enterprises) has no internal padding (wadding) where the cover overhangs the cap band on each side. Without such padding, the cap has a somewhat flimsy appearance and is very lightweight. The piping color is actually the Bundeswehr shade of Panzer pink – not a match with the original wartime Panzer branch piping.

Reproduction caps have been around as long as collecting caps has been a hobby. Though quality varies greatly, some of the older reproductions were made using original wartime materials. These are generally fairly good copies; their principal weakness lies in the fact that they were not necessarily made by individuals with extensive training and experience in the art of cap-making. Though convincing at first glance, they carry flaws which eventually give them away, whether these be errors in cut, poor construction techniques, or the like. Many modern reproductions on the other hand, whether individually-made or mass-produced, are of very poor quality and can be identified almost immediately.

Some reproductions really are first-class copies, however, both in terms of the raw materials used as well as in techniques of construction (that is, in their faithfulness to the originals). Consistency in the final product, however, exists only for caps made by the *same manufacturer* – not between reproduction makers as a group.

Top quality reproduction caps are a double-edged sword. On the one hand, they are suitably convincing when used by reenactors or by a collector as a filler for a hard-to-find cap, as they were meant to be. On the other hand, the very fact that they are so convincing makes them a danger in the wrong hands.

JANKE

Perhaps the finest-made and most technically accurate reproduction caps are those produced by a German firm named Janke Tailoring. These caps are sold by Janke through international distributors *without insignia of any kind* and are thus no more than military-style caps made to World War II specifications: they in no way transgress against the regulations/limitations imposed by Section 86a of the German Federal *Strafgesetzbuch* (StGB) [Penal Code]. The purpose of this section of the penal code is, in effect, to forbid the distribution of, or public display within a group of, "*symbols of any party or association declared unconstitutional by the Federal Constitutional Court.*" "symbols" includes any National Socialist period paraphernalia, "*...namely flags, emblems [or insignia], uniform items, slogans and greetings.*" This paragraph also forbids domestic production, foreign production (under German direction) of such symbols, as well as import or export thereof.

Thus, whatever reproduction insignia the cap buyer desires is added only after the caps reach the United States. Metal insignia is produced in the U.S. as well as in a number of other countries (including the U.K.), but there is no German involvement in any insignia manufacture. The quality of the insignia is quite good.

Janke Tailoring's owner has long held an interest in military history – particularly for headgear and uniforms. He had the good fortune to meet with and enlist the assistance of an elderly fellow who as a young man, worked as a cap maker for a German manufacturer during the war. This gentleman produces caps of all types for the Janke firm, all of which are sold specifically as reproductions. As of this writing, the fellow is reported to be training the company owner (Mr. Janke) and others in the old techniques so that Janke can continue the line after he retires.

One of the three distributors in the United States for *new* Janke caps is Mr. Bill Bureau, of Bill Bureau's Militaria. This very amenable businessman (himself a reenactor) sells reproduction German caps and uniforms on his Internet website – all clearly labeled as such. Most caps are Janke products, made individually to order in accordance with the buyer's choice of cap size, arm of service (Army, Luftwaffe, etc.), cap model visor cap, M43 cap, etc.) and branch (Infantry, Panzer, etc.).

A poor reproduction, which experienced collectors would recognize on sight. The crown has a weak frontal (interior) support, the cover material sags; the cap band is not stiffly vertical (it tapers inward toward the top), and the material is too dark. The eagle is also a poor reproduction (and the underlay color too light a green); it is applied much higher up on the cap front than specified in German Army regulations. Note the extremely wide brim ridge, which tapers off at each end – such a ridge is never found on authentic caps.

Interior: Though some makers used black for the visor reverse color, it is rather rare. On a cap already having so many irregularities, the black really stands out. Note also the way the cloth that forms the bottom cap edge actually *wraps over* each end of the visor: this was *not* a method used by German wartime (or pre-war) manufacturers for attaching visors (it *is* seen on some American military caps). The black sweatband is also uncommon (on an Army cap). The thickness of the bottom edge of the cap varies around its circumference. This item, offered on the Internet, was not labeled as a reproduction.

Janke-made Waffen-SS Panzer officer's Old-Style Field Cap (sometimes referred to as an M37), in fine quality field gray Trikot. The cap band's vertical height is correct. The material color, however, is too new (i.e. no age fading or wear from fifty years storage); another weakness is the Panzer Waffenfarbe (branch color), which is the current Bundeswehr shade rather than the slightly paler, warmer original wartime color. In addition, the brim of the flexible visor has a very slight ridge, whereas original visors did not. In most other respects – weight, feel, appearance, lining – the cap is superb. Cost: About $160 in the year 2000; the reproduction insignia cost an additional $40.

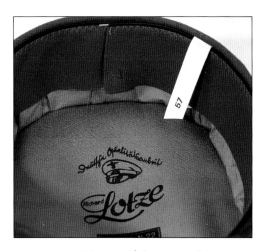

Looking toward the rear of the cap, the lining has no sweatshield (common on original examples from government stocks). The maker mark used here (*Lotze*), is not that of a known, confirmed WWII maker. Note the small reed rod at the rear of the sweatband (see below). The narrow white size tag is neither the correct shape for tags used during the war, nor is it attached to the cap in the same manner as an original.

View toward the front of the cap. Note the parallel fluting on the reverse of the flexible visor – a feature commonly found on authentic originals.

In Germany, Herr Janke managed to secure arrangements with several textile companies to produce field gray doeskin and Trikot fabrics, which – other than looking brand new (they are, of course) – differ little from original materials used during the war. The cost is high, since the quantities he orders, relatively speaking, are small. Hence, the reason why all Janke caps are hand-made individually to order and must be paid for in advance. Prices for these reproductions run about $160 per cap (in the year 2000) – no higher than the cheaper EREL lookalikes, and for that price comes an extremely well-made cap with nearly original fabrics – no synthetic materials or plastics are used in construction. The caps are solid, and well-padded (and thus, weighted) like the originals. In fact, they are in essence *modern-made* originals – a frightening concept for some dealers and collectors.

Over a period of several decades, Mr. Janke has diligently searched the newspapers for announcements of firm closings (original makers who survived the war but went out of business in the postwar years). When found, he would visit these companies and buy the stamp dies or transfer patterns for that firm's maker mark/logo. As a result, Janke caps include many original marks for quite a few of the original wartime manufacturers that are now no longer in existence (Christian Haug of Berlin, for example). He was also able to secure a stock of original Luftwaffe visors for Other Ranks caps which had been discovered somewhere in Germany. Janke reproductions of Luftwaffe visor caps thus come with original visors while the supply lasts.

VISOR CAPS

Janke products include both visor caps and field caps for all the armed services, as well as non-military headgear. Despite the superior quality and excellent materials, there are still a few minor points that allow Janke-made visor caps – new examples in particular – to be identified as reproductions without considerable difficulty. These are:

1. **Piping color:** On Panzer caps, the original pink Panzer branch Waffenfarbe simply does not exist anymore; there is no choice for the company but to make use of the current Bundeswehr shade, which tends more toward a cooler violet than the warm, pale pink found on authentic originals.

2. **Reed shaping rod:** A characteristic in the construction of all Janke-made visor caps is a thin black satiny material insert between the sweatband and the bottom of the cap. This material forms a loop at the base of the sweatband, in which a tiny reed rod is inserted to help the cap bottom maintain a round shape. A millimeter or so of the rod is usually visible at the rear of the cap, protruding from the material loop.

3. **Sweatband/stitching:** Janke also uses thin leather sweatbands, with a simple zigzag stitch for all its visor caps. The sweatband stitching in fact, is the only significant weak point on these caps – the threads are not quite as heavy as on originals, and thus can break more easily.

4. **Visor Ridge:** On reproductions of the *officer's old-style field cap*, the flexible leather visor is the same thickness as the originals, but has a very slight ridge on the visor brim (obverse) reminiscent of a Vulkanfiber visor. Original caps of this type have a smooth, even brim with no ridge.

The characteristic reed shaping rod found in most Janke-made caps, and the black material insert. Note the zigzag stitching.

Janke Kriegsmarine white top junior officer's cap with reproduction bullion wreath and cockade, but no national emblem (no National Socialist insignia is produced in Germany). Caps are shipped from Germany with *no* insignia other than wreath and cockade. The reproduction gold bullion Navy wreath on this example is very well done. The white cap material is the correct original waffle pattern, and the wire scallop embroidery on the visor is also correct. *Courtesy of Bill Bureau.*

nke reproduction Army infantry officer's visor cap. Aluminum reproduction eagle with reproduction hand-nbroidered bullion wreath and cockade. The wool is a somewhat coarse, average grade; the field gray color is ry good. *Courtesy of Bill Bureau.*

Reproduction Army infantry officer's visor cap. This is another mass-produced EREL-lookalike. Thin doeskin wool, no interior padding. Reproduction chromed metal eagle (extremely shiny), but an *authentic* oak leave wreath (applied after purchase). Thus, the eagle and wreath do not match, and are obviously not a set. The cockade is also a reproduction (made in England), with a pasted on red paper dot in the center. Although the cap band appears dark enough in this picture, it is actually too light a green.

Right: *Authentic* Army smoke troops (Nebelwerfertruppen) officer's cap for comparison. Thick, field gray "aligned" (Strichware) doeskin. The cap is fairly heavy, with a well-padded interior. *Courtesy of Gerard Stezelberger of Relic Hunter.*

Janke black Panzer M34 cap with soutache. In this case, a manufacturing date stamp on the cap lining later than 1942 would be incorrect since use of the soutache was discontinued in that year. This may provide another clue that the cap is a reproduction. The soutache is correct, if a bit thinner than an original.

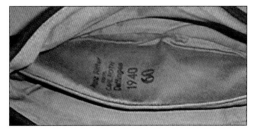

Interior. In this case, the manufacturing date stamp on the cap lining is suitable for a cap with a soutache. While lining markings in and of themselves can not be used to guarantee a cap's authenticity, they can sometimes provide a clue to help in identifying a reproduction. Note the correct interior ridge down the center of the cap top.

FIELD CAPS

Top quality reproduction field caps are much more difficult to identify but for their relative newness and the use of reproduction insignia (some owners apply original insignia, however). They are otherwise complete down to (in some cases) original maker stamps on the lining – which is why these can not be depended upon as an indicator of authenticity. Lining materials are correct, as is the interior construction.

The wool used for reproduction field gray caps usually has a heavily felted (coarse) finish more reminiscent of late war production. With authentic caps, this tends to be associated with the RB numbers used later in the war (instead of a named maker mark – though this is not a hard and fast rule. It is still wise to double check the type of wool finish against any lining marks – this may be of assistance in identifying a reproduction.

If one of these caps is worn for a period by a reenactor, however, the difficulty in discerning the difference increases.

Janke-made caps used by collectors and then resold as originals thus present far more of a problem – the field wear makes them that much more convincing. The reed shaping-rod on visor caps remains a giveaway, however, and the scent of recent wear (sweat, hair oil, etc.) will still be present. Any cap offered as an original on the militaria market has the potential to be a reproduction; while high quality reproduction caps can sometimes be a challenge to detect when offered as authentic, detection *is* usually possible given a careful inspection prior to any purchase. The need for these caps for legitimate purposes means that they will always be available; as a collector, you must remain perpetually on guard unless buying from a trusted, experienced source.

Left: Janke-made Luftwaffe Other Ranks reproduction Fliegermütze with reproduction machine-embroidered insignia. The workmanship and materials used are excellent. The insignia – while quite good – has some minor flaws: The tips of the eagle's wing feathers are a smidgen too wide, for example. Above: Interior: While the lining material is correct and correctly installed (complete with the central fore and aft ridge), original caps did not come with a narrow white size tag sewn to the bottom hem stitch line. *Courtesy of Bill Bureau.*

Army officer reproduction field gray M43 cap, as delivered (i.e. without insignia). Any insignia desired by the buyer can be applied *after* the cap arrives in the U.S. The wool quality is comparable to that found on authentic officer's caps originating from government stocks. The field-gray color tends to the green (partly due to the lighting), but is within limits. *Courtesy of Bill Bureau.*

Chapter 15

The Cap-Making Industry and the "Extra"-Mütze

By the outbreak of war in 1939 there were more than 90 individual cap makers in cities and towns throughout Germany, ranging from very small companies with only one workroom (classed as a Handwerk – handcraft business) to large operations with more than twenty five workers and over 200 square meters of floor space (classed as an *industrial* concern). Indeed, the years immediately preceding the Second World War were a heady time for this industry in general. The business climate had improved dramatically and by the mid-1930s many firms were enjoying the most solid financial footing that they had experienced in many years. The possibility of war still remained somewhat of an abstract concept prior to 1939; nor does it seem to have caused any undue concern among cap makers. That something would need to be done to keep the more than two million men in uniform occupied was not a topic they wasted time thinking about.

No, the main point of interest was the large military force in and of itself – the bigger the better! A larger Wehrmacht meant excellent business. Articles in the clothing industry's technical trade paper *Uniformen-Markt* (*UM*) waxed philosophic at times suggesting that Germans were simply "naturals" in uniform, while soldiers in many foreign armies did not take their uniforms seriously enough.

With several million men under arms by the end of the 1930s and at least two visor caps permitted or required for every officer and one for every soldier, airman, Navy NCO and Waffen-SS trooper, demand had increased dramatically. The industry experienced a surge in growth. Part of this included new manufacturers, among which were certain firms which possessed the needed start-up (or expansion) capital, but had little or no production experience in cap-making. Their primary interest was in grabbing as large a piece of the action as possible. A few of them produced and sold visor caps that were not at all in accord with official military regulations; some of them, amazingly enough, *had not even bothered to acquire and study those regulations (cap specifications) prior to commencing production.* As if this were not enough, there were also a few companies which apparently had no qualms about using inferior materials in their caps, then slapping on a generic lining logo such as *Deutsche Wertarbeit* [German Craftsmanship] or *Deutsche Qualitätsarbeit* [German Quality Craftsmanship] and then passing off these products as Extramützen (the German term for private purchase caps) at the high prices that such caps commanded.

An officer and an NCO (sergeant) enjoy a relaxed discussion on a wharf. The officer is wearing an Officer's old-style field cap (crush cap), the NCO a standard service visor cap. *Courtesy of Eric Mueller.*

What militaria collectors and dealers refer to today as a *'private purchase'* cap, the Germans called simply *'die "Extra"- Mütze'* – the 'extra', or 'special' cap. The price of a basic Extramütze started at around 6.90 Reichsmark (RM), and escalated from there – some to as high as RM 22.00 or RM 24.00. One former Luftwaffe pilot (officer) questioned about his first cap purchase, recalled the price paid at somewhere around RM20.00. By comparison, consider that a private in the German military received basic pay (*Wehrsold*) in the sum of RM13.30 *per month* in 1940 (not including any field or ration allowance). In foreign exchange values of 1939, one Reichsmark was equivalent to U.S. 40 cents (in 1990s dollars, roughly $3.50).

The potential for trouble with military authorities over inferior quality products fraudulently sold as *high quality* goods certainly did not escape the notice of industry leaders, most of who were from long-established companies (or who had worked in the trade for many years). A number of articles appeared in the *Uniformen-Markt* which, among other things, strongly criticized deceitful firms that purposefully cheated soldiers by misrepresenting inferior quality goods. These articles also included remarks about letters to the editor received from several unnamed manufacturers (who were already in production) that had actually written to ask about proper military specifications (after the fact, so to speak). The *U-M* response – with no little sarcasm – suggested contacting the military directly to ask for the latest regulations and found it very disturbing that this common sense action had not been taken by these firms *before* beginning production.

Time and again, *U-M* articles warned that one bad apple could easily spoil the barrel for all, if the military were to take notice of these transgressions and begin cracking down on any and all non-regulation headgear. Remember, that for the Army, the Waffen-SS and the Luftwaffe the regulation wool was Trikot (tricot) – but that 65% to

'Extra-Klasse' generic logo, with eagle. The small print in the space at the bottom says *Marke Standard* (standard brand). Nothing in the cap gives any indication of the true maker's identity. This particular generic logo has been appearing quite frequently in recent years – perhaps *too* frequently, in fact – on a wide variety of Army and Waffen-SS visor caps. Note the torn sweatband made of low quality, pressed paper material. *Courtesy of Bill Shea.*

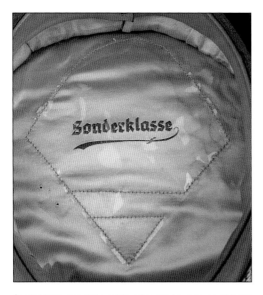

Sonderklasse generic logo. No other marks appear in this cap – the maker can not be determined. *Courtesy of Gerard Stezelberger of Relic Hunter.*

70% of all Extramütze visor caps were made of *non-regulation* wool (i.e., something other than Trikot). By their very nature, they were non-regulation, and the military would be quite within its legal rights to forbid all purchases of these caps by the troops. The main concern was that *too much* divergence from official standards by a few might draw unwanted attention and dire consequences for the industry as a whole – not just for those companies responsible for causing the problem.

DIE EXTRAMÜTZE – QUALITY WORKMANSHIP AND MATERIALS

With the privately purchased visor cap, *die "Extramütze"*, the principal meaning of the word Extra for cap makers and customers alike was first-class, or special, workmanship. The generic logos mentioned earlier stress exactly this point. Other common generic marks which do not identify the actual maker, but do tout quality, are:

1. *Erstklassig* (first class)
2. *Extraklasse* (extra class, best class)
3. *Sonderklasse* (special class)
4. *Sonderkonstruktion* (special construction)
5. *Sonderausführung* (special model)
6. *Sonderform* (special form)
7. *Meisterklasse* (master class)
8. *Luxus* (deluxe)
9. *Modern* (modern)
10. *die Deutsche Qualitätsmütze* (the German Quality Cap)

These terms and others are also used quite often in conjunction *with* a named maker. The correct brand name for the famous EREL caps, for example, is not EREL itself; Robert Lubstein, the maker of these caps, used EREL-*Sonderklasse* as his trademark name.

The German term *sonder* (special) used in such slogans specifically implies *hervorragende Qualität* – exceptional quality.

A further implication of the Extramütze concept was that only the highest quality materials were used in construction. After all, this was one of the main reasons for the expensive pricing of these caps. The industry demanded that every cap maker use nothing but the best materials available; in other words, no skimping on quality in order to pocket an extra Reichsmark or two! These high-quality materials, however, were often non-regulation.

As things stood, German military authorities more or less ignored the fact that most Extramützen were made with non-regulation wool. On the other hand, aberrations from the regulation field gray color that went beyond minor shade differences between textile manufacturers *did* draw attention during the pre-war years. Cap-making industry leaders strongly denounced any failure to ensure that the correct color was being used, and specifically addressed this issue in several *U-M* articles. The cap on page 213 provides an excellent example of an unauthorized color. This Army officer's cap is made of a very high quality doeskin – but in a completely non-regulation color. It is indisputably gray, rather than the correct, more greenish-hued field gray. This kind of obvious transgression was sure to be noticed during the pre-war years, but once the conflict was underway such fine points had a low priority and for the most part, were simply ignored.

Another common complaint was the appearance of visor caps with front crown heights clearly exceeding regulation measurements, as well as the growing tendency of some companies to increase the side panel measurements on Luftwaffe [removable] white top summer caps. With much of the side panel material overhanging the cap band in a nearly horizontal position, any increase in measurements made the overhang wider and more bulky – giving the caps a definite similarity to the normal Kriegsmarine white service visor. The following are comments from a *U-M* article on this theme:

> …every recruit goes around with a sort of 'generals' cap', which can eventually lead to limiting regulations for Extra-Mützen as well – and then a large part of the caps presently on hand will no longer be able to be worn. By instructing his customers, the seller must see to it that the line is drawn. Otherwise it can end up like the white removable cover [summer cap] for the Luftwaffe. In imitation of the Naval officer's cap, the cover simply could not be made large enough – until in many garrisons, apparently, bans were issued and a fundamental change in taste trends subsequently came about.

Each of these concerns and many others, were forcefully addressed in *U-M* articles. For the most part it appears that the general warning – *adhere to regulations as much as possible* – was understood by the majority (if not all) makers. Nonetheless, the industry was always trying to put out fires. As soon as one error was corrected, another appeared.

PIPING ISSUES

Piping on visor caps was, by regulation, not to exceed a maximum diameter of 0.2 cm. In addition, all three piping rings, the two around the top and bottom of the cap band and the upper ring around the cap crown, were to be the same diameter. The popular

Meisterklasse generic logo, in Fraktur font (German gothic). This mark also bears the phrase *ohne Schirmdruck* – (without visor pressure), and the notice 'D.R.G.M.'. The patent office-registered D.R.G.M. is probably for the system used to prevent the visor from pressing against the forehead. Which company held the D.R.G.M. rights is impossible to tell, however, without knowing the actual number. *Courtesy of Gerard Stezelberger of Relic Hunter.*

Army Gebirgsjäger (mountain troops) officer's visor cap made of high quality, extremely soft doeskin – but in a non-regulation color. Matching Leichtmetall insignia, including the Edelweiss cap badge authorized for wear by members of mountain troop units. Thick-style crown piping. Surprisingly enough, despite the superb exterior, the *interior* has a lesser quality composition sweatband. Maker: *Hans Schiederer*, of Landshutt (Bavaria).

belief held by many contemporary headgear collectors and militaria dealers that thicker crown piping appeared only after the war was well underway, is not correct. As early as 1936, in fact, some manufacturers had begun using *Biesen* (piping) for the crown of the visor cap that was slightly thicker than regulation specifications, while the cap band top and bottom piping remained the correct diameter. This thicker crown piping was apparently more aesthetically pleasing to many soldiers, and as a result its popularity grew quickly. Recognizing a good thing, some manufacturers decided that if more was good, then a lot more might be even better still. The diameter of the cap crown piping continued to increase until, once again, industry leaders felt compelled to step in with the warning that too much of a good thing was, in fact, not a very good thing at all. Extra-thick crown piping was becoming "entirely too conspicuous", they objected. In particular, it was glaringly obvious with a light-colored Waffenfarbe (branch color) such as the lemon yellow of Signals (*Nachrichtentruppen*), or the pink of the Panzer

Crown piping in regulation thickness matching the diameter of the cap band piping.

Thick-style crown piping. The diameter is over the regulation 0.2 cm, obviously thicker than the piping on the cap band.

branch. Upper limits on the piping diameter at a point just above that specified by regulations, were strongly recommended. Moderation was constantly urged as the key to success for everyone: this was acceptable, blatant excesses were not. The line, however, was a very fine one.

THE QUESTION OF CAP CLOTH

A customer shopping for an Extramütze could buy a cap right off the shelf or, in some cases, order one made to specifications. Material options in this case might include:

a. Trikot wool (regulation)
b. Doeskin (various grades)
c. Eskimo

In keeping with the concept of the Extramütze, the grade of Trikot used for a private purchase cap was generally of much higher quality than that used for caps produced for government stocks (though this was not necessarily true of late-war private purchase caps). A very high-grade regulation Trikot was just as much Extra in terms of quality, as either doeskin or Eskimo – both of which were non-regulation. An officer cap purchased from government uniform stocks on the other hand, was usually made of a lesser (coarser) grade of Trikot with a much more visible, heavier rib pattern.

GOVERNMENT ISSUE vs. THE EXTRAMÜTZE – *The Selbst-Einkleider*

In the field of soft (cloth) headgear collecting there is a great deal of confusion and a lack of consistency among dealers and collectors alike regarding the term *issue*. The word is often used (incorrectly) to distinguish officer caps made of *any grade* of Trikot, from a cap made of doeskin (or Eskimo) wool. The visual distinction between Trikot and doeskin is obvious, and so the assumption is that any visor cap made from doeskin is private purchase, that is, an Extramütze (true), while any visor cap or other type cap made from Trikot must come from government issue stocks (false). This usage is very misleading, as officers *were never issued* any uniform items (i.e. for free), and nearly all makers produced caps made of Trikot material. What then, is the real situation here?

The Wehrmacht recognized three main classes of personnel (as does the Bundeswehr today) in terms of uniform procurement:

1. The **Selbst-Einkleider:** All military officers and Wehrmachtsbeamter (Wehrmacht officials) with officer's rank. These individuals were expected to furnish their own uniforms and headgear themselves (Selbst), at their own expense.
2. The **Teil-Einkleider:** Included certain NCO ranks and other officials who were required to furnish specified uniform items (but not all) at their own expense – that is, they were partly (Teils) responsible for self supply.
3. **Government Issue:** The rank and file, who had all of their uniforms and headgear issued to them out of government supplies, from a depot at muster, or from their unit.

Classed in the Selbst-Einkleider category, officers were never *issued* caps and using this word to describe *any* cap made of Trikot cloth is therefore inaccurate. A far better and more precise term is *government stock* – which refers only to the quality of

This NCO is definitely wearing what is unquestionably a non-regulation hat in this unusual candid photo! He seems to be impersonating a magician, or so the "presto!"-style gesture (together with the eggs) suggests... No doubt he did not allow his subordinates to see this picture. *Courtesy of Eric Mueller.*

the wool used in the cap's construction. A cap (Army, Waffen-SS, or Luftwaffe) was either regulation i.e. Trikot wool (of any quality grade) or it was not: it was made of doeskin or some other material. A Trikot cap was thus automatically regulation – and could, at the same time, also be a private-purchase item (an Extramütze), if the material grade was of higher than average quality. Conversely, if made of only *average* quality Trikot (but better than that used for enlisted caps), with no maker mark (or with only a basic, generic maker mark), then it was usually a cap secured by the military on a contract, and made available for officers to purchase. These were *government stock* caps – the caps currently referred to by most collectors and dealers as "issue."

The dashing Oberstleutnant (LTC) Günter Goebel (as a Major, in this picture) who wears a fine example of an *Extramütze* (private purchase cap) with Leichtmetall eagle and hand-embroidered bullion wreath combination in this studio-posed award shot. Goebel, holder of the Oak Leaves to the Knight's Cross, served as a Staff Officer for the L Army Corps. The cap cover (top) is made of fine grade Eskimo. *Author's collection.*

German officers could and certainly did, sometimes buy caps from government stocks. They were never *issued* to an officer at no charge – he always had to pay – though he most likely enjoyed a government-subsidized bargain price. These comparatively inexpensive caps bought from government stocks were suitable for use in the field, while the much nicer, expensive Extramütze (private purchase cap) was left at home, or packed in the officer's belongings.

THE KLEIDERKASSE SYSTEM

The requirement for officers to provide their own uniforms is common to many western armies and dates as far back as Imperial Rome. U.S. military officers are also required to supply their own uniform needs at personal expense. So it was with the German Wehrmacht in the Second World War and continues in the modern Bundeswehr (Germany's Federal Armed Forces) today.

The German military recognized even before the First World War that the financial drain on young officers trying to provide for their own uniform requirements with expensive, privately purchased items might present problems. The solution was to create the [*Heeres*] Offizier Kleiderkasse, which was in effect, an [Army] Officer's Clothing Sales System. This institution, operated as a non-profit business and managed by a civilian director, established a system of financial accounts for each and every officer or official with officer status – in short, for everyone in the Selbst-Einkleider category. A mandatory monthly salary deduction was instituted, with proscribed amounts according to rank taken from each individual's pay and deposited in a personal, named Kleiderkasse account. In addition, the military services occasionally added a special sum as a clothing allowance to the officer's pay. Over time the deposit balances grew, managed by the Kleiderkasse on behalf of each and every member.

This clothing account helped German officers – especially the young junior officers – to build up the money needed to finance their uniform and headgear purchases. The officer himself did not really need do anything at all except monitor his monthly payroll deductions and account balance. The Heeres Kleiderkasse was first located at Berlin W 8, Wilhelmstraße 82-87 until May 8, 1935, when it relocated to Berlin W 62, Budapester Straße 9. In September of the same year, the building number was changed to 28. The Army's Kleiderkasse address never appears in any caps – only the *Offizier Kleiderkasse* etiquette itself.

The Kleiderkasse also published a catalog offering a large assortment of uniform items (caps, complete uniforms, insignia, tress, shoulderboards, uncut fabrics etc.), at prices lower than available on the commercial market. There were four options an officer could choose from when purchasing his uniforms, headgear and accessories:

1. **Kleiderkasse catalog sales:** Items could be ordered directly from the Kleiderkasse through its catalog, these were then delivered by mail, or could be picked up by the individual. In this sense, the Kleiderkasse functioned in a fashion very similar to the U.S. Military's Exchange Catalog system. (See Appendix G)

2. **Direct purchase:** Items could be purchased in person at the Kleiderkasse sales office in Berlin, or at one of several branch offices (called *Filliale*) located in different cities. In the modern Bundeswehr Kleiderkasse, the system functions in the same fashion. The U.S. Military's Clothing Sales Store, at which personnel can buy all manner of uniform items at their own expense (including certain items of field gear), is again similar.

3. **Outside sellers:** Items could be purchased from an outside seller, that is, a seller not affiliated in any way with the Kleiderkasse such as a haberdasher, a military goods (*Militäreffekten*) shop, or a manufacturer's outlet.

4. **Government stocks:** Items could be purchased from official government stocks. Such items were not free, but were most likely offered either at cost, or slightly above cost. Officer caps from government stocks were not equal in quality to caps offered through the Kleiderkasse or by an outside vendor – but met government standards and were fully serviceable.

DIE VERKAUFSABTEILUNG DER LUFTWAFFE – THE AIR FORCE [CLOTHING] SALES SECTION

The Army was not the only service that managed a Kleiderkasse. The Luftwaffe, ever intent on maintaining its very independent position relative to the Army, operated its own version of the system under the name die Verkaufsabteilung der Luftwaffe. As far as can be determined, the Verkaufsabteilung moved once in its history, relocating from its initial address to the final location of Berlin SW 68, Puttkamerstraße 16-18. The original address appears but rarely and only on early Luftwaffe caps, with the majority of Luftwaffe *Extramützen* sporting the later Verkaufsabteilung etiquette listing the Puttkamer Straße address.

DIE OFFIZIER KLEIDERKASSE DER KRIEGSMARINE – THE NAVY CLOTHING SALES STORE

The Navy likewise, had a similar system usually identified by the abbreviation O.K.K. in visor caps – when it appears at all (which is rarely – and only on EREL-Sonderklasse caps). The street address of the Navy's clothing sales office remains to be discovered.

MAKER ACRONYMS

German cap makers had a penchant for using acronyms formed from the first syllables of either the company owner's name, the town where the business was located, or a combination of both as trademark (logo) names. Thus the well-known firm of Peter Küpper in Wuppertal-Ronsdorf, for example, used the brand name *Peküro* representing **Pe + Kü + Ro** ('Ro' from Ronsdorf); Albert Kempf, also of Wuppertal-Ronsdorf, went by *Alkero*. Leonhard Paulig of Rothenburg-Oder (now part of Poland) went by **Leparo**, and so on (both of the first two companies survived the war, see their individual entries in Chapter 16).

Robert Lubstein (EREL-*Sonderklasse*) did not quite follow the usual pattern, using instead only the first letters of his name. In German, the letter *R* is pronounced 'eR', and *L* is pronounced 'eL', hence, ER/EL. To make certain the meaning is clear, the logo emphasizes the name initials by presenting them slightly larger and canted, while the two E's remain nearly vertical. EREL caps are the only ones to display either the Offizier Kleiderkasse or the O.K.K. mark as an integral part of the makers information.

Of the two, the O.K.K. mark is the rarer; in fact, the majority of visor caps made for Navy personnel were unmarked.

Another common form of brand mark was simply the first letters of the maker's name, with the first letter of the city where the company was located. Thus, HPC fo Hermann Potthoff of Coesfeld, for example.

DISTRIBUTORS/COTTAGE INDUSTRY

The relationship between major makers and distributors is akin to an iceberg – that is, a great deal is hidden below the surface. *Hermann Schellhorn*, for example, was a small company in the city of Offenbach am Main which manufactured high quality headgear and protective equipment for fire departments, but also distributed military caps made by the much larger August Schellenberg company of Berlin, and Peter Küpper of Wuppertal-Ronsdorf. Why was this? Did Schellhorn merely distribute these caps, or, did he actually produce them under license in his own workroom? Schellhorn's maker's mark also appears on the linings of these caps. Was he simply trying to increase his sales by using this distributorship to expand into a larger market?

Another question concerns the relationship of large makers to LAGO or AGO associations (small producers working together in a form of cottage industry). Perhaps these big companies had arrangements with a LAGO or AGO to help provide extra production capacity when this was needed to fill particularly large contracts. The firm of Clemens Wagner, for example, seems to have had such a relationship with an AGO located in central Germany; the AGO stamp appears beside the Clemens Wagner name on the sweatband reverse of some contract visor caps.

THE BUNDESWEHR KLEIDERKASSE AND POSTWAR MANUFACTURE

Die Kleiderkasse der Bundeswehr (the Bundeswehr's Kleiderklasse) is set up along the same format and structure as its Second World War predecessor – that is, as a non-profit, civilian-run business operation. After the creation of the German Federal Armed Forces in 1955 (and with it, the Bundeswehr Kleiderkasse), one or two of the larger cap-making companies seem to have secured similar (though not exclusive) supplier relationships with the new Kleiderkasse institution.

One of these firms was actually the former wartime maker **Peküro** (Peter Küpper) of Wuppertal-Ronsdorf, whose maker mark included the phrase '*Kleiderkasse der Bundeswehr*'. The firm was later forced to drop its military cap production due to growing economic pressure.

The ten year long postwar period without a military, from 1945 to 1955, had, in fact, been harrowing for German military cap makers in general. Many of those that had survived the war were simply unable to make ends meet through these extremely lean years – at least, not as cap makers. Peküro managed to stay afloat, but found itself in bad shape. Its Bundeswehr contract was not enough to support the business, and the company decided to make a major shift in product to sports wear and sport headgear; military cap production lines were closed and re-tooled.

Today, the only former wartime military cap maker still in business as a major manufacturer of *military* caps is Alkero (Albert Kempf), formerly of Wuppertal-Ronsdorf, now located in Teunz. From information kindly provided by the Albert Kempf firm, it is clear that postwar price pressures, production cost increases and a limited market necessitated difficult choices on the part of surviving cap makers. Albert Kempf took the unusual step of moving most of its military cap manufacturing activities to other countries (where production costs could be kept down). This decision ultimately proved a winner as, one by one, other companies like the once famous Peküro and Clemens Wagner had to drop put of the race. Today, according to the business office of the Bundeswehr Kleiderkasse, the main (and nearly exclusive) supplier of military caps to the Bundeswehr is, in fact, Albert Kempf.

INDIVIDUAL MAKER DATA

THE 'FANTASTIC FOUR'

If a list had to be compiled of the four greatest – and at the same time most famous – makers of private purchase caps (Extramützen) from wartime Germany, that list would be:

1. Robert Lubstein (EREL-*Sonderklasse*) [Berlin]
2. Clemens Wagner [Braunschweig]
3. Peter Küpper [Wuppertal-Ronsdorf]
4. August Schellenberg [Berlin]

Caps from all four of these well-respected cap makers are consistently excellent in both materials and construction, and are relatively common.

There were many other makers besides those four, however, including little-known companies that produced some excellent caps. Researching the many companies presented in the next chapter proved to be a very complicated and daunting task. The search for detailed records and other information on old German companies was frequently hindered by a host of inter-linked problems, some of which are noted here:

1. **Time periods:** Some German government agencies (particularly the Gewerbeamt offices) do not maintain records for any longer than a certain, specified period of time. Thus, after this period has elapsed (frequently the case given the age of these particular records), all older and inactive files are purged and destroyed, or sometimes shipped off to a larger, regional archive where they are buried amongst an ever-growing stock of such files.

2. **War damage:** Record availability is also greatly influenced by war damage. Many German cities were pulverized by allied bombing, particularly during the final months of the war, and although attempts were made to safeguard documents, paper was never completely bomb-proof. The Industrie und Handelskammer zu Berlin (Berlin Chamber of Commerce and Industry) for example, lost all of its pre-1945 company records in the last months of the war (February or March of 1945) when bombs hit the building in which it was located. The ensuing fires destroyed all of the IHK files. The Berlin Handwerkskammer likewise lost all of its records of small to mid-size companies. Thus, requests made to these agencies for information on Berlin-based companies dating prior to May of 1945 simply could not be fulfilled – the records don't exist. The city of Würzburg likewise lost all of its archival records in bombing fires; Dresden lost the majority of its civil records when the city was firebombed in 1944, and so on.

Occasionally, luck did manage to lend a hand. Documentation received from certain Berlin Amtsgericht (District Court) offices – the organization that maintains the local Handelsregister (commercial register) – included pre-war letters sent to the Amtsgericht by the IHK Berlin with detailed information on the companies in question. The IHK copies of these letters had of course, all been destroyed but the originals still existed in the Amtsgericht files!

3. **Company size:** Many firms were too small to ever qualify for the city Handelsregister (commercial registry). This leaves only city archives (Address Books) and the Business Register at the Gewerbeamt as possible sources – but

here we often arrive back at point Number One with regard to the Gewerbeamt's business register records.

4. **Political Realignments:** Political realities also impact research. For those firms that ended up on the East German side of the postwar border, even if the company continued the name usually did not since nearly every business was taken over and managed by the East German state. Former names were changed to accommodate a more socialist image. In some cases, old records were scrapped – the new socialist companies were to start afresh. The result of all this maneuvering was that any pre-war cap makers that were located in East Germany simply vanish from existence after 1951, like magic.

To make matters worse, several companies, such as Leparo (the firm of Leonhard Paulig), for example, simply vanished in 1945 along with the province in which they were located. This was the result of postwar border realignment. Leonhard Paulig was located in the town of Rothenberg-Oder, hence the maker name **Le+Pa+Ro**. This small town in the district of Grünewald in Schlesien (Silesia), lay east of the Oder-Neisse river line. After 1945, any land to the east of this line became part of Poland and the German town thus disappeared.

5. **Non-availability of actual examples:** Unless one is quite wealthy, it is simply impossible to own an example of a cap from every single maker – even if these could all be found. At other times a cap is available, but not at the time it happens to be needed. Not all maker entries in the following chapter include a photographic example of the makers mark, as desirable as that would be. An authentic cap may have been seen, but the item was often sold later by its owner before details could be noted or the mark photographed – nor could another example be readily found. Efforts were made (and continue) to secure as many examples of authentic makers marks as possible.

The individual company entries that follow in the next chapter are compiled alphabetically by city: Augsburg, Berlin, etc. In some cases, when no physical records confirming the company's existence have been found, data entries for these firms will be mostly blank. It is important to remember that a lack of confirming records does not necessarily mean that a given company did not exist, particularly if authentic original examples of caps made by that company have been confirmed. (Note: An alphabetical listing *by company name* is provided in Appendix B)

A WORD ON "EXCEPTIONS TO THE RULE", AND SOME NEEDED VOCABULARY

Every maker was different: Though their products were all made to a standard, there was some room for individuality. Given this variety, collectors need to expect that there will always be exceptions to the rule. Any exceptions that may be encountered should be studied that much more carefully, however, before coming to a decision to purchase.

One tendency to note while reading through the information and histories on these companies, is that the larger makers frequently used the rhomboid-style sweatshield (large, diamond with the top rounded off) on their *military* caps, while less well-known (usually smaller) makers used a pure diamond-shape. Some companies, on the other hand, used both: Officer old-style field caps, made by the major maker August Schellenberg, for example, use a diamond-shaped celluloid sweatshield, while all of the firm's private purchase [military] service visor caps have the rhomboid.

It should be noted here that the term "rhomboid", used to mean the rounded-end sweatshield diamond is, actually, not the correct name for this geometric shape. Since it has already been used for many years by many sources, however, I have used it in this book in order to avoid confusion.

VOCABULARY

Important Terms:

Certain German language terms need to be explained before moving on to the slice of history that follows in the next chapter, since they are used regularly in the individual maker data entries, and providing the English translation every time requires too much space. Note that der (masculine words), die (feminine words) or das (neuter words) merely represent the English article *the*. These articles may change form, depending on how the associated noun is used in a sentence (grammatical case).

das **Amtsgericht** – District Court/Records office, which maintains the commercial registry for businesses within its jurisdiction.

der **Betrieb** – [The] business (c.f. das Geschäft).

C O / N O etc. – Prior to 1945, Berlin was divided up into geographical sections identified by letter. *O* was Ost [east], for example; *N O* for Nordost [north-east], and so on. Thus, a Berlin address would always begin with 'Berlin', followed by the section of the city: N O 20 (Berlin Northeast [section] 20), for example, followed by the street. This system was discontinued in the early postwar years.

Deutsches Reichsgebrauchsmuster, abbreviated **D.R.G.M.** A *Gebrauchsmuster* is a listing of an individual invention recorded at the German patent office. It is not a patent, however, nor does it enjoy the full protection of a patent in either legal rights or period of validity. Instead, it provides limited protection and limited legal rights, while also being fairly quick to acquire (patent applications take a long time).

The presence of D.R.G.M. on a cap (without or without the accompanying number) *does not* necessarily mean that the particular cap maker held the rights to the invention. Many companies, in fact, bought licenses to use the invention from the actual D.R.G.M. holder. (For more information see Appendix C)

D.R.P. – **D**eutsches **R**eichs**p**atent, German National Patent. *Das* **Patentamt** means patent office. (See Appendix C)

D.R.P. angem. – Stands for Deutsches Reichspatent *angemeldet*, and means *patent pending* – in other words, a patent application has been submitted to the patent office for processing and (hopefully) award.

der **Einzelhandel** – Retail trade, retail.

das **Ersatzleder** – Substitute (replacement) leather, i.e. composition sweatband material.

eröffnet (past participle) – Opened, initiated.

erlöschen (past participle) – Dissolved; applies to a company's management structure. 'Die Gesellschaft ist erlöschen' means that the current company has been dissolved. In general, it means that the company has, in fact, been closed. (See: Handelsregister)

die "Extra"-Mütze / Extramütze – German term for a high quality, private purchase cap.

die Fabrik – Factory.

der Fabrikant – Manufacturer, factory owner.

das Fabrikat – Product, brand.

Gebr. – Abbreviation of *Gebrüder*, brothers.

Gegr. – Abbreviation of *gegründet*, founded.

das Geschäft – Business. (c.f. *der Betrieb*).

die Gesellschaft – Company. German firms changed their business form very often, depending on which system at the time offered the best tax (or executive pay) advantages. There are benefits and liabilities to each form:

- *das Aktiengesellschaft* – stock company (*Aktien* – 'stocks'). Abbreviation: **AG**
- *das Handelsgesellschaft* – trading company
- *offene Handelsgesellschaft* – General partnership company. Abbreviation: **OHG**
- *Gesellschaft mit beschränkter Haftung* – Limited liability corporation (similar to English 'Ltd., or L.L.C.'). Abbreviation: **GmbH**
- *das Kommanditgesellschaft* – Limited partnership company. Abbreviation: **KG**

Ges. Gesch. – Abbreviation of *Gesetzlich Geschutzt*, protected by law, patented. This can be used to refer to a D.R.G.M. item, an already patented item or a trademark name (in which case it would mean *registered trademark*).

das Gewerbe – Business (also 'das Betrieb')

das Gewerbeamt – The city (or district) Business Bureau that maintains records of all businesses operating within its jurisdiction. Any company, regardless of size or type, must register at this office (at least according to current regulations). The Gewerbeamt often has a statue of limitations of sorts, on old records. If the file is not active, and the specified time period has elapsed, records are either destroyed or, in some cases, relocated to a district archive elsewhere (usually at the State level).

das Gewerberegister – Business Registry (registration required for all commercial trade businesses) das ***Handelsregister*** – A city's commercial registery (only covers businesses in certain classes, particularly large firms considered to be of industrial level, and generally of the same type as those covered by the Chamber of Commerce and Industry). Any changes in a company's principal personnel (owner/director, chief clerk, or the company management structure must be reported (and the documents notarized). Companies entered in the Handelsregister must also continue to meet the qualifying requirements. If not, their registration may be terminated and the entry closed. This does *not* necessarily mean that the company has gone out of business, however. The Handelsregister is maintained by the local Amtsgericht, which also has the authority to initiate insolvency proceedings. If found insolvent, the firm is legally closed/dissolved.

die ***Handwerkskammer*** – Chamber of Trade, in the sense of handcraft/handtrade (*voluntary* registration, and open only to business of small and medium-size whose products classify as a Handcraft).

Hochachtungsvoll – respectfully yours; formal closing form for a business letter.

der ***Hut*** – Hat.

die ***Hüte – Mützen*** – Hats and caps.

die ***Industrie- und Handelskammer*** [IHK] – Chamber of Commerce and Industry; again, voluntary registration and open only to larger companies and factories which manage their accounting and inventory records in a specific manner. The IHK usually conducts a company inspection/audit when a firm announces its desire to be entered in the Handelsregister, in order to ensure qualification requirements are met; it then recommends the company to the Amtsgericht for registration (the IHK and the government office of the Amtsgericht have a close-knit relationship).

die ***Insolvenz*** – Insolvency.

das ***Insolvenz Verfahren*** – Insolvency proceeding.

der ***Kommanditist*** – Limited partner (in a limited partnership company). This position often led to company ownership. (c.f. *Prokurist* below).

das ***Konkursverfahren*** – Bankruptcy proceeding.

der ***Lieferant*** – Supplier, contractor.

das ***Lieferjahr*** – Year of delivery.

die ***Mütze*** (Mützen) – Cap (caps).

die **Mützenfabrik** – Cap factory.

das ***Mützenfabrikation*** – Cap manufacturing.

Ohne Schirmdruck – Without forehead pressure.

*die **Prokura** / der **Prokurist*** – proxy (position name)/ confidential or head clerk, deputy. The Prokurist was an individual authorized by the company owner to represent the firm in making official, legally binding business decisions. The person could be given full rights to act individually, or be limited to acting jointly with one or more other *Prokuristen*. This position was frequently a stepping stone to ownership of the company. An individual could not hold the positions of owner or chief executive and Prokurist (or Kommanditist) at the same time.

*die **Qualitätsarbeit*** – Quality Craftsmanship.

*die **Qualitätsmütze*** – [High] Quality cap.

*das **Patentamt** –* (see entry under DRP)

'*Sehr geehrter Herr…* – 'Dear Mr…. ' standard greeting for a business letter.

***Stirndruckfrei** –* Free of forehead pressure (compare *ohne Schirmdruck* – without forehead pressure).

Stirn- und Schläfendruckfrei – Free of pressure on the forehead and temples.

Stirnschutz – Forehead protection [i.e. from pressure caused by the edge of the visor].

*die **Uniform-Mützenfabrik*** – Uniform cap factory (i.e. making caps strictly for uniformed personnel, rather than civilian use).

*die **Verhinderung*** – prevention (i.e. sweat bleed-through).

*die **Wertarbeit*** – Value Craftsmanship.

Chapter 16

Individual Maker Entries

AUGSBURG

COMPANY: *Josef Weithmann Militäreffekten*

BUSINESS ADDRESS:

WWII period:
Josef Weithmann Militäreffekten
Kayerstrasse 14 V. Langemarkstrasse
Augsburg

Postwar:
Josef Weithmann
Mützenfabrikation, Orden, Studentenartikel, Vereinsbedarf, Pokale*
(street address not yet known)
Augsburg

* *Cap manufacture, Decorations, Articles for Students, Association requirements, Pokals*

OWNER or BUSINESS LEADER:
1. Josef Weithmann
2. Olga Weithmann (daughter), after Josef's death in the fall of 1940.

CURRENT STATUS:
Out of business. Olga Weithmann herself initiated the company liquidation procedures in 1969, and the Handelsregister entry was officially closed on March 26, 1970, after her death.

FOUNDED:
May 1,1889. Augsburg's trade register and associated records up to and through the war years were all destroyed in Allied attacks on the city [presumably air raids] that occurred on February 25 and 26, 1944. (The city's destroyed Handelsregister was rewritten in 1946.)

TRADE NAME if any:
None

MAKERS MARK, WITH LOCATION:
No photo (or other historical information, or descriptions of headgear bearing the Weithmann logo) is currently available.

Confirmation Source:
1. *Uniformen-Markt*
2. Staatsarchiv Augsburg
3. Authentic example

BAD-AACHEN

COMPANY: *Hut Kelg*

BUSINESS ADDRESS:

WWII period:
Hut Kelg
Bad-Aachen, Kramerstr. 2, Ecke Katschof

OWNER or BUSINESS LEADER:
Unknown

CURRENT STATUS:
Unknown. No current records for the firm have been found. The German city no longer exists and may be part of present day France.

FOUNDED:
Unknown

TRADENAME If Any:
None

SWEATSHIELD SHAPE:
On Extramützen: *Rhomboid*

MAKERS MARK, WITH LOCATION:
First line: **Hut Kelg**
Second line: "Kramerstr. 2 Ecke Katschof" printed on the cap lining.
No authentic example of the Hut Kelg makers mark is currently available.

HISTORICAL BACKGROUND:
None yet available.

Confirmation Source:
Authentic example

BAD CANNSTATT

COMPANY: *Alfred Valet Mützenfabrik*

BUSINESS ADDRESS:

WWII period:
Alfred Valet Mützenfabrik
(Address unknown)
Stuttgart-Bad Cannstatt*

* A suburb of Stuttgart

OWNER or BUSINESS LEADER:
1. Alfred Valet [Sr.]
2. Alfred Valet [Jr] and Eugen Valet (from 1912) – merchant
3. Kurt Valet, Manufacturer (from 1957)
4. Kurt Valet and Eugen Valet (1960)
5. Helmut Müller, merchant (from 1961)

CURRENT STATUS:
No longer exists. The company was closed and its assets liquidated on August 25, 1966.

FOUNDED:
Before 1887

TRADE NAME If Any:
None

SWEATSHIELD SHAPE:
On Extramütze visor caps (private purchase): *Rhomboid*

MAKERS MARKS, WITH LOCATION

Alfred Valet Mützenfabrik Stuttgart-Bad Cannstatt stamped in ink on the sweatband reverse (contract caps – Government Issue).

COLORS:

Field Caps: M34	*Material:*
Army: Gray	Cotton (Trikot weave)

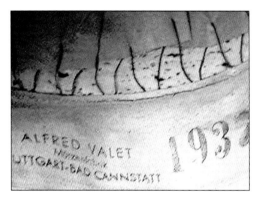

Ink stamp makers mark on the reverse of an Army Other Ranks visor cap (cavalry piped). The cap is a standard Government Issue [contract] cap, and thus has a sweatshield but no mark on the lining (other than head size).

Alfred Valet makers mark stamped in ink on the lining of an Other Ranks M34 field cap. *Courtesy of Dale Paul.*

HISTORICAL BACKGROUND:
Alfred Valet (senior) probably opened his business in the late 1870s or early 1880s; the exact date can no longer be determined. His company was first listed on the city's Handelsregister roll on May 11, 1887, and this serves as evidence that the firm had done quite well – it would not have been able to qualify, otherwise. Alfred continued to develop his business and clientele, until handing the company over to his two sons, Alfred Jr. and Eugen. This transfer of the business is recorded in a Handelsregister entry dated April 10, 1912. The firm's business field at that time was identified as "Wholesale business in hats, caps and fur goods with manufacturing." The company passed through the First World War and survived the difficult economic period of the following decade. With German remilitarization under the Third Reich, the Valet brothers began production of Wehrmacht visor caps and field caps on government contract (issue caps). It is not certain whether the company also produced high-end Extramützen (private purchase caps) as well; to date, no authentic examples of these have been seen. Alfred Valet Mützenfabrik once again survived the end of a major conflict, and struggled

through the early postwar years; somehow it managed to continue meeting the Handelsregister qualification requirements and was not removed from the roll.

In 1957, Kurt Valet (possibly Alfred's son) briefly became the sole owner as Alfred left the business – perhaps to retire. By 1960, Eugen reappears as part owner, but he departed the company for good in the following year, 1961. The company's deputy (head clerk), a fellow named Helmut Müller (married to a Valet daughter), took charge of the firm on August 17, 1961, but its fortunes were already on the wane. It continued to decline until closing permanently on or around August 26, 1966.

Confirmation Source:
1. Amtsgericht Stuttgart (Registergericht)
2. Authentic example

COMPANY: *Erich Weiblen*

BUSINESS ADDRESS:

WWII period:
Erich Weiblen
(Address unknown)
Stuttgart-Bad Cannstatt

OWNER:
Unknown

CURRENT STATUS:
Presumed no longer in existence. Records for the firm have not yet been located, and it never qualified for Handelsregister entry (thus a very small firm).

TRADE NAME if any:
None

MAKERS MARKS, WITH LOCATION:
No example of an authentic Weiblen makers mark is currently available.

HISTORICAL BACKGROUND:
None yet available, research continues.

Confirmation Source:
Authentic example

COMPANY: *Paul Wagenmann*

BUSINESS ADDRESS:

WWII period:
Paul Wagenmann
(Street address not known)
Stuttgart-Bad Canstatt

OWNER:
Mr. Hehner (likely following on Paul Wagenmann himself)

CURRENT STATUS:
Unknown; presumed no longer in existence.

FOUNDED:
Unknown

TRADE NAME If Any:
None

MAKERS MARKS, WITH LOCATION

**"Paul Wagenmann
Inh: Hehner"***
Appears printed directly on the lining material. No authentic example of the Wagenmann makers mark is currently available.

* Inhaber – owner

HISTORICAL BACKGROUND:
No information of any value has yet turned up for the Wagenmann company. It was never entered in the city's Handelsregister, indicating that it never qualified (the company was either too small, or did not use the required bookkeeping/business system). Research continues.

Confirmation Source:
Uniformen-Markt

BAD-KISSINGEN

COMPANY: *Anton Baumgart*

BUSINESS ADDRESS:

WWII period:
Anton und Willi Baumgart Mützenfabrikation
Bad Kissingen, Bachgasse 8

Postwar:
Anton und Willi Baumgart Herren- und Damen Schneiderei*
Bad Kissingen, Bachgasse 8

* Men and Women's Tailor

OWNER or BUSINESS LEADER:
Anton Baumgart

CURRENT STATUS:
The Handelsregister entry for the company was not closed until December 10, 1978 however the firm apparently went out of business with the death of Herr Anton Baumgar on January 20, 1955. No records for the firm date after 1955 (except for the 197 closing entry in the Handelsregister); nor do records for any descendant company exis in city files.

FOUNDED:
1918

MAKERS MARKS, WITH LOCATION:
No authentic example of the Baumgart makers mark is currently available.

CONFIRMATION SOURCE:
1. Amtsgericht Bad-Kissingen
2. Authentic example

J.B. Holzinger makers mark. Though Holzinger pointedly advertises the Frischluft ventilation system, this invention was actually licensed from the Coesfeld firm of Hermann Potthoff (familiar to collectors only by its logo of *HPC*, since the actual name *Hermann Potthoff* never appears on HPC military caps).

BERCHTESGADEN

COMPANY: *J.B. Holzinger*

BUSINESS ADDRESS:

WWII period:
J.B. Holzinger
(Address unknown)
Berchtesgaden

OWNER or BUSINESS LEADER:
Unknown

CURRENT STATUS:
Unknown (records have yet to be located).

FOUNDED:
Unknown

TRADE NAME If Any:
None

SWEATSHIELD SHAPE:
On Extramützen (private purchase caps): *Rhomboid*

MAKERS MARKS, WITH LOCATION
In red ink, usually printed on the sweatshield reverse.
First line, in a large Fraktur [German gothic] font – note in the photographic example the initial *J* is slightly larger than the *B*: "**J.B. Holzinger**"
Second line, in a vertical semi-script (the initial 'U' is vertically centered on the rest of the word): "Uniformen"
Third line, in Fraktur: "Berchtesgaden"
Fourth line, in italics, with the initial letter in cursive script: "Frischluft"
Fifth line: "D.R.G.M. 1419757"

PATENTS or D.R.G.M.:
None, however Holzinger used HPC company's Frischluft system under license. The HPC- registered D.R.G.M. number therefore appears in Holzinger caps.

COMPANY DATA:
No specific data available.

HISTORICAL BACKGROUND:
None yet available.

Confirmation Source:
Authentic examples

BERLIN

COMPANY: *August Mordasch Herstellung von Uniformmützen*
(Manufacture of Uniform Caps)

BUSINESS ADDRESS

WWII period:
August Mordasch
Berlin SO 36, Schlesischestrasse 39/40

OWNER or BUSINESS LEADER:
August Mordasch, with son Wilhelm Mordasch

CURRENT STATUS:
No longer exists.

FOUNDED:
Unknown

TRADE NAME If Any:
None

MAKERS MARK(S), WITH LOCATION:
Unknown; no authentic example of an August Mordasch makers mark is currently available.

COMPANY DATA: (As of 1938)

Business Facilities:
Total Space: Approximately 95 square meters

Employees:
2 sales personnel
14 skilled employees

Financial Data:
Sales: Annual sales of 81,000 RM was achieved in 1937; in 1938 the figure was 56,617 RM, and by the end of September 1939, sales were at 59,153 RM.

HISTORICAL BACKGROUND:
By the mid-1930s the August Mordasch company was fully active (if not earlier) – the actual founding date can not be determined. It was a fairly small size firm in terms of floor space at 95 square meters and only sixteen employees. While its actual production capacity is unknown, the company's annual sales figures compare very favorably against those of larger manufacturers.

In 1938, Herr Mordasch felt ready to apply for entry in the city's Handelsregister. This was normally done with the assistance of the [Berlin] Chamber of Commerce and Industry (***IHK***), which submitted letters of recommendation for registration to the Amtsgericht on behalf of qualified companies. The initial **IHK** Berlin letter was submitted by a Herr Grasshof – a fellow who appears a number of times in the same capacity in the files of other companies – on July 5, 1939. The Amtsgericht seems not to have responded as expected. Mordasch, on his own, decided to submit a notarized document reporting the company for registration (dated September 20, 1939), but the attempt had instead an altogether negative effect which led to some sort of penalty.

Things were serious enough that August's son, Wilhelm Mordasch, paid a visit to the IHK (or was called there), where he discussed the situation with Mr. Grasshof, including mention of his father's independent submittal. Grasshof then wrote a second letter to the Amtsgericht on November 3, 1939, in which he noted having been informed by Wilhelm Mordasch of the notarized submittal. He explained also that the results of an additional investigation of the Mordasch company (by the IHK) supported his earlier recommendation. He reiterated Mordasch's qualification for registration as an upstanding company, and he makes a very interesting comment at the end of the letter, writing: "*Die fruchtlose Pfändung ist unverständlich.*" [The fruitless seizure is incomprehensible].

Seizure of what? The word *Pfändung* can also mean garnishment, but again, garnishment of what? A question for which, unfortunately, there is no answer: This second IHK letter is the last record in the Mordasch file – there is nothing else. The company finally achieved its Handelsregister entry, but the government's seemingly unwarranted action may eventually have forced the company to close, perhaps sometime in 1940; in any case, the August Mordasch firm does not seem to have survived the war.

The Mordasch notarized document which reported his company for entry in the Handelsregister.

Confirmation Source:

1. Amtsgericht Charlottenburg files
2. *Uniformen-Markt*

COMPANY: August Schellenberg Uniformmützen-Fabrik

BUSINESS ADDRESS:

WWII period:
August Schellenberg
Berlin O 27, Alexanderstr. 40

Postwar:
August Schellenberg
Berlin N 20 Wriezener Str. 7

OWNER or BUSINESS LEADER:
1. August Schellenberg
2. Heinz Schellenberg (son), part owner
3. Minna (neé Maron) and Käthe (neé Milde) Schellenberg

CURRENT STATUS:
No longer in existence.
The last record of an action in the Handelsregister files is a request for a register excerpt dated April 6, 1955. The company's registration was never officially closed out, but there is no evidence of the firm's existence later than 1955 (regardless of source).

Schellenberg logo, close-up. The company did not limit itself to one color of ink (or lining color); Schellenberg logos appear in black, silver, or red ink, with the latter color usually appearing on gold or gray colored linings.

Schellenberg logo mark on a dark colored lining.

Schellenberg diamond-style makers logo in silver ink. The sweatshield is cut in a vertical diamond, while the logo itself is horizontal. The label reads "Special [extra] Class, forehead pressure free D.R.G.M.." At the bottom is 'Ges. Gesch.'(registered).

FOUNDED:
October 1, 1930. First entry in the Handelsregister (Amtsgericht Charlottenburg): 1934

TRADE NAME or LOGO If Any:
Schellenberg bear logo (see below)

SWEATSHIELD SHAPE:
1. Extramützen *(Rhomboid)*
2. Officer's old-style field cap: *Diamond (vertical)*

MAKERS MARKS, WITH LOCATION
Makers mark colors: Red, black, silver (rare)
Schellenberg bear logo:
For the rhomboid sweatshield pattern, the mark appears either on the lining material, or directly applied to the underside of the sweatshield. **Schellenberg** with the word **Original** just above the name. Below the name, a circle enclosing a bear (the bear sometimes appears alone i.e., without any name mark). Muzzle facing left, standing with legs and arms outstretched to the left and right, paws face up: The left paw (as viewed) supports the letter *A*; the right paw the letter *S*. On the bear's head is a visor cap. Below the bear, following the bottom of the circle in some examples, the word "*Stirndruckfrei*"; beneath this a long box enclosing the word "SONDERKLASSE" in capital letters.

VERIFIED PRODUCTS

1) *Schirmmützen*:
• Army Officer and Other Ranks Extramütze/Other Ranks contract (Government Issue) caps
• Luftwaffe Officer, Other Ranks Extramütze
• Allgemeine-SS and other government agency Extramütze

Fine example of a Schellenberg Army officer's Extramütze. The cap piping color is for Panzer troops. *Courtesy of Gary Baier.*

2) Officer old-style field cap: Army

COLORS:

On private purchase grade officer/Other Ranks caps:

Visor Cap Lining:	*Material*
Army: Gold, Bluish-gray	Silk or rayon
Luftwaffe: Gold	Sweatband: Leather; composition (Ersatz)

Officer old-style field cap lining:
Army: Gold

Sweatband colors for Visor caps:
Army: Tan
Luftwaffe: Tan

Visor reverse colors:
Army: Tan; reddish tan
Luftwaffe: Green (standard)

GRADES:

Schellenberg "**Rekord**" (appears primarily on caps sold by distributors), Schellenberg "**Original**" (no indication which was higher).

PATENTS or D.R.G.M.:

D.R.G.M.: Number: **1 383 756.** Date of D.R.G.M. registration•: August 27, 1936 "Device for the Prevention of Forehead Pressure in Caps" [*Vorrichtung zur Verhinderung von Stirndruck bei Mützen*]

•with the Reichspatentamt [National Patent Office] in Berlin

COMPANY DATA:

Business Facilities:
1 office
1 work floor (room)
1 storage room
Total Space: Approximately 200 square meters

Employees:
2 sales people
5 cap makers
9 trained seamstresses
1 [female] 'preparer'
2 management interns

Equipment:
1 riveting machine
1 pasteboard cutter
1 stamping machine
15 special sewing machines

Production capacity:
8000 caps **per month**, average estimated capacity (1934)

Financial Data:

1. Business capital: Approximately 30,000 ReichMarks

2. Sales: Yearly sales valued at 30,000 RM was expected for 1933; for the period January 1, 1934 to November of 1934, the company looked to take in 68,000 RM, and forecast a minimum of 10,000 RM per month thereafter.

DISTRIBUTORS:
Hermann Schellhorn, Offenbach am Main (see Schellhorn entry under *Offenbach*)

HISTORICAL BACKGROUND:
One of the more well known German cap manufacturers from World War II was the firm of August Schellenberg. Something of a latecomer to the scene, the company was founded in 1930 with a capital sum that allowed it to open right from the start with a medium to large-sized, industrial-rated production capacity. Sometime in 1934 Schellenberg applied for registration with the Industrie und Handelskammer [IHK] zu Berlin [Berlin Chamber of Commerce and Industry], which subsequently forwarded its recommendation for Schellenberg's official Handelsregister [Commercial Registry] entry to the responsible Amtsgericht, in this case, the Amtsgericht Charlottenburg.

The Amtsgericht obliged. Interestingly enough, Herr Schellenberg was not at all pleased with the business title used for the company's Commercial Register entry. In a forcefully written objection submitted to the Amtsgericht Charlottenburg and dated 1935, he states:

> "I applied with the Chamber of Trade and Industry for entry under the company name of August Schellenberg Uniform Cap Factory. The Chamber then investigated whether my firm can be regarded as a uniform cap factory and confirmed this to me in writing, at the same time informing me that entry in the Handelsregister has been initiated. At my registration with the Amtsgericht Charlottenburg I distinctly stressed that the firm be entered under the name 'August Schellenberg Uniformmützenfabrik'. I would like to mention that the designation 'Mützenfabrikation' [cap manufacture], as well as 'Herstellung' [production] and 'Vertrieb von Mützen' [sale of caps] is completely inaccurate, since I produce only uniform caps and these are made in my factory.
>
> I therefore request that you set the entry right, and sign with

Heil Hitler!
August Schellenberg
Uniformmützen-Fabrik

By the end of 1936, Schellenberg's company had been awarded a Gebrauchsmuster (August 27th of that year) which gave him limited, protected rights to his registered invention: A forehead pressure prevention system for visor caps. Any company that desired to use this system in its products had to pay a license fee to Schellenberg. The licensee was also required to note somewhere in the cap that a protected item was being used. Stating 'D.R.G.M.' or 'gesetzlich geschutzt ' [protected by law] somewhere on the cap was usually sufficient as a notice. Sometimes, the actual number was also given. Caps by quite a few different makers, in fact, appear with Schellenberg's D.R.G.M. number.

With business going so well, Herr Schellenberg decided (also perhaps in 1936) that it was time to bring son Heinz into the business. He officially registered this intention in December of that year, with the effective date set for January 1, 1937. From this point on, there are no further records for the company in official Amtsgericht files unt

---

Uniformmützen aller Art

liefert günstig

August Schellenberg
Uniformmützenfabrik
Berlin O 27, Alexanderstr. 40
Spezialität: Reichswehr - Extramützen

Schellenberg advertisement from the July 1, 1935 issue of *Uniformen-Markt*. Text reads:

Uniform caps of all types
delivered at favorable terms
August Schellenberg
Uniformmützenfabrik
Berlin O 27, Alexanderstraße 40
Specialty: Reichswehr Extramützen

1942, when it seems tragedy struck: Both August and son Heinz Schellenburg died (sometime in late 1941 or early 1942).

It is not clear whether father and son died at the same time, or of the same cause, nor is the actual cause given. August's widow, Minna Schellenberg, submitted an official notice to the Amtsgericht in June of 1942 that both husband August and son Heinz were dead. The submission also noted that, as the legally designated heiress, she would continue to run the company under the same name together with her daughter-in-law Käthe Schellenberg (the beneficiary of Heinz' share in the business). Thus, the two women took over the reins from mid 1942 and ran the business thereafter – though it appears Minna was the one who actually managed things.

What might have claimed the lives of father and son so close together, if not the same time? If Heinz had died later – that is, separately from his father, his wife Käthe would probably have had to register her inheritance and intention to continue on with the company (in Heinz' stead) in a separate application to the Amtsgericht. That this was not done, however, suggests that August and Heinz died together – perhaps they were caught in an Allied bombing raid.

The end of the war in 1945 was not the end of the August Schellenberg company; it still had one last gasp. The firm's business premises was relocated to Berlin N 20, Wriezenerstraße 7. The Landesarchiv Berlin shows a 1949 city Address Book entry for Schellenberg already at the Wriezenerstraße 7 address.

Documentation on the company for this period is practically non-existent, with the only other post war entry in Amtsgericht Charlottenburg records being a 1955 request for an extract of the Schellenberg Handelsregister registration. This was submitted by a different Amtsgericht office, perhaps the office that had jurisdiction over the company's new location. With this, however, all documentation stops.

The Handelsregister entry for Schellenberg appears to never have been officially closed; there is no August Schellenberg company (or any variation thereof) listed in contemporary Berlin records. Presumably, this well-respected cap maker went under at some time during the mid-1950s, if not before – perhaps unable to stand against competition from its surviving wartime peers such as Carl Halfar. In any case, the founder himself, August Schellenberg, was simply one more casualty among the many dead in World War II.

Confirmation Source:

1. Amtsgericht Charlottenburg
2. Landesarchiv Berlin, *Herr Wulf-Ekkehard Lücke*
3. *Uniformen-Markt*
4. Authentic Examples

COMPANY: *Carl Derwig Uniformfabrik*

BUSINESS ADDRESS

WWII period:
Carl Derwig Uniformfabrik
Berlin NW 40, Alt Moabit Nr. 10 b

Postwar:
Unknown

OWNER or BUSINESS LEADER:
Carl Derwig (residence: Berlin N 113, Wichertstr. 6)

CURRENT STATUS:
Uncertain, but presumed no longer in existence.

FOUNDED:
Unknown

TRADE NAME If Any:
None

SWEATSHIELD SHAPE:
On Extramützen: *Rhomboid*

MAKERS MARK(S), WITH LOCATION
Printed on the sweatshield reverse in gold ink.
First (top) line in large-sized, arcing capital letters "**CARL DERWIG**"
Second line (capitals): "UNIFORMEN"
Third line (capitals): "MILITÄREFFEKTEN"
Fourth line: "Berlin N.W. 40"
Fifth line: "Alt-Moabit 10b"
Sixth line: "Müllerstr. 77"
Seventh line: "Friedrichstr. 97"
Eighth line: "am Bahnhoff F[illegible]str."
Beneath these eight lines, within a long oval, in script: *Derwig Sonderform* [Derwig special model]
Beneath the oval in fine print capitals, "GES. GESCH. Nr. 51803"
Finally, in upward angled (left to right), large Fraktur (German Gothic font) letters: "**LUXUS**" (Deluxe)

GRADES:
Luxus

PATENTS or D.R.G.M.:
Possibly a patent, Nr. 51803 (not yet researched)

PECULIARITIES OF DESIGN (If Any):
None noted

HISTORICAL BACKGROUND:
No historical information is available. It could be inferred from Derwig's business title *Militäreffekten* [military goods], that the company was no more than a distributor or uniform tailor selling caps made by another company under its own name. That Derwig may have held a patent (not a currently identified patent number) suggests that his company actually did manufacture caps – presumably only private purchase pieces. The multiple addresses also show that Derwig had several shops, perhaps run by his brothers, or his sons.

Confirmation Source:
1. Landesarchiv Berlin, *Herr Wulf-Ekkehard Lücke*
2. Authentic example

COMPANY: *Carl Halfar Uniform-Mützenfabrik*

BUSINESS ADDRESS:

WWII period:

Carl Halfar Uniform-Mützenfabrik
Berlin N 20, Prinzenallee 74

Postwar:
Carl Halfar Uniform-Mützenfabrik
Berlin N 20, Prinzenallee 74

OWNER or BUSINESS LEADER:
1. Carl Halfar
2. Dr. Anton Viktor Halfar (son) [born December 13, 1901], from 1936
3. Elisabeth Halfar (neé Haßelberg, born August 30, 1913), wife and heir of Anton. From 1981

FOUNDED:
April 1, 1890 in Mörchingen (a town in the former province of Lothringen, now part of France).

CURRENT STATUS:
Closed in 1982, inactive perhaps as early as 1969.

TRADE NAME If Any:
None

MAKERS MARKS, WITH LOCATION:
On the lining, the name **CARL HALFAR** in capital letters printed in silver ink.
On the next line, in capitals: "UNIFORM-MÜTZENFABRIK"
Third line: "Berlin N 20, Prinzenallee 74", followed by the cap size.

On field caps, the company name appears in the usual fashion as an ink stamp, usually in black ink.

VERIFIED PRODUCTS:

1. Schirmmützen:
 Army Officer/Other Ranks Extramütze, Other Ranks contract caps (Government Issue)
 Luftwaffe Officer/Other Ranks Extramütze

. Tropical Field Cap (garrison cap): Army. **Lining color:** Red

. M43 Einheitsfeldmütze:
Army Other Ranks
Luftwaffe Other Ranks

. Helmets: (unknown whether fiber or steel)

COLORS
In private purchase grade officer/Other Ranks caps:

Interior of a Carl Halfar Army Panzer officer's Extramütze (private purchase) visor cap. No mark other than the company name, address and size normally appears on Halfar caps. Note the somewhat coarse, linen-like, magenta colored lining material, typical on Halfar officer caps.

A Halfar officer's Extramütze showing thick-style crown piping and nicely matching Leichtmetall (aluminum) insignia. Fine doeskin wool. Waffengattung [branch]: Panzer.

Makers mark on a Carl Halfar Luftwaffe Other Ranks Extramütze (private purchase) visor cap. Orange-rust colored waterproof chintz lining – typical of Halfar caps for enlisted personnel. The sweatband ends overlap on the side (left) of the cap, rather than at the rear. *Courtesy of Bill Shea.*

Lining:
Army: Magenta
Luftwaffe: Orange rust

On Government Contract caps
Army (Other Ranks): Orange rust
Luftwaffe (Other Ranks): Orange rust

Sweatband Colors
For visor caps:
Light tan (Army, Luftwaffe)

Visor Reverse Colors:
Army: Crème
Luftwaffe: Green (standard)

Material:
Cotton (incl. Officer caps) [linen weave]
Chintz (Other Ranks)

Chintz
Chintz

GRADES:
None

PATENTS or D.R.G.M.:
None

PECULIARITIES OF DESIGN (If Any)
None noted.

DISTRIBUTORS:
Unknown

COMPANY DATA:
Primary business as a supplier of uniform caps to military, state and city authorities and in much lesser scope, also some retail business.

Business Facilities (main building with courtyard):
1 office
2 work rooms (production rooms)
1 storage room
1 material locker
Total Space: Approximately 330 square meters

Employees:
1 Business director (not a "Master Cap Maker")
1 sales person
34 trained workers
5 Home workers *(worked out of their homes)*

Equipment:
1 Thread stitching machine *(or, staple machine)*
1 [Gas driven] steam oven
1 cutting machine
1 steel cutting machine
2 balancers
1 stamping machine
1 steam system (gas driven)
2 pasteboard cutters
18 electric [driven] sewing machines
20 electric motors

Financial Data:

. Business capital: Approximately 26,900 RM

. Rent for business facilities: 1200 RM *(presumably per month)*

. Sales: Yearly sales valued at 61,637 RM in 1934.

HISTORICAL BACKGROUND:
Founded by Carl Halfar in Mörchingen in the region of Lothringen in 1890, Halfar moved the firm bearing his name to Berlin seventeen years later, in 1907. With access to a much bigger customer base in the Prussian capitol, the company grew quickly. Upon Carl Halfar's death in 1936, son Dr. Anton Halfar assumed the direction of the company. Under Anton's leadership, the firm gained registry in the Handelsregister of the Amtsgericht Charlottenburg. Halfar's business was a *Uniform-Mützenfabrik* – a uniform cap factory. The company did not involve itself in any civilian (commercial) cap production. With remilitarization under the Third Reich, business boomed and Halfar, in fact, had one of the largest facilities in Berlin in terms of actual floor space at 330 square meters (one hundred square meters larger than August Schellenburg's company, for example). Nor, it seems did the war dampen the company's successful operations. The only available records pertaining to this period from the Handelsregister files show that Halfar opened a branch company under the same name in the city of Gnesen on February 16, 1942 (Richard-Wagner Straße). He apparently shifted 80% of his Berlin production to this facility, in order to safeguard it from the ever increasing Allied air attacks on the German capitol. According to Halfar's son, Hans-Jürgen Halfar (see

further down), this move to Gnesen was a direct response to an instruction from the government, though this has not been confirmed.

Just over a year later, in May of 1943, the branch company was converted into an independent (wholly-owned) subsidiary, at which point it ceases to be recorded under Halfar's own (parent company) Handelsregister file. The fate of the subsidiary is unknown, but it was most likely destroyed when overrun by the Soviet Army's advance on Berlin.

In May of 1945, as the dust of battle slowly settled around the shattered ruins of Berlin, Carl Halfar Mützenfabrik found itself still standing amidst the rubble of the capitol, at the same address, Prinzen Allee 74 (or what was left of it). Prinzen Allee was situated in the Western-occupied sector of the city, within the French Zone.

The early postwar years were extremely difficult. Berlin offered only a limited number of customers, and competition for the available business was ferocious among the surviving Berlin cap makers. Carl Halfar sold its products to every possible government agency in Berlin: The postal service [Bundespost], the national railway [Bundesbahn] (once this was up and running again); the Berlin police forces; in short, any government organization (with a few exceptions) that used caps was courted by Carl Halfar, often with success. This business alone, however, was simply not enough to support the company; the cap makers who had made it through the war now found themselves on the economic chopping block, quickly being whittled away to less than a handful as the war-shattered economy slowly recovered and prices for raw materials, labor and overhead began to climb.

Something had to be done to lesson the ever-tightening financial squeeze. In desperation, perhaps, Anton Halfar began looking outside of Berlin in the new Federal Republic (West Germany) for a solution and, as luck would have it, he found one. The fledgling West German federal government had just instituted a federal border patrol force (a paramilitary police agency) known as the Bundesgrenzschutz (BGS), and the organization was in need of a cap supplier. The key to success for Anton Halfar in this case was the "national" character of this agency: BGS personnel were stationed throughout West Germany. A significant new customer, indeed, and Halfar managed to secure a long-term, exclusive contract to supply caps to the BGS force. The caps were produced in Halfar's Berlin factory, then shipped to West Germany (through East Germany). It was a major coup for the company and it secured Halfar's financial health for several years to come.

While the few remaining Berlin makers were fighting over the local customer base, Halfar had gone national, or rather, international (since the caps had to pass through East Germany and Soviet Army – later East German – customs inspectors on their way to the West). No objection of any kind to these cap shipments was raised either by the Soviets, or later the East Germans; they seemed uninterested in the matter. The company's financial picture was rosy. Anton Halfar did not waste much time searching for other organizations and it was not long before Carl Halfar Mützenfabrik came to depend heavily on the BGS as its primary customer – a fatal flaw, as it turns out.

Halfar had either unwittingly or knowingly overlooked a very important fact: regulation imposed years earlier under the Four Powers Agreement of the Allied military authorities specified that *no company in Berlin, whether in the Western or Eastern Sector of the city, was permitted to sell weapons or equipment (including uniforms and uniform caps) to a military organization* – i.e., to any organization that would be classified as a combatant in the case of a war. The paramilitary BGS in fact, was classified as a combatant force. Still, there were no objections, no sign of any problem.

But that was to change. Carl Halfar Mützenfabrik was located in the French occupied zone of West Berlin. In 1967, things took an alarming and totally unexpected turn when French Army officials paid an unannounced visit to Anton Halfar at the company office. They informed him that they had learned of his BGS contract (already seven years in operation!) and explained further that it was forbidden for any Berlin-based company to supply a combatant organization. He must therefore, terminate his contract with the BGS immediately. It was a command, not a request: Halfar had no choice but to comply.

With the loss of its major customer, the company could no longer afford to stay in business and was forced to close down, possibly that very year. Anton Halfar passed away on February 27, 1968, perhaps as a result of the shock caused by the sudden demise of the company he had pulled back from the brink of ruin and then built into a thriving business.

After Anton's death, his wife Elisabeth assumed control, but the company was essentially closed; there was little or no active operation. She made no official report to either the IHK or the Amtsgericht concerning the situation. Perhaps she wished to confound the company's competitors? In any case, it was not until 1981 that she took any action when she submitted a document to the Amtsgericht Charlottenburg officially proclaiming her inheritance of the firm (some fifteen years after the fact), and her intention to continue the business under the same name. Less than a year later, she submitted another document reporting the company's dissolution. In truth, there had been no active company for years.

Meanwhile, the Amtsgericht was apparently unaware of the firm's demise in 1967, and did not close Halfar's Handelsregister file. While reviewing and updating its records in 1981, the Amtsgericht decided to inquire of the Berlin IHK (Chamber of Commerce and Industry) concerning what the IHK might know about the status of Carl Halfar Mützenfabrik. The IHK had in fact, already begun its own investigation – with unsatisfactory results. It reported that there was little or no business activity going on at Carl Halfar. The IHK had even advised the company about converting to a G.m.b.H. (limited liability company) – since this would allow it to maintain its Handelsregister qualifications. In response, the Halfar company had asked for several time extensions, which were granted. The extensions subsequently ran out, however, with no response from the company to any further IHK inquiries.

The IHK thus informed the Amtsgericht that it had no objection to the company's removal from the Handelsregister – and the Amtsgericht concurred: the *Carl Halfar Uniform-Mützenfabrik* entry was officially closed in 1982. The 1981 document from Elisabeth Halfar was probably prompted by the IHK's inquiries (Elisabeth must have decided that some action had to be taken), but it did no more than buy a little time – it appears she did not want to see the company officially closed, even though physically it no longer existed.

The son of Anton Halfar, Herr [Mr.] Karl-Jürgen Halfar, provided an interesting twist to the Carl Halfar Uniform-Mützenfabrik story. In a personal interview conducted by Herr Klaus-Peter Merta (Director of the Militaria Collection of the German Historical Museum, Berlin), Mr. Halfar explained that his family was of the opinion that a zealous competitor had "blown the whistle", so to speak, on the BGS contract. The unnamed competitor had made French military authorities aware (either directly or anonymously) that the Carl Halfar contract transgressed against a clause in the Four Powers Agreement. While there is no actual proof to support this supposition at this time, circumstantial evidence suggests that this may, in fact, have been the truth. Unquestionably, competition for customers was fierce and the BGS represented a very lucrative contract; Halfar's competitors were no doubt envious. But which company might have been the one to tip off the French? An intriguing coincidence appears in a BGS officer's cap dating from the late 1960s or very beginning of the 1970s, i.e. the period immediately following Halfar's demise, yet prior to the BGS's conversion to a darker moss green uniform color. The cap bears the logo Peküro (trademark of the Peter Küpper firm in Wuppertal-Ronsdorf) and further proclaims *Spezialanfertigung für den B.G.S.* [Special Manufacture for the BGS]. With Carl Halfar now completely out of the picture, Peter Küpper, it seems, was one of the very first companies to take advantage of the new situation. Perhaps, it had a head start?

According to Mr. Halfar, his mother Elisabeth also requested that he maintain the company's phone book and business listing active – something he faithfully continue to do even after her passing. In fact, Carl Halfar Uniform-Mützenfabrik appeared first place (on a *very* short list) in an Internet search of German cap makers conducte by the author in the year 2000 using a German online business directory listing firm by industry branch. One letter sent to the company address returned as undeliverabl two subsequent letters vanished into limbo (nothing came back) – much to the conste nation of the author. Informed of this situation, Herr Merta investigated and discovere the current Mr. Halfar, still living in the same building which once housed his father factory. It appears that after Mr. Merta's contact and interview, Mr. Halfar decided best to let the company take its place in history, and has since removed Carl Half Uniform-Mützenfabrik from all public listings.

Confirmation Source:
1. Amtsgericht Charlottenburg
2. Landesarchiv Berlin, *Herr Wulf-Ekkehard Lücke*
3. Interview with Herr Karl-Jürgen Halfar
4. Authentic examples

COMPANY: *Christian Haug Mützenfabrik*

BUSINESS ADDRESS:

WWII period:
Christian Haug Mützenfabrikation
Berlin N O 18, Höchste Straße 29
CRIHA-Haus

Postwar:
Christian Haug
1. Berlin No. 55, Bötzowstraße 8 (Haug residence)
2. Berlin Charlottenburg 9, Neidenburger Allee 40

OWNER or BUSINESS LEADER:
1. Christian Haug
2. Wolfgang Bremer

CURRENT STATUS:
Business closed presumably in the early to mid 1950s. No current active records for th firm or for any descendant company exist.

FOUNDED:
1904

TRADE NAME If Any:
CRIHA **Gut und dauerhaft**

CERTIFICATIONS:
RZM authorized cap maker: RZM No. 48

SWEATSHIELD SHAPE:
On Extramützen: *Rhomboid*

MAKERS MARKS, WITH LOCATION

Printed directly on the cap lining in silver ink:
"Deutsche Qualitätsarbeit" [German quality work], Below this, a circle with a silver center bar and the letters CRIHA (this likely represents *C[h]ri* from Christian – less the *h* – and the *Ha* from Haug). Within the circle, "*gut und*" (above the name bar), "*dauerhaft*" (below the bar): *good and long-lasting*. Below the circle, "Christian Haug BERLIN NO 18 Höchstestr. 29".

VERIFIED PRODUCTS

Schirmmützen:
• Army Other Ranks Extramütze
• Luftwaffe Other Ranks Extramütze
• Kyffhäuserbund caps (non-military)
• Ordnungspolizei caps (non-military)

COLORS

On private purchase grade officer/Other Ranks caps:

Visor Cap Lining:
Army: Reddish brown (rust)
Luftwaffe: Reddish tan (light); gray

Material:
Cotton
Cotton (linen weave)

Christian Haug makers mark in a private purchase Luftwaffe Other Ranks cap. Cotton lining in a linen weave.

Sweatband colors for Visor caps:
Army: Reddish brown
Luftwaffe: Dark tan; gray

Visor reverse colors:
Army: Reddish brown
Luftwaffe: Green (standard)

GRADES:
No specific grade names

PATENTS or D.R.G.M.:
None

PECULIARITIES OF DESIGN (If Any):
None noted.

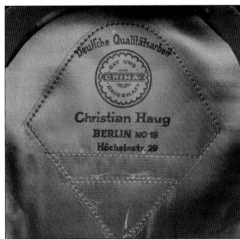

Reproduction Christian Haug makers mark for comparative purposes. The cap is made by Janke. In fact, the makers mark is produced using one of Haug's original die stamps – thus, in a certain respect it is original. *Courtesy of Bill Bureau.*

Christian Haug ink stamp on the sweatband reverse of a contract (Government Issue) Luftwaffe Other Ranks cap. The name itself is nearly illegible, but the Höchstestraße address suffices to identify the company as Christian Haug. The unit stamp is for the elite General Göring Regiment, later to become a Luftwaffe field Panzer division. *Courtesy of Dale Paul.*

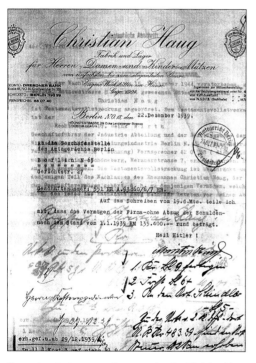

Christian Haug advertisement from the May 1, 1935 issue of *Uniformen-Markt.* The text reads:

Complete
Uniform Caps
Any quantity delivered quickly
Christian Haug, Cap Factory
Berlin No 18, Höchste Straße 29

The only extant pre-*war* document for the Christian Haug company, dated December 22, 1939.

HISTORICAL BACKGROUND:

The firm Christian Haug *Mützenfabrikation* was founded by the gentleman of the same name in 1904. The company's letterhead from December of 1939 gives the business address at Höchstestrasse 29, corner of Landsberg Strasse, CRIHA-HAUS. The firm survived the lean years following the First World War, struggling through the difficult economic conditions of the Weimar period and on into the National Socialist era. By 1939, a relative named Gottlob Haug was also involved in the business, as the chief financial officer – though it is not clear whether he was son, or brother to Christian. This position he shared with another fellow named Walter Bremer; both were authorized to represent the company and make independent executive decisions in the company's interest (this position is called *Einzelprokura* in German).

The only pre-war document still in existence is a letter from the company on its own letterhead: "*Christian Haug, Factory and Warehouse for Gentlemen, Ladies and Children's Caps, from the simplest to the most elegant Genre, Own workshop on the premises.*"

The letter reports, in response to a request for an assessment of the firm's worth, that as of January 1, 1939 the company – before subtraction of debts – was valued at approx. 135,400 Reichs Marks.

No further records turn up until 1946, at which time a document appears concerning the disposition of the assets of founder Christian Haug, who passed away before the end of the war on September 15, 1944. From this point on, the business was apparently run by Wolfgang Bremer, a relative of Walter Bremer, the financial officer who already worked for the company. Haug's personal property outside of the business was left to his widow, Frau Wilhelmine Haug (neé Audersch), the company to Wolfgang Bremer. As fate would have it, the company found itself at war's end – like Robert Lubstein – situated in what would become the Soviet occupied Eastern Sector. Wolfgang, at least, was wise enough to see the danger in this, and apparently fled to the Western Sector taking the company's assets with him before they could be confiscated (confiscation/sequester of assets in the Soviet Zone began in October of 1945). It is not clear whether Walter Bremer, or any of the Haugs followed him.

In 1950, someone tried to have the Christian Haug company's Handelsregister records transferred to the Amtsgericht Berlin-Mitte in the Eastern Sector (by this time, the Amtsgericht was fully under Soviet control). One of the Bremers responded to this attempt (probably Wolfgang), by writing a memo to the Amtsgericht Charlottenburg (on new company letterhead), asking that these records not be released. This is the last official entry in the company's files. It is not certain whether the new Christian Haug company ever went fully into business, or simply faded into oblivion. It never again qualified for entry in the Handelsregister, in any case, and if indeed it opened, it most likely did not survive beyond 1955.

Confirmation Source:
1. Amtsgericht Charlottenburg
2. *Uniformen-Markt*
3. Authentic examples

COMPANY: *Erwin Freudemann*

BUSINESS ADDRESS:

WWII period:
Erwin Freudemann
Berlin Steglitz, Florastr.

Postwar:
Erwin Freudemann
Berlin Zehlendorf, Kleinaustr. 27

OWNER or BUSINESS LEADER:
Unknown, presumably Erwin Freudemann.

CURRENT STATUS:
Presumed out of business. No residency records currently in existence for this company.

MAKERS MARK(S), WITH LOCATION
No authentic example of a Freudemann makers mark is currently available.

Confirmation Source:
Uniformen-Markt

COMPANY: *Emil Schebeler, Mützen- & Pelzwaren-fabrik Hut-Lager engros*

BUSINESS ADDRESS (primary)

WWII period:
Emil Schebeler
Berlin N O 55, Immanuelkirchstr. 6

Postwar:
Emil Schebeler
Berlin 55, Immanuelkirchstr. 6

OWNER or BUSINESS LEADER:
1. Emil Schebeler
2. Edmund Schebeler, businessman, after Emil's death
3. Maria Schebeler (widow of Edmund), from August of 1942.
4. Willi Schreiber

CURRENT STATUS:
Out of Business. Last file record dated 1957.

FOUNDED:
1870. Legal documentation gives a change in company management method as of July 9, 1907. First listing in the Handelsregister of the Amtsgericht Charlottenburg on December 19, 1937.

TRADE NAME If Any:
None

SWEATSHIELD SHAPE:
On Extramützen: *Rhomboid.*

MAKERS MARKS, WITH LOCATION:
In silver ink, printed on the lining material in stylized, canted letters, "**Emil Schebeler**".
Second line: (The company address).
The cap size is usually stamped on the lining in black ink above the makers mark.

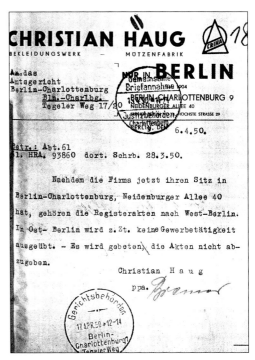

Final record in the Christian Haug file. Note the new triangle logo and the new address (upper right) of Berlin, Charlottenburg 9, Neidenburger Allee (in the Western Sector). The text, dated April 6, 1950, reads:

Since the firm now has its premises in Berlin-Charlottenburg, Neidenburger Allee 40, the Register files belong in West-Berlin. In East-Berlin, no business activity is taking place at this time. – request the files not be handed over.

Christian Haug
ppa. *Bremer*

Complete interior view of an Emil Schebeler Other Ranks Extramütze (Luftwaffe). Despite being a private purchase cap (thus with some flexibility in interior colors), it conforms more or less to regulations with a standard reddish brown (rust) colored cotton lining (linen weave) and tan sweatband. The visor reverse is the usual Luftwaffe green. Branch piping is for flight or paratrooper (Fallschirmjäger) personnel.

COLORS:

On private purchase grade Other Ranks caps:

Visor Cap Lining:
Army: Reddish brown (rust)
Luftwaffe: Reddish brown

Material:
Cotton
Cotton, linen weave

Sweatband colors for Visor caps:
Luftwaffe: Tan, reddish brown

Natural leather; composition

Visor reverse colors:
Army: Tan
Luftwaffe: Green (standard)

GRADES:
None specified

PATENTS or D.R.G.M.:
None

COMPANY DATA:
None available

HISTORICAL BACKGROUND:
Founded by Emil Schebeler in 1870, with its first place of business at Prenzlauer Alle 36, the company started with cap manufacture and used only this in its business title. Later, it added fur products to its product line, and renamed itself to include this business (*Kürschner*). By 1937, the company had relocated to Immanuelkirchstraße 6, where it was to remain.

The only source for the company's founding date is found on the cap linings of private purchase visor caps, and on the company's letterhead – there is no official notation on any government documents.

Emil's son, Edmund Schebeler (businessman) was running the company as early as 1907. In a letter to the Imperial Amtsgericht (Prussia) written in 1910, he inquired about the possibility of starting another business named 'Berlin Hat-Bazaar, owner Edmund Schebeler' [*Berliner Hut-Bazar, Inh. Edmund Schebeler*], and also inquired about the necessary opening procedures. Nothing seems to have come of this particular idea; Herr Schebeler however did open a subsidiary company in the town of Soldin at some point during the following years. Records are incomplete, and the start date of the subsidiary is impossible to determine. Its size also remains a mystery, as does the reason for its eventual closure on June 9, 1934 (most likely for financial reasons).

During the early Third Reich years and through the start of the Second World War, Herr Schebeler quietly ran his business until he passed away on October 19, 1941. His wife, Maria Schebeler (neé Ganzlin), took direction of the firm as the sole heiress and officially announced her intention to continue the business with no change in the company name. The Amtsgericht had actually reviewed the inheritance specifications in the Schebeler Will and Testament, on November 11, 1941 and determined that Frau Schebeler was indeed the inheritor. For some reason, however, the bureaucratic actions needed to complete the process were not discharged until July 27, 1942.

Despite the challenges of running a business in a war torn city, Maria found enough time to secure a new husband. In the letter to the Amtsgericht reproduced on the next page, she explains that due to her remarriage (to a fellow named Oechsner), the company's registration information in the Handelsregister needed to be changed. She continued to run the company on her own.

At war's end, the Emil Schebeler company was among the survivors, with its business premises unchanged at the same address. The building that housed the company, if damaged, was apparently salvageable.

Communications with the Amtsgericht begin once again from May of 1951, with Maria reporting the company's restructuring and the addition of a new partner, a businessman by the name of Willi Schreiber. The company's line of business was identified as, "the manufacture and sale/distribution of caps and also hats." In an interesting coincidence, the street address given for this Mr. Schreiber was Heinrich-Roller-Straße 25 – just a bit farther down the street from Robert Lubstein's ERELHAUS building at Heinrich-Roller-Str.16-17 in the Eastern Zone!

Business continued, apparently in a manner sufficient to maintain the company's place in the Handelsregister for at least the next few years. On January 4, 1957, Maria Oechsner retired from the company, leaving Herr Schreiber to continue on in her stead as sole owner.

Just how long Schreiber managed to carry on the business is not known: no further records exist, nor was the company's Handelsregister entry ever closed out. There is no contemporary evidence of the firm (or any possible offspring) to be found in Berlin public records. *Emil Schebeler Mützen-Fabrik*, like so many of its peers, slipped quietly into history.

Confirmation Source:
1. Amtsgericht Charlottenburg
2. Authentic examples
3. *Uniformen-Markt*

Copy of a letter dated July 23, 1943 to the Amtsgericht from Maria Oechsner (widow and heiress of Edmund Schebeler). The company letterhead provides the firm's founding date (1870). The signature block is interesting, with Maria Oechsner's name appearing at the bottom. Another letter dating from 1951 (not shown) has the exact same company letterhead – but in Roman font. The signature block is also the same but without the "Heil Hitler!" The letterhead text reads:

Emil Schebeler owner: Edmund Schebeler, Cap Factory, Hat Warehouse engros, founded 1870 Manufacture of first class blue Yacht Club and civil caps, Special section for Uniform Caps

COMPANY: *G.A. Hoffman/BEROLINA-Mützenfabrik Hoffmann & Co. KG*

BUSINESS ADDRESS:

WWII period:
G.A. Hoffmann
Berlin SW 29, Gneisenau-Straße 33

Berolina Mützen- und Sporthutfabrik
Berlin C 2, Maßmannstr. 12

Postwar:
G.A. Hoffmann
Berlin SW 29, Gneisenau-Straße 33

OWNER or BUSINESS LEADER:
1. C. Georg A. Hoffmann
2. Ernst Hoffmann
3. Gert Hoffmann

CURRENT STATUS:
No longer exists. City Handelsregister (business registry) entry closed on March 15, 1973.

FOUNDED:
1871 by Carl Georg Albrecht Hoffmann

Crisp example of the G.A. Hoffman firm logo. From an Army private purchase visor cap for Other Ranks – that is, an Extramütze for an enlisted soldier or NCO. The lining is rust colored cotton. This mint, unissued cap still retains the original stuffing paper used for storage.

G.A. Hoffman ink stamp logo from an Other Ranks tropical cap. "G.A. Hoffman Berlin SW 29 1942". Standard red tropical cap lining (in comparison, see below for the Berolina logo and lining color).

TRADE NAME If Any:
GAH

SWEATSHIELD SHAPE:
Visor caps: For G.A. Hoffmann: *Rhomboid;* For BEROLINA: *Diamond*

MAKERS MARKS, WITH LOCATION:
For *G.A. Hoffmann*:
Company logo in silver ink, printed on the cap lining under the sweatshield. The logo is an oval, in which is a silver circle. Inside the circle are three staggered cubes, each block (in the lining material color) containing one silver capital letter, the three letters together forming **GAH**. Outside the center of the circle and to the left, is the word "**Fabrik-**"; to the right of the circle is the word "**Marke**" [Fabrik Marke – factory trade mark]. Above the oval are the words **Die Deutsche Qualitätsmütze** [The German Quality Cap], below the oval appears:

First line: "**G.A. Hoffmann**"
Second line: "BERLIN S.W. 29"
Third line: "Gneisenaustr. 33"

VERIFIED PRODUCTS

Under the *G.A. Hoffman* Logo:
1) *Schirmmützen:*
• Army Officer/Other Ranks Extramütze
• Luftwaffe Other Ranks Extramütze
2) *Einheitsfeldmützen (Tropical):* Army Other Ranks (visored) field cap
3) *Einheitsmützen (Tropical):* Army Other Ranks
4) *Pith Helmets:* Army and Luftwaffe
5) *Non-military visor caps*
6) *Panzer Crash Helmets/Berets*

Under the *Berolina* Logo:
Berolina did business at least initially with only [commercial] associations, wholesale merchants, and distributors. High-grade Extramützen would not have fit into this picture; High quality private purchase caps with a **Berolina** lining makers mark – though extremely rare – have, in fact, been seen however. The company produced large quantities of field caps, as well as government contract visor caps.

1. *Schirmmützen:* Army officer Extramütze [very rare] /Other Ranks contract caps

2. *Einheitsmützen (Tropical):* Army Other Ranks garrison caps

MAKERS MARK:
In silver ink on the cap lining, a large triangle. Within the triangle is a circle, the edge of which touches the triangle sides. At the top center of the circle (within the circumference), a thick capital letter *G*. Below this, within the circle, a tall, outlined letter *A*. The crossbar for the *A* is shaped like a *V*. In the center of the letter is a small mark, which may be a *D*, an *H* or an unknown symbol. Outside of, and below the triangle, in lower case letters, the name: "Berolina".

COLORS:
On private purchase grade officer/Other Ranks caps:

Visor Cap Lining: *Material:*

Army:

G.A. Hoffmann: Light reddish brown (rust)   Rayon, silk, or cotton (Other Ranks)

Berolina: Gray   Rayon or silk

Sweatband colors for Visor caps:

Army: *G.A. Hoffmann* – Reddish-brown; tan Leather

      Berolina – Reddish tan (dark)   Leather

Visor reverse (underside) colors:

Army: Reddish brown; dark tan (both companies)

Tropical Field Cap Linings:

Army: *G.A. Hoffman* – Red; *Berolina* – Khaki

GRADES:
No specific grade names

CERTIFICATIONS:
RZM license No. **10**

PATENTS or D.R.G.M.:
None noted

PECULIARITIES OF DESIGN (If Any):
None noted

COMPANY DATA:
No technical data

HISTORICAL BACKGROUND:
The G. A. Hoffman company was opened in 1871 by the Berlin merchant *Carl Georg Albrecht Hoffmann.* Preferring not to use "Carl", he instead gave the new company his own two middle initials and family name as its official title. The business did well, and by April 22, 1901, it was entered for the first time in the Charlottenburg district's Handelsregister. Entry in the Handelsregister was considered by German companies to be something of a milestone on the road to success, but it did not come without a price: the qualification requirements had to be maintained indefinitely. If the company failed to meet these requirements at any time, its trade register entry could be closed.

This was not a problem: the company's success in the Prussian capitol, Berlin, was sufficient that in 1905 the company even went far afield and opened a subsidiary in Vienna, Austria. Also in that same year, the founder's son, Ernst Hoffmann, entered the firm as chief financial officer. This man, a Berlin merchant by profession, was the personality under which the G.A. Hoffmann company would later achieve its period of greatest growth and success. Over the next decade there would be significant changes. At some point the elder Hoffmann died, leaving his wife Auguste in control of the company until she, too, departed the firm with her passing in 1913. Prior to this, in 1909, the Vienna subsidiary – apparently no great success – had been closed.

With the death of Auguste Hoffmann, son Ernst effectively became the head of the company with brother Hans Hoffmann also employed. By 1919, the first of many bitter postwar years following on the German defeat in World War I, Ernst Hoffmann had become sole owner of the firm. Brother Hans had departed the business (not without first signing an official notarized "blessing" on the new chief, which was submitted to the Amtsgericht). The company's future was now solely in Ernst Hoffmann's hands.

The Berolina makers mark as it appears on high-end Extramützen. Only two of these, in fact, have ever been seen (to the author's knowledge). *Courtesy of Gerard Stezelberger of Relic Hunter.*

Die anerkann, :n Qualitätsfabrikate in
Uniformmützen jeder Art
Soldatenbundmützen, NSDAP.-Mützen (Lizenz Nr. 10 der RZM.), Kyffhäuserbundmützen
G. A. HOFFMANN · Uniformmützen-Fabrik · Berlin SW 29, Gneisenaustr. 33
Gegr. 1871

For several years he was content to simply weather the difficult economic situation in Germany and quietly run the business, but by the end of the 1920s he began to make a name for himself within the industry. He sought out responsible positions which would eventually present him the opportunity to have a major impact on the cap-making industry's development and direction.

With the rise of Hitler and the National Socialist bureaucracy, Ernst Hoffmann saw both the company and his own fortunes begin to turn. He had suffered through the economic turmoil of the late 1920s and early 1930s, which saw crisis after crisis. The company had posted significant losses in 1931 and 1932, but the tide had begun to turn and the company showed positive growth by late 1933. Military expansion under Hitler meant significant increases in the demand for military headgear production and major business for the company. Ernst, now in his mid-fifties, had by this time also become the head of the Union of German Cap Manufacturers. In 1934 he also became leader of the technical sub-group '*cap industry*' (within the economic group "clothing industry"), a position which handled technical issues, intra-industry relations, industry development and oversight, and gave him considerable influence. Ernst Hoffmann had certainly become a very big fish within the cap-making industry ocean.

In 1938, Ernst brought his son, Dr. Gert Hoffmann, into the company as deputy (head clerk) – the same position in which he himself had started years before. Gert was a promising young fellow who had done well academically – in fact, he held a law degree. Also in 1938, the company (together with some unknown backers), took over a pre-existing Berlin cap maker known as Mützenfabrik L. Thorner. The original owners of L. Thorner were apparently not of the correct "Aryan heritage": a newspaper notice recording the takeover mentions that, *"L. Thorner has been transferred to Aryan hands."* The new company became a G.A. Hoffmann subsidiary and was renamed BEROLINA Mützenfabrik Hoffmann & Co., K.G., located on Maßmannstr. (Massmann street)

Headgear seen to date with the *Berolina* ink stamped makers mark is tropical field caps, contract visor caps (ink stamp mark on the sweat band reverse) as well as two officer Extramützen. It is not certain what other types of headgear the firm may also have manufactured.

Young Dr. Gert Hoffman assumed the helm of the new business, which at first dealt only with wholesalers and large concerns. Berolina most likely serviced military and combat police field cap needs and mass-produced contract visor caps without getting involved in the more complex (and less cost-effective) production of Extramützen [private purchase] visor caps until sometime later.

With the start of the war, Ernst Hoffmann earned another feather for his cap as the new Director of the Reichs Clothing Distribution Bureau. By now, he had become one of the most influential (and powerful) people within his industry. He turned 60 years of age on September 6, 1941, and his birthday was celebrated with a half-page article in the *Uniformen-Markt* trade paper. The article mentions the many contributions he had made to the industry over the years, including ironing out a smoother relationship be

G. A. HOFFMANN Uniformmützenfabrik
Berlin SW 29, Gneisenaustraße 33
Herstellung von Uniformmützen jeder Art und Menge
Erstklassige Werkmannsarbeit seit 1871
Sämtliche Mützen für die Parteiformationen der NSDAP.
Lizenz Nr. 10 d. R.-Z.-M.

tween military and civilian cap makers, and setting pricing and minimum quality standards for mass-produced caps. The article praised his contributions to the *U-M* paper itself, as well.

During the war, no major changes were entered in official records. Both G.A. Hoffmann and Berolina continued to operate successfully despite increasing air raids and first a faltering, then a failing, economy. In January of 1945, Ernst Hoffmann revised his Will and Testament (for the final time). He named son Gert as his successor; not, he said, because of their relationship, but rather because he had been impressed by Gert's dedication to the business. Ernst also made sure to leave a position in the company (as part owner) open for his son-in-law, Kurt Müller, to be, *"filled upon the young man's return from the field."* Müller, however, is never mentioned again except in an oblique reference through a document signed by his wife: "...the widow Müller, born Hoffmann" – which suggests that he became, instead, a battlefield casualty. Ernst himself passed away in the early postwar years.

In May 1945, amid the ruins of Berlin, Allied occupation forces installed an interim military government and assumed control of the German national and local economy within its sector of responsibility. Amidst the slowly clearing rubble, Gert Hoffmann set about salvaging what he could of the original G.A. Hoffmann business in hopes of building a new one. It was a daunting task, with difficult times facing the shattered German economy that would take nearly a decade to overcome. In 1949, the Allied military government closed itself down (though the Occupation remained in place), and began transitioning power back to German national control with the fledgling Federal Republic of Germany. As a part of this transition, the tight grip on the German economy was released somewhat.

The G.A. Hoffman company, still struggling in business, once more held its future in its own hands – but business for a cap maker in Berlin was a brutal proposition (the Berolina company was never reopened).

No longer hampered by the military government's rigidly applied fiscal control policy, competition between surviving firms for Berlin's limited customers grew fierce. Gert Hoffmann managed to hold the G.A. Hoffmann company together until 1973, but in that year it failed to meet qualification requirements for the Handelsregister and was officially removed from the active registry.

Whether the firm found its end in that same year, or continued on for a while longer as an inconsequential small business (before finally succumbing to financial failure), is a question for which the answer is lost in the sands of time.

Confirmation Source:
1. Amtsgericht Charlottenburg files
2. *Uniformen-Markt*
3. Authentic examples

Berolina ink stamp makers mark on a tropical cap lining. *Courtesy of Bill Shea*

Ernst Hoffmann, owner of the G.A. Hoffmann firm and leader of the Clothing Industry's technical subgroup "Cap Industry", in a photo that appeared in the *Uniformen-Markt* on the occasion of Hoffmann's 60th birthday.

COMPANY: *Gebrüder Alm* **Uniform-Mützen-Fabrik.**

BUSINESS ADDRESS:

WWII period:
Gebr. Alm Uniform-Mützen-Fabrik
Berlin C 2 Münzstr. 15

OWNER or BUSINESS LEADER:
Otto and Wilhelm Alm

CURRENT STATUS:
No longer exists. No current records for Alm or any descendant company exist in official Berlin records.

FOUNDED:
1918

TRADE NAME If Any:
None

SWEATSHIELD SHAPE:
On Extramützen: *Rhomboid*

MAKERS MARKS, WITH LOCATION:
Printed on the lining in a flowing Fraktur (German Gothic) font, **Gebr. Alm** in silver ink.
Second line: "Uniform-Mützen-Fabrik"
Third line: "Berlin C2"
Fourth line: "Münzstr. 15"
Beneath this, the year of delivery (also silver).

VERIFIED PRODUCTS

Schirmmützen:
• Army Other Ranks Extramütze
• Luftwaffe Other Ranks Extramütze

To date, nearly every cap seen with the Gebr. Alm logo has been a Luftwaffe cap for Other Ranks – most likely purchased by NonComs (Non Commissioned Officers: individuals holding sergeant's rank). Only one Army cap has been reported.

COLORS:
On private purchase grade Other Ranks caps:

Gebr. Alm makers mark as it appears on visor cap linings.

Visor Cap Lining:	*Material:*
Luftwaffe: Reddish brown ('havana brown'/rust)	Chintz (waterproof)

Sweatband colors for Visor caps:
Luftwaffe: Reddish tan or tan

Visor underside colors:
Luftwaffe: Green (standard)

GRADES:
No specific grade names.

PATENTS or D.R.G.M.:
None noted.

PECULIARITIES OF DESIGN (If any):
None noted.

HISTORICAL BACKGROUND:
This company was founded by two brothers from the family Alm in 1918 (according to records with the Landesarchiv Berlin), which is particularly interesting since 1918 wa

not a very auspicious year for German companies – particularly those in the cap making business. The company location was initially at Schöneberg, Brunhildstr. 5 (the Alm residence), but a 1937 ad in *Uniformenmarkt* places the firm at Berlin SW 61, Kreuzbergstraße 30. Between 1937 and 1939 it moved once again, relocating this time to its final address at Münzstraße, where it remained. In its advertisement, the company states: "Special uniform caps".

An entry for the Gebr. Alm company has not been found in any post-war era Berlin Address Books, nor is there any other documentation available for the firm. It appears that Gebr. Alm was one of the Berlin-based cap makers that did not survive the end of the war.

Confirmation Source:
1. Landesarchiv Berlin, *Herr Wulf-Ekkehard Lücke*
2. *Uniformen-Markt*
3. Authentic Examples

COMPANY: *Herbert Grell Mützenfabrikation*

BUSINESS ADDRESS:

WWII period:
Herbert Grell Mützenfabrikation
Berlin C 2, Gollnowstr. 28

OWNER or BUSINESS LEADER:
Herbert Grell, Mützenmachermeister (Master cap maker)

CURRENT STATUS:
No longer exists

FOUNDED:
Unknown

TRADE NAME If Any:
None

MAKERS MARKS, WITH LOCATION:
No authentic example of a Herbert Grell makers mark is currently available.

VERIFIED PRODUCTS:
Military caps

COMPANY DATA:

Business Facilities:
 at Gollnowstr.28)
 office
 storage rooms

 at Gollnowstr.19)
 Production office
 Work floor (room)
Total Space: Approximately 240 square meters

Advertisement for Herbert Grell from the *Uniformen-Markt* trade paper, Issue 17, September 1, 1941. The text reads:

Caps for all uniformed personnel Herbert Grell Berlin C 2 Gollnowstraße 19

(The word *Uniformenträger* in literal translation, is 'wearers of uniforms').

Letter from company owner Herbert Grell to the Amtsgericht Charlottenburg. He boldly advertises his title *Mützenmachermeister* on the company letterhead. Within the document he mentions both Wehrmacht contracts, and colonial work.

Employees:
35

Equipment:
41 "modern" special machines with individual motors

Production capacity:
2000 to **2500** visor caps *per week*, available for delivery (1941)

Financial Data:
1. Business capital:
Approximately 20,000 ReichMarks
2. Sales:
Annual sales volume at a value of 96,000 RM for 1940; for the last quarter, 20,000 RM monthly.

HISTORICAL BACKGROUND:

A letter from the Industrie und Handelskammer zu Berlin [Chamber of Trade and Industry] dated June 27, 1941 was sent to the Amtsgericht Charlottenburg recommending that the company Herbert Grell Mützenfabrikation be officially entered in the Handelsregister (commercial register). The company was already registered with the Berlin Handwerkskammer [HWK] and had been removed from the HWK roll on June 19, 1941 (since a company could not be simultaneously registered on both the HWK roll and the Handelsregister). The HWK register was for small to medium-sized firms only – which by this time Herbert Grell certainly was not, if the firm's impressive production capacity and sales volume for 1940 is any indication.

Unfortunately, there are no other details for the Herbert Grell company in the Amtsgericht files. It is not known whether the company survived the war; it is unlikely that a company with such a large production capability and considerable contract work simply went out of business. The lack of any further documentation suggests that it was probably destroyed in bombing raids sometime prior to 1945.

Confirmation Source:
1. Amtsgericht Charlottenburg
2. *Uniformen-Markt*

COMPANY: *Holters Uniformen*

BUSINESS ADDRESS:

WWII period:

Holters Herrenmoden Uniformen*
Berlin W. 50 Tauentzienstr. 16, Eingang Marburger Str.

Post WWII:
Wilhelm Holters, Schneidermeister (Master tailor)
Berlin Lichterfelde, Berliner Str. 30

* *Men's fashions, Uniforms*

OWNER or BUSINESS LEADER:
Wilhelm Holters

CURRENT STATUS:
Uncertain; presumably out of business. The firm is registered in the city Handelsregister as early as 1930 (HR-Nr. 29 046), but detailed records have yet to be located.

FOUNDED:
Unknown

TRADE NAME If Any:
None

SWEATSHIELD SHAPE:
On Extramützen: *Diamond*

MAKERS MARKS, WITH LOCATION:
Within an oval cameo, a soldier seen from the collar up in three-quarter view, facing to his left, wearing a uniform and a visor cap. Printed in black ink on a light colored lining; on special-order (custom-made) caps with a *black* lining, the logo may appear reversed, in white.

VERIFIED PRODUCTS

Schirmmützen:
• Army Officer/Other Ranks Extramütze (also, Other Ranks caps per government contract)
• Waffen-SS General/Officer/ Other Ranks Extramütze

COLORS
Private purchase Officer Grade:

Visor Cap Lining:	Material:
Army and Waffen-SS: Creamy white or gold	Silk or rayon
Special order (custom): black	Silk or rayon

COLORS
On private purchase grade officer caps:

Sweatband Colors
a. for visor caps:
Tan (Army, Waffen-SS) [most common]
Black (*Special order* [custom made])

Visor Reverse Colors:
Army and Waffen-SS: Light tan
Black (special order)

GRADES:
No specific grade names.

PATENTS or D.R.G.M.:
None

PECULIARITIES OF DESIGN (If Any):
None noted.

The normal Holters mark was in black ink on a creamy white or pale gold lining. This example is a Holters makers mark on a private purchase, custom-made cap: white ink on a black lining. (This cap belonged to Joachim von Ribbentrop, Reichs Foreign Minister). *Courtesy of Bill Shea.*

DISTRIBUTORS:
Unknown

HISTORICAL BACKGROUND:
None yet available. That the firm is out of business, or at the very least no longer making military caps, is near certain due to the fact that Holters' original dies for its makers mark are apparently in the possession of the reproduction cap maker, Janke Tailoring. Janke-made reproduction visor caps, therefore, may bear one of several makers marks, including that of Holters Uniformen, as well as Christian Haug, Otto Schlientz and others, on its cap linings. The mark itself is, in effect, original.

Confirmation Source:
1. Authentic examples
2. Landesarchiv Berlin, *Herr Wulf-Ekkehard Lücke*

COMPANY: *Knuth & Wiese Mützenfabrikation*

BUSINESS ADDRESS

WWII period:
Knuth & Wiese Mützenfabrikation
(Address unknown)

OWNER or BUSINESS LEADER:
Max Wiese and Wilhelm Knuth, merchants

CURRENT STATUS:
Unknown; no records have yet been located for this firm.

HISTORICAL BACKGROUND:
No data yet available.

Confirmation Source:
Uniformen-Markt

COMPANY: *ROBERT LUBSTEIN (EREL-Sonderklasse)*

BUSINESS ADDRESS:

WWII Period:

1. **Robert Lubstein**
Berlin NW 21, Alt Moabit 105 (until 1939) [cap production only]

2. **Robert Lubstein**
Berlin NO 55, Heinrich-Roller-Straße 16-17
(all production operations were combined at this location from 1939)

Post-War:
Robert Lubstein
Berlin NO 55, Heinrich-Roller-Straße 16-17

OWNER or BUSINESS LEADER:
1. Robert Lubstein
2. Margarete Lubstein (wife and heiress), from 1948

FOUNDED:
1902 in Berlin

TRADE NAME (LOGO) If Any:
E*REL-Sonderklasse* (Ges,Gesch.)

PATENTS or D.R.G.M.:

Deutsches Reichspatent Nr. 694 529, Class 41b Group 2
This patent was for a sweatband modification that provided an *air cushion*, as a means of relieving pressure from the visor on the forehead (officially announced on July 4, 1940). *"Robert Lubstein in Berlin, Kopfbedeckung mit Schweißband"* [Headgear with Sweatband]. See Appendix C.

    Robert Lubstein applied for and secured at least one – perhaps more – patents (evidenced by the phrase D.R.P. angem which appears in EREL caps (usually as part of a sweatband logo, but not always present). Also, according to his makers marks, Lubstein already held a Gebrauchsmuster as well, though like the patent number(s), the actual D.R.G.M. number itself *never actually appears on any EREL or Robert Lubstein-marked cap.*

    Unfortunately, the German Bundespatentamt (federal patent office) does not have the capability of searching for old patent records *by company name*; without a number to work with, things become exceedingly difficult and potentially very expensive. In this case, luck lent a hand.

    Lubstein's Reichspatent 694 529 turned up entirely by accident when the author requested a copy of a patent issued under this number – unaware at the time whose patent it was. No reference was given in the article where it appeared as to what the patent was about (other than headgear-related), nor was there any name associated with it. When the requested patent copy was finally received, however it turned out to be for Lubstein!

SWEATSHIELD TYPE:
On Lubstein military caps with visors (including the officer's old-style field cap): *Rhomboid*
Older caps however, and non-military headgear, may use a *Diamond* sweatshield form.

MAKERS MARKS, WITH LOCATION

. *"Robert Lubstein*
Grösste Berliner Militär-Mützen-Fabrik
Berlin N.W. 21 Alt Moabit 105
Lieferjahr"

    "Robert Lubstein, Berlin's Largest Military Cap Factory, Berlin N(orth) W(est) 21 Alt Moabit [avenue] 105. Delivery year" Above this information is a [family?] crest.

    The Robert Lubstein cap line (and with it the makers mark) was discontinued after 1939 (visor caps only). Until that time, Lubstein sold two separate lines of military caps: one under his full name (Robert Lubstein line), and the other under his trademark alphabetized initials, EREL-*Sonderklasse*.

. **EREL-*Sonderklasse*** – Horizontal diamond, longer in the horizontal axis, inside of which is an oval with the name EREL centered in it. The two *E*s are positioned verti-

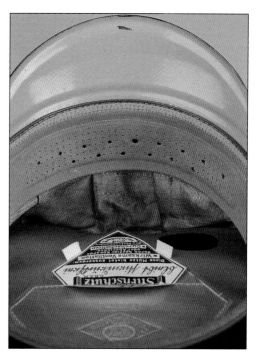

The patented EREL-*Sonderklasse* air cushion sweatband. Note the extra piece of material stitched to the bottom at the front, tapering off at the temples. This extra piece extended up behind the main sweatband and, in effect, created three layers of material that trapped air between them – hence the air cushion concept.

Robert Lubstein makers logo. Lubstein manufactured caps under his own name, as well as under the EREL-*Sonderklasse* logo. This cap bears the *Alt-Moabit 105* address. Note that there was *no specific model grade* for caps with the Lubstein logo, as was the case with EREL-*Sonderklasse* caps (Privat, Extra, etc.). The makers mark does, however, specify *Sonderklasse*.

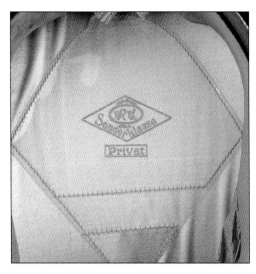

EREL-*Sonderklasse* logo. In this case, the EREL diamond logo appears alone, which means this cap was sold outside of the officer Kleiderkasse system. The grade is Privat – the "top of the line" model.

EREL-*Sonderklasse* logo accompanied by the Offizier Kleiderkasse *Berlin* etiquette. This cap would have been sold and purchased through the Kleiderkasse catalog, or at one of its branch offices. The EREL diamond is positioned *below* the Kleiderkasse mark on Army and Waffen-SS caps. This example is an Extra model. The model name always appears in Roman font.

cally but canted slightly to the left, the *R* and *L* are taller, and angle strongly to the right. Beneath this is "Berlin". The word "*Sonderklasse*" lines the inside bottom edge of the diamond. "Ges. Gesch." (registered – i.e. as in a "registered trademark") is printed just above the EREL oval, in very fine print. The correct trade name for these caps *includes* the word *SONDERKLASSE*. The EREL name alone, is technically incorrect.

3. OFFIZIER KLEIDERKASSE BERLIN – Army and Waffen-SS officer caps, printed on the lining material, on EREL caps sold through the Army's Kleiderkasse system (used also by the Waffen-SS).

4. Die Verkaufsabteilung der Luftwaffe – Luftwaffe caps, printed on the lining material above the logo, on caps sold through the Luftwaffe clothing sales system.

5. O.K.K. – [Offizier Kleiderkasse der Kriegsmarine] on the lining material, *below* the Erel logo, on caps sold through the Navy clothing sales system.

6. EREL/EREL Stirnschutz – embossed on the sweatband, usually left temple. Often silver colored on caps through 1942. Sometimes the words *Patent Stirnschutz* are included [patented forehead protection].

7. EREL Stirnschutz D.R.G.M., D.R.P. angem. – [Deutsches Reichsgebrauchsmuster, Deutsches Reichspatent angemeldet] "Gebrauchsmuster, Patent Pending" in ink, printed on the left side of light-colored sweatbands.

8) Paper or foil tag (***only on models with the vented cockade***) [see below].

9. On caps made under contract for Government Issue (*Lieferungsmützen*), an ink stamp makers mark with '**Robert Lubstein Berlin NW 21**' (and year of delivery), found on the sweatband reverse.

Left: EREL-*Sonderklasse* logo with the Luftwaffe's Kleiderkasse etiquette, *Die Verkaufsabteilung der Luftwaffe* at its early (original) address. Note the sweatshield is diam*ond-shaped* in this example of a very early Luftwaffe cap. *Courtesy of Gerard Stezelberger of Relic Hunter.*

Right: EREL-*Sonderklasse* logo with the final etiquette for the Luftwaffe Kleiderkasse at its *Puttkamerstraß* address. This later address is the one most often found on Luftwaffe visor caps. *Courtesy of Gerard Stezelberger of Relic Hunter.*

Kriegsmarine officer's cap interior with the EREL-*Sonderklasse* logo mark and the additional etiquette: *O.K.K.-Berlin* (Offizier Kleiderkasse der Kriegsmarine). Makers marks are infrequent on Kriegsmarine caps, and an EREL logo with O.K.K. is therefore quite a rare bird. On the Kriegsmarine EREL, the Kleiderkasse logo appears *below* the logo diamond, rather than above it as on other service caps. EREL Kriegsmarine caps do *not* normally indicate the model grade (Extra, Privat, etc.). *Courtesy of Gerard Stezelberger of Relic Hunter.*

EREL-Sonderklasse sweatband logo, in black ink applied at an oblique angle on the left temple of the sweatband. The cap model is an Extra, offered through the Kleiderkasse. *Courtesy of Gerard Stezelberger of Relic Hunter.*

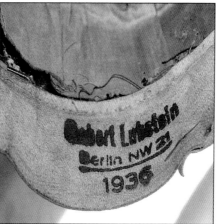

Left: Horizontally embossed sweatband logo in Fraktur (German Gothic font), filled with silver ink, on the left temple of the sweatband. The cap model is an Extra, offered through the Kleiderkasse. *Courtesy of Bill Shea.*

Right: Robert Lubstein ink makers stamp from an Other Ranks visor cap manufactured under a government contract. "Berlin N W 21" is the city sector, followed by the year of delivery. No markings appear on the cap lining itself except for the size. This stamp is on the sweatband reverse at the rear of the cap.

VERIFIED PRODUCTS

1. *Schirmmützen:*
• Army General/Officer/Other Ranks Extramütze (also, Other Ranks caps per government contract)
• Waffen-SS General/Officer/ Other Ranks Extramütze
• Luftwaffe General/Officer/Other Ranks Extramütze
• Kriegsmarine Admiral/Officer/NCO Extramütze
• Polizei officer/Other Ranks Extramütze
• Allgemeine-SS (reportedly seen on an authentic example) Extramütze

2) *Sommermützen*:
• Luftwaffe white top [summer] **Lining:** Pale crème, off-white.
• Polizei [white top] visor

3) *Officer Old-Style Field Cap:*
• Army General/Officer
• Waffen-SS Officer

4) Tropical Field Cap: Army Officer **Lining color:** Red

5) *'Blue' Cap: (Kriegsmarine)*
• Blue top (*non-removable*) visor for Admiral/Officer, NCO/OR
also: **White top** (*removable*) visor: Admiral/Officer, NCO/OR

6) *Fiber-Tschakos* (Polizei – Police)

7) *Fiber-Helmets* (for parade)

COLORS:
On private purchase grade officer/Other Ranks caps:

Visor Cap Lining:	EREL Model	Material:
Army and Waffen-SS		
Gold (champagne).	Extra, Standard, no mark	Silk or rayon
Pale bluish-green.	Privat	Silk or rayon
Luftwaffe:		
Gold (champagne)	Extra, Standard, no mark	Silk or rayon
Pale bluish green	Privat	
Kriegsmarine:		
Dark blue, white.	No model name used	Silk or Rayon
On Government contract caps		
Army (Other Ranks)	Reddish brown	Cotton

Sweatband Colors
a. for visor caps:
Light tan (Army, Waffen-SS) [most common]
Reddish tan (Army, Waffen-SS, Luftwaffe, Kriegsmarine)
Dark brown (late war Luftwaffe; rarely Army)
Dark bluish green [pebbled] (usually late war Luftwaffe)
Gray (Luftwaffe; rare color)

b. for the Officer's old-style field cap:
Crème/light tan
Reddish tan

U-M (Uniformen-Markt) ad, October 15, 1938. This Robert Lubstein advertisement for his Erel-Sonderklasse cap presents the EREL model grades that Lubstein offered. Many people may be surprised to note the 140 class, since no definitive example of this heretofore unknown grade has ever been seen (as far as can be determined), nor has any documentation yet been located for it. The text reads (same order as the advertisement):

With the motto Erel it's the Erel-Sonderklasse that matters!
The elegant and modern officer's cap
Robert Lubstein, Berlin NW21
Largest Berlin uniform cap factory.

Visor Reverse Colors:

Crème [Army; Waffen-SS]
Tan [Army; Waffen-SS]
Reddish-tan [Army; Waffen-SS]
Green [Luftwaffe]
Black [Kriegsmarine]

GRADES: (Models) Listed from highest to lowest
1. **Privat**
2. **Extra**
3. **Standard**
4. **140**

DISTRIBUTORS
1. Militäreffekten **Sperling** (Dresden)
2. Militärwarenhaus **Hans Dürbeck** (Wien)
3) **Wilhelm Voigt** Magedeburg [Otto von ?]
4) **Thiele & Edellinger**, Darmstadt
5) **V. Osterwalder**, Lüneburg

A sweatband EREL mark (in any style, whether embossed or printed) like this one, combined with an EREL-*Sonderklasse* logo on the lining, gives the so-called *double EREL*.

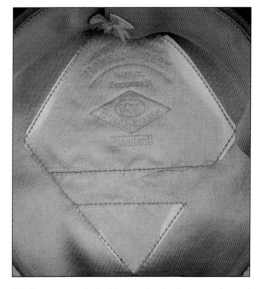

Distributor example. In this case, the distributor was located as far away as Wien (Vienna): "MILITÄRWAREN-HAUS Hans Dürbeck, Wien IX, Berggasse 31" [Military Store Hans Dürbeck, Vienna IX, Berg St. 31]. This cap happens to be a Standard model. *Courtesy of Gerard Stezelberger of Relic Hunter.*

SIGNIFICANT EREL CHARACTERISTICS:
One of the most distinctive features of certain EREL caps is the vented cockade. This air circulation system was an option available to any purchaser for an additional fee or, if the cap was from stock, then at a higher price. There is no indication that the vent system was confined to any particular EREL model though it is most often associated with the *Extra* model, it could appear on a *Privat* model as well – yet never on the *Standard*. Perhaps, if an individual could only afford a Standard in any case, it was not likely he could afford the extra cost for the vent option.

The concept for the ventilating system was simple: in place of the usual solid, red felt-covered cockade center, Lubstein's vented caps used a special cockade with a hole in its center and a metal tube behind it. The hole served as an air vent and was covered with a fine, red painted wire mesh that looked solid from a distance. Directly behind the cockade, the cap band was also pierced through to the cap's interior, allowing air from the vent to pass into the cap where it could then circulate about the wearer's head. This cap band hole held a hollow, metal shaft. On the inside of the cap, a rectangular gasket panel made of leather or cardboard – likewise with a central hole – was secured to the back of the pasteboard behind the sweatband and lining cloth. The panel hole was ringed by a washer through which the end of the metal airshaft tube protruded. The shaft itself was secured in place by splitting the end to create several tongues, which were then crimped back over the washer rim. See page 264 for a picture and Appendix E.

Another distributor's logo "Militäreffekten Sperling, Dresden N" located above the EREL logo. *Sperling* is the company name; the first word means *military effects* (goods), i.e. uniform items, insignia, etc. Dresden is the city; 'N' is *Nord* (north). Sold outside of the Kleiderkasse system, this cap does not have the Kleiderkasse etiquette. The cap model is a Standard. *Courtesy of Gerard Stezelberger of Relic Hunter.*

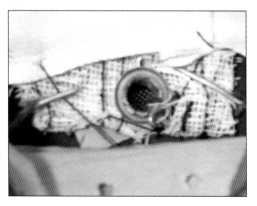

View of the air vent hole through the cap band. The vent shaft is located just above the forehead area (behind the sweatband). There is a rim around the vent hole; the end of the airshaft is split and the wedges are crimped back over a fixed washer. (Sweatband has been pulled down).

UNMARKED CAP LININGS

Many mid-war and later private purchase EREL caps have the impression of the makers mark on the cap lining – but it is not inked in. Some caps do not even bear the impression; the lining has no mark at all. These styles are particularly prevalent from 1943 onward, which can be determined due to the fact that many EREL-*Sonderklasse* Extramützen (private purchase caps) actually have a date stamped on the sweatband reverse (left side). The date gives the month, day and year that the cap was manufactured. Though full (inked) lining marks did not disappear completely, they become much less common. Exactly what lay behind Lubstein's shift to this "blank" lining is not known. Perhaps it was a cost cutting measure.

Embossed Gothic font and printed EREL logos on the sweatband remained; however the embossed type – which on early caps was filled in with silver ink – was now left bare.

Paper or Foil Tags:

Some EREL caps offer the bonus of a printed paper or foil tag, the top half of which is sewn to the lower section of the celluloid sweatshield. The presence of one of these advertisement tags is no guarantee of authenticity in itself, however. Many reproduction caps are EREL imitations – right down to the paper tag. These are usually quite easy to spot as the reproduction tags are printed in black ink only, and the paper is too thin. Original EREL paper tags were printed in blue, red, and black ink on a coarse (high fiber content), thick paper. A few reproductions have been seen with colored tags, but these are still rare and in any case, the vent system itself will be absent from the reproduction. Adding such a system would be not be cost effective for a manufacturer of reproduction caps, most of whom would prefer to stay with less expensive copies which generate better profit margins. Of more likely concern to collectors is someone taking the vented cockade (only) and cap lining from a real, but damaged EREL, and attaching them to an original cap made by some other maker. The difference would secure an extra $200 perhaps, for a cap with one of the more desirable branch colors like the grass green of the Gebirgsjäger (mountain troop) troops. Such a transfer would be difficult, but certainly possible to carry out for anyone with a little patience and sewing skill. The sewing up inside the cap crown (along the line of the crown piping) would have to be done by hand, and this would be noticeable in any careful inspection. (Note that there are some war-vintage sewing machines still around, which have been used to reassemble dismantled original caps.)

If the cap has a vented cockade, the interior pasteboard cap band should be checked to ensure the correct interior vent set-up is indeed present. If the pasteboard is blank, alarm bells should ring! If the lining bears the EREL-*Sonderklasse* mark, check it very carefully: the tension across the lining surface should be even all around, with the sweatshield correctly centered. From outside, use your fingers to push one of the side panels in slightly, in order to expose the line of stitching that secures the lining to the cap top. Look for any large and/or uneven stitches, or anything that seems out of place.

The image of the advertisement tag on page 265 is taken from a sample printed in black ink on a clear, sweatshield-shaped (the full size sweatshield) acetate sheet. Though a reproduction, the typeface is perfect. Also present is the EREL logo and the cap model designation (in this case, Extra). The logo and model name, however, in actual use would not normally have been black, nor would the paper tag text appear anywhere but on the paper tag itself, sewn to the sweatshield. In any case, the actual EREL logo was always located on the lining fabric *under* (not on) the celluloid sweatshield. Other difficulties in making exact reproductions lies in the paper texture and the ink colors used for the originals, both of which are very hard to duplicate. In terms of printing, even the old-style script on the tag obviously presents no hindrance to reproduction.

The two subsequent photographs show *authentic* EREL tags, one foil, one paper.

Double EREL:

As mentioned earlier, the phrase "double EREL" means an EREL cap that has both the normal logo mark on the lining, plus an embossed or printed EREL mark on the sweatband (usually the left side). The latter frequently appears together with the words D.R.P. angem.' (see Appendix C for an explanation of this term)

HISTORICAL BACKGROUND:

Robert Lubstein and his company have long been something of a mystery. The Robert Lubstein Uniform Cap, Fiber Helmet and Tschako Factory was the firm that owned the *EREL-Sonderklasse* trademark name (logo). The correct title for these famous caps is not the EREL name alone, but rather the combination EREL-*Sonderklasse*. This was a trademark name, *not* an independent or subsidiary company. Until 1939, in fact, the Robert Lubstein company marketed caps for two distinct lines: the "Robert Lubstein" line, and the "EREL-*Sonderklasse*" line. It is probable that the latter name had its origin with caps marketed exclusively through the Army's clothing sales office (Kleiderkasse), and was later extended to caps offered through other distributors, first displacing and then eventually replacing the Robert Lubstein line altogether in 1939.

After that year the Robert Lubstein brand cap disappears from view in favor of the *EREL-Sonderklasse* line (though not for other types of headgear, such as tschakos). *EREL-Sonderklasse*, however, remained no more than a legally registered product trademark, and no search through still extant Berlin records will ever turn up any evidence of a separate company with this name.

Considered by most militaria collectors to be the finest military cap maker in pre- and wartime Germany, the Berlin merchant Robert Lubstein opened the Robert Lubstein company in 1902 at Berlin NW 5, Rathenowstr. 59, where it remained for some years.

Lubstein started from the outset with two main goals: 1) To concentrate on only uniform headgear, and 2) To offer the best quality product possible.

He was successful, and in a hand written registration document from January 26, 1912, Lubstein records that the company employed a workforce of some 40 odd people in the "production of goods", with a customer base of 300. This document is reproduced on page 266, and includes more interesting information. The business card, attached at the top, shows the company's original (first) address and touts "Berlin's largest Military Cap Factory, en gros, export". (No records exist to show whether he did indeed export, and if so, to which countries.)

A reproduction of the EREL paper tag, as a transfer on a sheet of clear acetate. The script used on the third line is closest to *Offenbacher Schrift* (there were several styles of old German *script*, including Sütterlin). The text reads, starting from the top:

"EREL"
Forehead protection D.R.G.M.
Remains free from forehead pressure [script reads: "bleibt stirndruckfrei"]
In addition this cap offers
* *effective ventilation* by means of a special cockade or insignia
hinders sweat from *soaking-through* the cap band.

Courtesy of Bill Bureau.

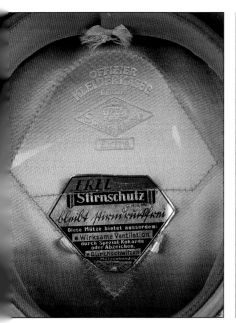 

Far left: EREL tags appear in both foil and paper, here the foil variety. The ink colors used are the same on both types, and the tags are sewn to the celluloid sweatshield in the exact same fashion. *Courtesy of Gerard Stezelberger of Relic Hunter.*

Left: EREL tag made of paper. Any tag, whether paper or (especially) foil, is a rare item since the tags usually tore off at one point or another during normal wear, or were removed by the cap owner after purchase. They are in effect an advertisement for a cap with the cockade vent system. *Courtesy of Gerard Stezelberger of Relic Hunter.*

Reproduction cap with its EREL 'lookalike' tag, for comparison. Unlike the multi-colored original tags, most reproductions are black and white only, and the caps have no actual air vent.

Partial translation of *selected* items:

2. Yearly sales: 160,000 Marks
7. Number of people employed in an average year:
   a. *Sales personel:* 1
   b. *Counter and register* personnel:* 2
   c. People involved in the manufacture or processing of goods, 30 to 35
   d. Storehouse worker, driver [*coach*], service personnel, [illegible] 2 people

   Section I. Handwork for the finest caps
   Section II. Electric power for contract caps (Lieferungsmützen)

13. Number of suppliers and customers: 300 customers
14. Does the employer himself work in the business? *"[I] personally lead my business."*

* Not necessarily a cash register; the word can also mean "pay office."

The reference to *Lieferungsmützen* (caps delivered on contract to suppliers) suggests that as early as 1912, Lubstein was supplying headgear contracts either for the military or for some other government agencies (or both). It is also interesting to note the remark about counter and register personnel – perhaps the company also sold its caps directly to customers through a small showroom?

By the time of the Third Reich, Lubstein's principal cap making facility had long been reestablished at Berlin NW 21, Alt-Moabit 105. This workshop was separate from a larger facility operating at another location, which housed Lubstein's fiber helmet and tschako production as well as the firm's business offices. Lubstein manufactured caps for both his Robert Lubstein line, and the EREL-*Sonderklasse* line at the Alt-Moabit address. It seems logical that the EREL-Sonderklasse concept, born from the two letters *R* and *L* (pronounced in German as *eR* and *eL*), was actually created by Lubstein for caps sold exclusively through the Army's clothing sales system (the Offizier Kleiderkasse), but the EREL brand became so well established and had so overshadowed the other, that the decision was made to completely discontinue the Robert Lubstein line.

Contrary to a common belief within the collecting community, the EREL-*Sonderklasse* line was *not* in fact, a Third Reich era creation. Lubstein had first begun using this logo as early as 1921 or 1922. His advertisements for EREL-*Sonderklasse* caps in the *Uniformen-Markt* trade paper, like the one presented on the next page, specifically mention them as having been, *"favorites of officers and officials...for about 15 years."* In this advertisement, Lubstein was apparently responding to an increase in companies that were trying to capitalize on his business by using the term *Sonderklasse*.

In 1939, Lubstein closed out the Alt-Moabit facility and consolidated all of his cap production into the larger building already housing his business offices and manufacturing operations for the company's fiber helmet and police Tschako production. This quadrangular structure, with its bigger rooms and greater floor space than the Alt-Moabit facility, was located at Berlin N O 55, Heinrich-Roller-Straße 16-17. It was referred to as the ERELHAUS – a name that also figures in the company's telegraph address.

The building was what Germans refer to as *prächtig* [grand], and the Lubstein company owned at least half – if not all – of the structure. In an advertisement that

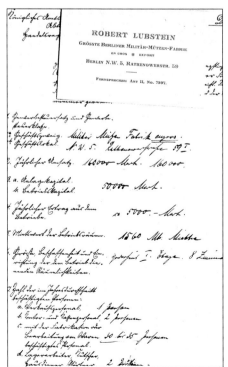

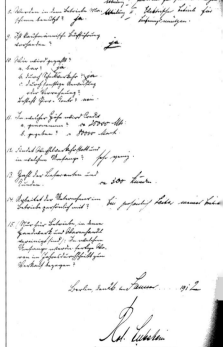

From a 1936 issue of *Uniformen-Markt*. The text reads:

Only uniform caps displaying the company logo shown [to the right] guarantee the well established and sought after products of my firm known as "Erel-*Sonderklasse*", favorites of officers and officials for about 15 years!

Caps identified merely as "sonderklasse" but with no connection to my patented trademark 'Erel' are not my products! Demand only uniform caps with the lining mark "Erel-*Sonderklasse*".

Rare example of a Reichswehr period officer's cap manufactured with the EREL-*Sonderklasse* logo. This cap dates to some time after February 1927, since the use of twisted silver chin cords (the cords are of silver, not aluminum) had not been officially instituted prior to that date. The cockade color (green and white) indicates that the officer was from Saxony. Internet offering.

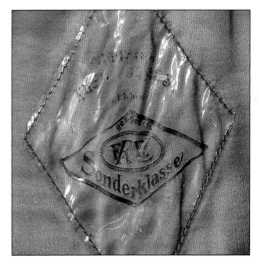

Interior, with the EREL logo. Although partially worn off, the *Offizier Kleiderkasse* (Army clothing sales store) etiquette (and 'Berlin'), is still present – which indicates that Lubstein held a contract to supply the Army Kleiderkasse well prior to the start of the Third Reich.

appeared in the April 15, 1939 issue of *Uniformen-Markt* [*U-M*], Lubstein publicly announced the move. It appears that he also used this opportunity to close out his Robert Lubstein product line. This ad is reproduced on page 268.

The quadrangle section of the building at least, belonged to Lubstein (the separate front wing appears grayed out in the ad). Cap production took place in the rear-most wing of the building behind the inner courtyard. Note the square tower that forms the left rear building corner (bottom left) in the illustration, and the Robert Lubstein sign panels atop the wings.

Other Lubstein advertisements give an idea of some of the different headgear manufactured by the company. All of the advertisements on pages 268-269 are taken from the *Uniformen-Markt* (*U-M*) trade paper.

Except for paid advertisements such as these, no other mention of the Robert Lubstein company – or the man himself – ever appears in any *U-M* articles. Within the German cap-making industry, the individual who headed the technical sub-group for the cap industry was Ernst Hoffman – *not* Robert Lubstein, as one might have expected given his apparent status as a leading cap manufacturer.

That Lubstein was a major player is indisputable. The company's exclusive (or nearly so) military supply contract to produce officer's caps for the Heeres [Army] Offizier Kleiderkasse system (as early as the Weimar period) indicates that Lubstein had developed significant connections within the Army from early on. Later, during the early years of the Third Reich, he was also able to contract with the Kriegsmarine and Luftwaffe clothing sales organizations as well, and these contracts allowed Lubstein to advertise the relationship on his cap linings. Though the Luftwaffe contract at least, was not exclusive (many caps by other makers also appear with the *die Verkaufsabteilung der Luftwaffe* logo etiquette), only EREL caps display the Navy's O.K.K. etiquette, and the Offizier Kleiderkasse etiquette on Army and Waffen-SS caps (that is, only on Robert Lubstein products). It should be mentioned that the Koblenz-based firm of Carl Bangert may also have supplied caps locally to at least the Koblenz branch of the Army Kleiderkasse at some time during the mid 1930s. Beyond this, no caps with the *Offizier Kleiderkasse – Berlin* lining etiquette appear under the name of any other company.

Lubstein probably established his contracts with the Luftwaffe Verkaufsabteilung and the Kriegsmarine O.K.K. around 1935, which appear to have remained active throughout the war. His long-standing, exclusive arrangement with the Army must certainly have been a major frustration to his competitors.

Not content with this business alone, however, Robert Lubstein also provided caps to distributors – whether haberdashers or "sellers of military goods" (Militäreffekten-Händler/Militärwarenhaus). The EREL-*Sonderklasse* trademark in this case appeared without any other mark, though the distributor's name might appear on the lining *above* the EREL diamond.

The last document from pre-war time period files relating to the Robert Lubstein company was a confirmation letter to the Amtsgericht regarding Lubstein's qualifications for retaining his [current] Handelsregister listing. This letter was written in 1938 by Herr Grasshof of the Berlin Chamber of Commerce and Industry (IHK). No other documents or entries are noted in the records during the war years.

In April and May of 1945, the company survived the Soviet assault on and capture of Berlin. Despite its exceedingly good fortune up to that point (it had also survived the Allied bombings), at the close of hostilities it found itself situated in what was to become the Soviet occupied Eastern Zone – the "ehemaliger Ost-Sektor" (former Eastern Sector, as the Germans now refer to it). Lubstein did not escape completely unscathed, however. The next entry in the Handelsregister files is a memo from September 1946 on company letterhead, addressed to the newly created (Soviet controlled) Amtsgericht-Mitte. In the memo, Lubstein requests ("*äusserst dringend*" – extremely urgent) a sum-

U-M, April 15, 1939. Lubstein's consolidation of the Alt-Moabit cap-making operation at his larger Heinrich-Roller-Straße facility is announced here:

I have moved my entire production into the larger facilities
of my business premises at
Berlin NO 55,
Heinrich-Roller-Straße 16-17
(near Alexanderplatz)
Robert Lubstein
Special Factory for
"Erel-Sonderklasse" brand uniform caps,
* Blue caps * Sport caps
* Contract caps * Helmets *Tschakos
Equipment for Fire Department and Air Defense.

May 15, 1939 *U-M* advertisement for white summer caps for Luftwaffe and Polizei (police) forces. The illustration on the left shows a police officer's cap in white waffled cotton fabric "conforming to the latest regulations." At the bottom of the ad is Lubstein's claim of being "Berlin's largest Uniform cap factory."

mary (excerpt) of his company's file from the trade register, needed for submission to his bank. His own records, he explains, were lost "during combat actions" (*…bei den Kampfhandlungen…*). This document appears on the following page.

For reasons unknown, Lubstein apparently did not try to move either his company – or his capital – to the West (some of his competitors did exactly this). Perhaps he was too old by this time to bother? Instead, he began rebuilding his production anew, with his main customer now the Soviet Army.

The four photographs on pages 270-271 offer solid evidence that the Robert Lubstein company did, in fact, survive the war – though the EREL-Sonderklasse pr*oduct line did not*. Things in the new, postwar world were not as they had been, however; whether due to lack of quality materials, or adequate facilities, Lubstein's product went from the superb caps made for German Wehrmacht personnel to a much lower quality headgear. The company was now back in business, despite the fact that its assets had been confiscated (along with those of other businesses) by the Soviets in the fall of 1945. The confiscation took place under authority of Order No.124 of the Soviet Military Government. Perhaps, this is why Robert Lubstein remained in the East – hoping for an eventual return of his assets? In accordance with Order No.27 of the Chief of Garrison

Advertisement for fiber helmets and tschakos, *U-M*, January 15, 1939. The text reads:

> outstanding as are all Erel-Products…
> *Erel – Fiber tschakos* for municipal police, local police and rural constabulary
> *Erel – Fiber helmets* extra light for Army and Air Force
> Robert Lubstein Berlin NW 21, Alt-Moabit 105
> *Largest German fiber helmet and tschako factory*
> *Largest Berlin uniform cap factory*

Note that the claim of being the largest factory has changed slightly from the 1912 version.

U-M, January 1, 1943. This advertisement shows another Lubstein logo, this one for fiber helmets with the intitials 'RL' creating a face. Center text reads:

> Special factory for very fine
> "Erel-Sonderklasse" brand uniform caps.

The right side of the ad displays a 40 year anniversary wreath.

Worth noting on this memo is the fact that the company letterhead has not yet been changed: it still reads "Lieferant der Wehrmacht und Staatsbehörden" (Supplier for the Wehrmacht and State Agencies). The EREL-*Sonderklasse* mark is still present, as well.

Right: Soviet (motorized) infantry Other Ranks daily service visor cap circa 1947/48, front view. *Courtesy of Herr Klaus-Peter Merta.*

Below: Interior lining of the same cap. In Cyrillic print within the box, the maker's name appears: "Robert Lubstein, Berlin". The headsize is 56; the number 47 may represent the year of manufacture (1947). *Courtesy of Herr Klaus-Peter Merta.*

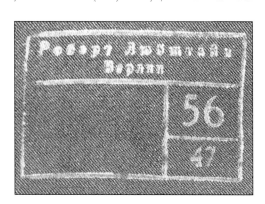

and Military Commander of the Soviet Occupation Sector, the confiscation order was lifted in April of 1947, though it was still not until May of 1948 that Lubstein finally regained control of the company's assets. Just before this in April, an independent trustee (fiduciary) known as a Treuhänder, was officially assigned to oversee the return of the company assets. This was completed by June, and the trusteeship was ended.

On May 25, 1948, Robert Lubstein appeared at the Amtsgericht Berlin-Mitte office – his final appearance in fact – to document this fact for the company's commercial register file. The document reads:

Then appeared
the merchant Robert Lubstein, Berlin-Grunewald,
Auguste Viktoriastraße 24,

authenticated by photographic I.D.

presented the accompanying copy of the dissolution decision from 5.12.1948 by the German Trustee Administration
and declared:

As sole owner of the firm entered in the Handelsregister of the Amtsgericht Berlin-Mitte in Section A under No. 94278:

Robert L u b s t e i n

I report for entry in the Handelsregister:

a. Due to rescinding of the confiscation [order] the appointment of the trustee Friedrich Wilhelm Groß is dissolved,
b. the appointment of the salesman Oskar Lubstein, Berlin NO 55, Heinrich Roller-Straße 16
I have assigned as
sole procurator.
The procurator • will appear shortly at the legal office for signing.

[signed] Robert Lubstein

Soviet cavalry/motorized and technical troops (includes Armor, Artillery) Other Ranks daily service visor cap circa 1948/50, front view. Note the finer quality wool used in this cap, compared to the previous example. *Courtesy of Herr Klaus-Peter Merta.*

Cap lining. Again, the Cyrillic letters within the box name the maker:"Robert Lubstein, Berlin". *Courtesy of Herr Klaus-Peter Merta.*

• equivalent to a chief financial officer, with (in this case, sole) authority to make official decisions regarding the company.

This was the last time Robert Lubstein personally conducted any business with the Amtsgericht. With this action, a great chapter in German cap-making history neared its inevitable end. Robert Lubstein was hospitalized that same year at the St. Hedwig Hospital in [East] Berlin, where he passed away on November 8, 1948.

The company, however, continued on – but its days were numbered. Lubstein's wife, Margarete Lubstein (neé Fischer, born in Berlin in 1883) as the heiress became sole owner and signed all official documents from that point on. She also elected to keep the business name unchanged, as "Robert Lubstein." In a memo on company letterhead dated November 23, 1949, she wrote to the Amtsgericht Charlottenburg (the Amtsgericht with original jurisdiction over Lubstein), which in 1949 was safely located in the western sector of Berlin. She requested that office transfer the company's trade register file to the Eastern Zone:

Regarding: HR 94 278
Since the seat of my company lies in the Eastern Sector, I request that the Handelsregister files pertaining to my firm be transferred to the Amtsgericht Berlin-Mitte... (i.e. to the Soviet zone).

The EREL logo still appears on the letterhead.

The famous Robert Lubstein company had indeed survived the war and then, as a manufacturer of cloth headgear for local Soviet military forces, had also survived the first six postwar years in relative safety. This state of affairs simply could not last, however: Communism was the new order of the day for the Eastern Zone and with the dawn of the DDR in 1949, it was no more than a matter of time until private ownership of companies was forbidden by the new state. All businesses were, in fact, eventually taken over by the East German government and renamed *Volkseigene Betriebe* (VEB) – People's [own] Companies. With this move, the great firm of Robert Lubstein as we know it, effectively passed forever into history – with the admiration of all serious collectors of German cloth headgear. The last recorded official document concerning the Robert Lubstein company dates to September 30, 1952, and was actually submitted by another officially appointed trustee, presumably the fellow who oversaw final confiscation and reallocation of company assets to the new East German state.

Front of the ERELHAUS. The archway opens into the inner courtyard. This particular building face forms the front arm of the square quadrangle that housed the Robert Lubstein offices, storerooms and production floors. The top of the archway can be seen in the illustration from the *U-M* advertisement shown earlier. Note the 'Trabant' automobile parked beneath the L4 mark. Nicknamed the *Trabbie*, it served as the East German "people's car." *Courtesy of Herr Klaus-Peter Merta.*

Right: Rear building of the quadrangle (on the opposite side of the inner courtyard from the archway seen in the previous picture), showing the entrance to the corner stairway shaft / tower that is also easily visible in the *U-M* illustration (left rear corner). In these rooms, in this rear building, the renowned EREL-*Sonderklasse* caps were made. *Courtesy of Herr Klaus-Peter Merta.*

Far right: The right corner stairway shaft in the same rear building. *Courtesy of Herr Klaus-Peter Merta.*

There is a remote possibility that cap production continued for a time under the new company name of VEB Perfekt, manufacturing caps for the equally new NVA (National People's Army) but this has not yet been confirmed; it matters not, however: Robert Lubstein was gone.

Interestingly enough, besides the many Lubstein and EREL-*Sonderklasse* caps that still exist, one other part of the Lubstein/EREL story still survives today, and that is the former *ERELHAUS*. Though having undergone some renovation/restoration, and with no signs remaining of what was once the great Robert Lubstein factory of course, the building itself still stands as silent testimony to the amazing company it once housed. The "Robert Lubstein" signs that once topped the building wings are long gone but with a little imagination, it still looks remarkably like the *ERELHAUS* of old as pictured in the *U-M* advertisement on page 268.

Confirmation Source:
1. Amtsgericht Charlottenberg [Handelsregister], *Frau Martine Pietsch*
2. Amtsgericht Berlin-Mitte (DDR) Handelsregister files
3. Investigation by the Director of the Militaria Collection of the German Historical Museum, Berlin, *Mr. Klaus-Peter Merta*
4. *Uniformen-Markt*
5. Authentic examples

COMPANY: *Wilhelm Schwarte Mützenfabrikation*

BUSINESS ADDRESS:

WWII period:
Wilhelm Schwarte Mützenfabrikation
Berlin C 2, Schillingstr. 22

OWNER or BUSINESS LEADER:
Unknown

CURRENT STATUS:
Unknown

FOUNDED:
Unknown

TRADE NAME If Any:
None

SWEATSHIELD SHAPE:
On Extramützen: *Rhomboid*

MAKERS MARKS, WITH LOCATION:

In silver ink, embossed into the sweat shield. A long oval with the word "Sonderklasse".
Beneath this, the first line in Fraktur font (German Gothic), **"Wilhelm Schwarte"**.
Second line: "Berlin C2"
Third line: Sch[illegible]gstr. 22

VERIFIED PRODUCTS

Schirmmützen: Police Officer Extramütze

COLORS:
On private purchase grade officer caps:

Visor Cap Lining:	*Material:*
Police: Bluish-gray	Rayon or silk

Sweatband colors for Visor caps:	
Police: Tan	Thin leather

Visor reverse colors:
Police: Tan

HISTORICAL BACKGROUND:
No data yet available

Confirmation Source:
Authentic examples

The *Wilhelm Schwarte* makers mark. *Courtesy of Gerard Stezelberger of 'Relic Hunter'.*

BIRKENFELD

COMPANY: *Christian Nagl & Sohn Mützenfabrikation u. Kürschnerei*

BUSINESS ADDRESS:

WWII period:
Christian Nagl & Sohn
Birkenfeld (street address unknown)

OWNER or BUSINESS LEADER:
Unknown

CURRENT STATUS:
Unknown

FOUNDED:
Unknown

TRADE NAME If Any:
None

MAKERS MARKS, WITH LOCATION:
First line: "**Christian Nagl & Sohn**"
Second line: "Mützenfabrikation u. Kürschnerei"*
Third line: "Birkenfeld"

* *Cap manufacturing and furrier shop*
No example of an authentic Christian Nagl makers mark is currently available.

VERIFIED PRODUCTS

Private Purchase Models:
Schirmmützen

GRADES:
No specific grades noted

PATENTS or D.R.G.M.:
None noted.

HISTORICAL BACKGROUND:
None yet available.

Confirmation Sources:
Authentic example; no official confirming documentation yet available

BRAUNSCHWEIG

COMPANY: *A. Bohnsack*

BUSINESS ADDRESS:

WWII period:
A. Bohnsack
Braunschweig, Friedrich-Wilhelm-Str. 30

OWNER or BUSINESS LEADER:
August Bohnsack

CURRENT STATUS:
No longer in existence.

FOUNDED:
The firm of August Bohnsack first appears in the Braunschweig city address books i
1898, however March 2, 1920 is the earliest official date in the city's Gewerbean
[Business Bureau] register.

TRADE NAME If Any:
None

SWEATSHIELD SHAPE:
On Extramützen: *Rhomboid*

MAKERS MARKS, WITH LOCATION:
No example of an authentic August Bohnsack cap is currently available

VERIFIED PRODUCTS

Schirmmützen: Army Officer/Other Ranks Extramütze

PATENTS or D.R.G.M.:
None

HISTORICAL BACKGROUND:
The August Bohnsack company was active in the business field of "Tailor, Commerce in Military Goods and Articles of Equipment" from March 2, 1920 until January 2, 1958 according to Braunschweig city Gewerbeamt records. It does not appear that the company ever reached a level that would enable it to qualify for entry in the city Handelsregister. No records therefore, are available from that institution. The Gewerbeamt records, however, show that the company did survive the war; though its postwar business activity is not specifically identified, it was probably limited to tailoring. There was no longer any customer for "military goods and articles of equipment" except, perhaps, for soldiers of the occupation forces. By the time the Bundeswehr was created it was too late for the company. Though August Bohnsack's Business Bureau register (not to be confused with the Handelsregister) entries ceased in 1958, the firm remained listed in the city Address Books until 1966.

Confirmation Sources:
1. Ordnungsamt Braunschweig [Business Section], *Frau Gelhard*
2. Amtsgericht Braunschweig [Handelsregister], *Frau Müller*
3. Stadt Braunschweig (Stadtarchiv) [City of Braunschweig Archiv], *Herr Dr. Garzmann (Archive Director)*
4. Authentic examples

COMPANY: *Carl Lippold*

BUSINESS ADDRESS:

WWII period:
Carl Lippold
Braunschweig, Bohlweg 14

OWNER or BUSINESS LEADER:
1. Carl Lippold
2. Hans Lippold, merchant
3. Hans and Hans-Eberhard Lippold, merchant; (presumably the sons of Carl Lippold)

CURRENT STATUS:
No longer exists, closed in 1989.

Makers mark for Carl Lippold.

FOUNDED:
Uncertain, perhaps earlier than 1885. The company, under Hans Lippold, was registered at the city Gewerbeamt on July 1, 1935.

TRADE NAME If Any:
None

SWEATSHIELD SHAPE:
On Extramützen: *Diamond*

MAKERS MARKS, WITH LOCATION
Printed directly on the cap lining in black ink:
"Carl Lippold" in large-sized script; below the name, a solid black bar.
Next line: "UNIFORMEN" in vertical capitals.
Next line: "*BRAUNSCHWEIG*" in italicized capitals.
Last line: "*BOHLWEG 14*" in italicized capitals.

VERIFIED PRODUCTS

Private Purchase Models:
Schirmmützen:
• Army Officer Extramütze
• Luftwaffe Officer white Sommermütze ([removable] white top summer cap)

COLORS
On private purchase grade officer/Other Ranks caps:

Visor Cap Lining:
Luftwaffe (summer): Pearl white

Material:
Rayon or silk

Sweatband Colors
On Private Purchase visor caps:
Luftwaffe: Pale gray

Visor Reverse Colors:
Luftwaffe: Green (standard)

GRADES:
None noted

PATENTS or D.R.G.M.:
None noted.

HISTORICAL BACKGROUND:
The firm of Carl Lippold was in operation some time prior to 1885, however the city's Gewerbeamt [Business Bureau] records show no registration until July 1, 1935. This is unusual, since German companies, regardless of size, capital, or operating form, are normally required to register with this office before they can open their doors.
The company was also entered in the city's Handelsregister in 1948 – but a notation in the file actually gives a "date of first entry on February 10, 1885." The Carl Lippold company's age therefore, seems far older than the Business Bureaus' records would indicate.
In 1935, the company business was identified as "Gentleman's and Ladies' Tailor, Uniforms and Commerce in Cloth and Confections." The owner at that time was Hans Lippold. Hans took advantage of the great increase in business created by Germany's

remilitarization and produced high quality uniform caps for private purchase sales (Extramützen). It is not certain if the company also got involved in government contract work, as no contract-type (Government Issue) caps with a Carl Lippold ink stamp on the sweatband reverse have yet been seen.

The Carl Lippold company survived the war, with Hans-Eberhard Lippold (presumably Hans' son) joining the firm in 1948. No extant records indicate the company's line of business from this point in time – though it is safe to say that military uniform caps were hardly likely a part of it. In January of 1974 Hans Lippold left the company, probably to retire, with Hans-Eberhard assuming control. There are no further entries in the Handelsregister file for the company (which was never officially closed out). Business Bureau records, however, show that the Carl Lippold company continued in business up until November 14, 1989, when it finally closed its doors for the last time.

Confirmation Sources:
1. Ordnungsamt Braunschweig [Business Section], *Frau Gelhard*
2. Amtsgericht Braunschweig [Handelsregister], *Frau Müller*
3. Authentic examples

COMPANY: *Clemens Wagner Mützenfabrikation*

BUSINESS ADDRESS:

WWII period:
Clemens Wagner Mützenfabrikation
Braunschweig 1, Bültenweg 72-73

Subsidiary factory:
Hamburg 6, Schanzenstr. 75-77

*Postwar:**
Firma Clemens Wagner
Braunschweig
(March 1, 1959 to July 19, 1977)

* as "Manufacturer of Men's Underwear and Caps, Wholesaler of Textiles, Raw Materials, Caps and Shirts"

OWNER or BUSINESS LEADER:
1. Clemens Wagner
2. Bernhard Wagner
3. Rolf Wagner

CURRENT STATUS:
The main company officially ceased operations in 1977 (last Address Book entry 1976), although the company's cap manufacturing operations may have ceased as early as 1973.

FOUNDED:
1898

TRADE NAME, If Any:
Wartime: *CW*
Postwar: **CleWa** [**CLE**mens **WA**gner]

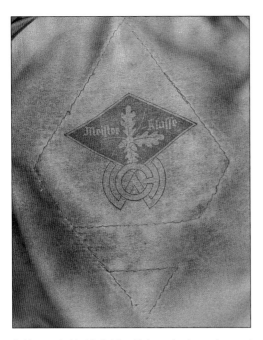

Oakleave triad in black ink with large-sized, superimposed initials "CW". The sweatshield is gone.

The CW sweatband logo, in this case gold ink on a dark reddish brown band. Note the extra section (gray color) added to the bottom of the sweatband at the forehead, which tapers away at the temple area.

SWEATSHIELD SHAPE:

On Extramützen: *Rhomboid.* Note, however, that Clemens Wagner caps for **non-military** organizations commonly appear with a diamond rather than the rhomboid.

MAKERS MARKS, WITH LOCATION:

1. *CW* in curved, superimposed capital letter (outlines), printed directly on the lining. (Usually in black ink).
2. Oakleaf triad and the word *Meisterklasse* within a diamond, printed directly on the lining material, usually in black or gold ink, with the CW logo beneath.
3. The word *Erstklassig* with a stylized spread-winged eagle, in red or brown ink on early military caps and on many caps for non-military organizations (printed directly on the lining). This mark was most likely used prior to 1939.
4. Sweatband: On some caps, the superimposed *CW* logo with *WIRKLICH* [authentic] in capitals printed below it. Beneath *Wirklich* is the phrase *Stirn- u. Schläfendruckfrei.* This arrangement is found on the left temple of the sweatband, usually in black or dark blue ink (on light colored bands) and in gold ink (on dark colored bands). When the sweatband printing appears, the lining makers mark itself is frequently without the CW letters.

Inks used on lining logos:
a. Black
b. Red (on pre-war cap models, particularly those with the phrase *Erstklassig*)
c. Gold (with oak leaf/Meisterklasse logo on wartime caps)

VERIFIED PRODUCTS

Private Purchase Models:
Schirmmützen: [visor caps]
• Army General/Officer/Other Ranks Extramütze
• Waffen-SS General/Officer/ Other Ranks Extramütze
• Luftwaffe General/Officer/Other Ranks Extramütze
• Polizei officer/Other Ranks (including Combat Police) Extramütze

2) **Sommermützen**: Luftwaffe officer/NCO white top [summer] visor. *Lining:* Pale crème, or off white

Contract Other Ranks Caps for Government Issue:
1) **Schirmmützen:** (visor caps)
• Army
• Luftwaffe (some made in consortium with an AGO)

3) **Tropical Field Cap:** Luftwaffe Other Ranks *Lining color:* Grayish tan

Einheitsfeldmützen (M43):
1) Army Other Ranks field caps (contract)
2) Luftwaffe Other Ranks field caps (contract)

Non-military caps: Caps for various government agencies and organizations

COLORS
On private purchase grade officer/Other Ranks caps:

Visor Cap Lining: *Material:*
Army and Waffen-SS: Steel gray; gold   Rayon or silk
Luftwaffe: Gray; gold (champagne)   Rayon or silk
       pale crème; natural (off-white)

Contract (Government Issue) Visor Cap Lining
Army: Reddish brown.   Material: Cotton
Luftwaffe: Reddish brown.   Material: Cotton

Sweatband Colors
a. for visor caps:
Light tan (Army, Waffen-SS)
Dark brown (Army)
Reddish tan (Army, Waffen-SS, Luftwaffe)
Gray (Army)
Pale creamy gray (Army and Luftwaffe generals)

Visor Reverse Colors:
Tan (Army, Waffen-SS)
Reddish tan (Army, Waffen-SS)
Gray (Army)
Green (Luftwaffe standard)

Postwar Bundeswehr Army Other Ranks visor cap lining with the "CleWa" logo printed on the sweatshield reverse.

GRADES: (Models) *Listed from highest to lowest*
It is not certain whether the terms *Meisterklasse* and *Erstklassig* represented actual model grades.

PATENTS or D.R.G.M.:
None noted

PECULIARITIES OF DESIGN (If Any):
Clemens Wagner caps tend to have very high (wide) sweatbands.

HISTORICAL BACKGROUND:
The Braunschweig merchant Clemens Wagner founded his company of the same name in 1898. It was a family-run business through most of its history, with family members of different generations involved with the company at one time or another – and another Clemens Wagner overseeing the company at its end. The business grew steadily, if slowly, during the difficult years between the two world wars and by 1922, the firm identified its business activity as, "Uniform Cap Maker, Commerce with Foreign Products." According to the Braunschweig Amtsgericht, Clemens Wagner qualified for entry in the Handelsregister [commercial registry] and was added on April 15, 1924.

     The company's success eventually led it to expanded to a second facility, a branch factory located in Hamburg, which was active throughout the war (though probably heavily damaged, if not destroyed, prior to 1945). This second factory greatly increased the firm's production capacity though it is not certain whether any Clemens Wagner Extramützen were manufactured there. Ink stamped makers marks inside contract caps (Government Issue) made by Clemens Wagner often list the Hamburg branch as well as the main Braunschweig location.

     Interestingly enough, CW seems to have had more orders at times than even its own large production capacity was able to handle (either in whole or in part). In such a case, CW contracted with a local AGO to do whatever work was required to fulfill the contract obligation in question. Or, perhaps a particular job was assigned a lower importance than another project already running at full production, and this job was farmed

The text reads:

Uniform caps of any kind
in best German craftsmanship and highly finished form
for all purposes
supplied by
Clemens Wagner, Braunschweig 1
SPECIAL UNIFORM CAP FACTORY
Phone 933 Since 1898 Government contractor

Well over 1000 caps deliverable daily

out to an AGO. The finished products bear *both* the Clemens Wagner maker stamp and an AGO stamp. (For an example, refer to Chapter 2).

In the Clemens Wagner advertisement above from the July 15, 1936 issue of *Uniformen-Markt*, the company's numerical figure for caps available for delivery on a daily basis bears noting. Wagner either had a very high production capacity, or a very large stockroom (or warehouse) – or both.

The excellent quality and sharp appearance of Clemens Wagner's Extramütze headgear was highly regarded, and the company certainly has to be ranked among the "Fantastic Four" makers: Robert Lubstein (EREL), Peter Küpper (Peküro), Clemens Wagner and August Schellenberg. On October 1, 1938, the company celebrated its 40-year anniversary. An industry news Chronicle entry in the *Uniformen-Markt* trade paper noted on this occasion:

> From small beginnings has emerged one of the largest uniform cap businesses of Greater Germany. The founder and current senior director presides over a firm in its full vigor, energetically supported for the past few years by his two sons. A branch factory exists in Hamburg, which supplies primarily the needs of Schleswig-Holstein. Top workmanship has characterized the endeavor. The *U-M* extends its congratulations.

With the end of the war, Clemens Wagner found itself still with a large production capacity – but with no main customer for its product. The once powerful Wehrmacht was no more than a memory and the new Federal Armed Forces, the Bundeswehr, would not be born for almost a decade. Clemens Wagner shifted its military production to civilian head wear while also focusing on caps for the country's newly reorganizing official agencies – police and fire departments, agencies responsible for highway and road maintenance and city reconstruction, and the postal service. The company delved into new areas as well, which were not directly connected with cap making. By 1959, the company's official business activity had changed to: "Production of men's underwear and caps; wholesale business in cloth, trimmings, caps and shirts."

With the creation of the Bundeswehr in 1965, Clemens Wagner saw an opportunity to again service the customer it had worked with for so many years in the past: the military. Part of Wagner's cap production was converted back to military caps and for some years the company supplied headgear to Bundeswehr personnel. Products included visor caps, field caps, berets and crash helmets (with beret covers) for Panzer crewmen. With this partial return to military cap production, CW apparently decided to create an acronym for its products (mentioned previously) – something which it had never done during the war (it had been content to remain simply CW). Bundeswehr

In this December 15, 1938 ad, Clemens Wagner mentions its *Erstklassig* model Extramütze. The text reads:

Uniform caps of every model and variety for any purpose,
at home and abroad
Specialty: „Erstklassige" Wehrmacht-Extra-Mützen
supplied by
Clemens Wagner
Special–Uniform-Cap-Factory since 1898
Main facility: Braunschweig 1, Bültenweg 72-73, Phone 897
Branch facility: Hamburg 6, Schanzenstr. 75-77, Phone 436862
Well over 1000 caps deliverable daily

caps produced by Clemens Wagner thus bear the makers mark CleWa. This mark is *incorrect* on authentic pre-1945 caps manufactured by Clemens Wagner.

The Bundeswehr, however, was established with a personnel strength that was no more than a shadow of the Wehrmacht. Germany was still struggling to overcome a major manpower shortage caused by the loss of a great percentage of its male population of military age. The math did not add up well for Clemens Wagner: Competition was increasing, yet the number of available customers remained essentially unchanged; with the improving economy came increases in raw material, overhead and labor costs.

Clemens Wagner's military cap production continued for a time, but the company was eventually forced to close down first this operation and then the other parts of its business as its financial situation deteriorated. Military production apparently ceased some time around 1973, but it was not until 1979 that the company actually flirted with bankruptcy proceedings. This was the end, and fittingly enough, it was another *Clemens Wagner* who was assigned as liquidator of the Clemens Wagner company assets. The Handelsregister (commercial registry) entry for the company was officially closed out several years later, on November 5, 1986.

With the loss of Clemens Wagner, only Peter Küpper (Peküro) of the "Fantastic Four" remained active – but its days, too, were numbered.

Confirmation Sources:
1. Stadtarchiv Braunschweig
2. Amtsgericht Braunschweig (Handelsregister), *Frau Müller*
3. Ordnungsamt Braunschweig (the Business Bureau registry section), *Frau Gelhard*
4. *Uniformen-Markt*
5. Authentic examples

COMPANY: *Georg Kurz*

BUSINESS ADDRESS:

WWII period:
Georg Kurz
Braunschweig, Hagenring 2

OWNER or BUSINESS LEADER:
Unknown

CURRENT STATUS:
Unknown

FOUNDED:
Unknown

TRADE NAME If Any:
None

SWEATSHIELD SHAPE:
On Extramützen: *Diamond*

MAKERS MARKS, WITH LOCATION
Printed on the lining material in black ink, a horizontally-oriented diamond, superimposed over the bottom half of a stylized eagle. Within the diamond is a wide oval bar (lining material color); inside the bar is the word:

Georg Kurz makers mark. *Courtesy of Gerard Stezelberger of Relic Hunter.*

Exterior view. Georg Kurz caps are finely crafted of high quality materials. Note the thick-type crown piping. *Courtesy of Gerard Stezelberger of Relic Hunter.*

"Erstklassig" in canted script, angling slightly upward on the right. At the bottom of the diamond, "GESETZL.GESCH."

Beneath the diamond, in script, the name "Georg Kurz".

Second line (beneath the name): "Braunschweig", also in script.

Final line, in fine print capitals: "HAGENRING 2"

VERIFIED PRODUCTS

Schirmmützen: Army Officer Extramütze

COLORS
On private purchase grade officer caps:

Visor Cap Lining:
Army: Gold

Material:
Rayon or silk

Sweatband Colors
On Private Purchase visor caps:
Army: Tan

Visor Reverse Colors:
Army: Tan

GRADES:
None noted

PATENTS or D.R.G.M.:
None noted.

HISTORICAL BACKGROUND:
With a business activity identified in Gewerbeamt [city Business Bureau] register records as "Tailor and Commerce with Fabrics", the Georg Kurz business was listed from February 10, 1931. The company had more than one business location; others included Friedrich-Wilheln-Str. 65 and Altewiekring 23. No examples of these two additional addresses have yet been identified on caps made by Georg Kurz, however.

The company never sought Handelsregister status, thus no records are available from that source, nor do any records exist pertaining to the wartime period (not unusual). A check of postwar records turned up nothing as well, which suggests that the company either went out of business during the war, or was destroyed.

Confirmation Sources:
1. Ordnungsamt Braunschweig [Business Bureau registry section], *Frau Gelhard*
2. Authentic examples

BREMEN

COMPANY: *Heinrich Balke Hut- und Mützenfabrik*

BUSINESS ADDRESS:

WWII period:
Heinrich Balke Hut- und Mützenfabrik
Bremen, Bachstr. 79-80

Postwar:
Heinrich Balke GmbH & Co.
Bauerland 19
D-28259 Bremen

OWNER or BUSINESS LEADER:
1. Heinrich Friedrich Christoph Conrad Balke, merchant (prior to 1900)
2. H.F.C.C. Balke and Johann Wilhelm Conrad Freyburg (from 1900)
3. J.W.C. Freyburg, H.F.C.C. Balke and Louise Sophie [neé Adler] (from 1912)
4. J.W.C. Freyburg and Carl Johann Heinrich Balke, merchant [presumably the son] (from 1919)
5. Carl Johann Heinrich Balke (from 1929)
6. C.J.H. Balke and Dieter Karl Balke, merchant [the latter presumably C.J.H. Balke's son] (from 1969)
7. Dieter Karl Balke (from 1971)

CURRENT STATUS:
Closed on August 1, 2000 after an insolvency investigation and associated proceedings (initiated by the city of Bremen).

FOUNDED:
1888

TRADE NAME If Any:
None

SWEATSHIELD SHAPE:
On Extramützen: *Rhomboid*

MAKERS MARKS, WITH LOCATION:
No authentic example of a Heinrich Balke makers mark is currently available.

VERIFIED PRODUCTS:

Schirmmützen: Army Officer Extramützen

GRADES:
No specific grades noted

PATENTS or D.R.G.M.
None

PECULIARITIES OF DESIGN (If Any):
None noted.

HISTORICAL BACKGROUND
The Heinrich Balke company was established at some time prior to 1900 by the Bremen merchant Heinrich Friedrich Christoph Conrad Balke. By 1900, the company was doing quite well enough to qualify for the city's Handelsregister (commercial registry) and was officially listed on November 1, 1900. The company remained in business throughout the next forty years, with Carl Johann Heinrich Balke (the son?) assuming leadership of the business from 1929. Carl Johann steered the company through the Third Reich period, producing headgear (private purchase visor caps) for the Armed Services and other government organizations.

May of 1945 found the company still in existence, but temporarily in limbo while the dust of war settled. Business began once again in 1946, with Carl Johann still at the helm and the company's name (and business structure) altered to GmbH (limited liability company). The postwar economic situation was difficult, but the company made it through the hard times and seems to have been doing fairly well at Carl Johann's death, on April 4, 1971.

Dieter Karl Balke (perhaps Carl Johann's son), took control and carried on the business with no change in name. Unfortunately, there is no documentation available that clearly identifies the company's products or activities during the postwar period – that is, whether it continued producing caps, or shifted to some other product or venture.

Whatever the Heinrich Balke postwar business might have been, in the long run it was unable to prevent the company from encountering financial difficulties. On August 1, 2000, the Bremen Amtsgericht initiated insolvency proceedings against the firm. Such proceedings automatically result in dissolution of the company involved, and with that, the more than one hundred-year old history of Heinrich Balke Hut- und Mützenfabrik came to an end.

Confirmation Source:
1. Amtsgericht Bremen [Handelsregister], *Frau Schönwälder, Justizhauptsekretärin*
2. *Uniformen-Markt*
3. Authentic example

BRESLAU

COMPANY: *Gustav Binner Mützenfabrik*

BUSINESS ADDRESS:

WWII period:
Gustav Binner
Breslau 1, Ohlauerstr. 42

OWNER or BUSINESS LEADER:
Unknown, but the original owner was presumably Gustav Binner.

CURRENT STATUS:
No longer exists.

FOUNDED:
Unknown

TRADE NAME If Any:
None

SWEATSHIELD SHAPE:
On Extramütze: *Rhomboid*

MAKERS MARKS, WITH LOCATION
Printed in silver ink on the lining material, "**Gustav Binner**" in a large, Fraktur font (German Gothic).
Second line: "Mützenfabrik"
Third line: "Breslau"
Fourth line: "Ohlauerstr. 42"
Fifth line: "Tel. 69180"
Sixth line: "Lieferjahr"

Gustav Binner makers mark in a Luftwaffe Other Ranks Extramütze.

VERIFIED PRODUCTS:

Private Purchase Models:
Schirmmützen: Luftwaffe Other Ranks Extramütze

COLORS:
On private purchase grade Other Ranks caps:

Visor Cap Lining:
Luftwaffe. Reddish tan (rust)

Material:
Cotton (linen weave)

Sweatband Colors
Luftwaffe: Tan

Visor Reverse Colors:
Luftwaffe: Green (standard)

GRADES:
No specific grades noted

PATENTS or D.R.G.M.:
None noted.

HISTORICAL BACKGROUND:
No information yet available. Research is hampered by the fact that half of Breslau is now the Polish city of *Wroclaw* (the city and territory became part of Poland after the war).

Confirmation Sources:
Authentic example

COMPANY: *Emil Wolsdorff*

BUSINESS ADDRESS:

WWII period:
Emil Wolsdorff
Breslau (street address unknown)

OWNER or BUSINESS LEADER:
Presumably Emil Walsdorf

CURRENT STATUS:
No longer exists.

FOUNDED:
Unknown

TRADE NAME If Any:
None

SWEATSHIELD SHAPE:
On Extramützen : *Rhomboid*

MAKERS MARKS, WITH LOCATION:
Printed on the cap lining in black ink within a large, toothed circle:
The name "**Emil Wolsdorf**"
Second line: "Nachfolger"*
Third line: "**Breslau**"
Fourth line: "Cottbus-Görlitz-Forst"

* *Successor (to what, is not known)*

VERIFIED PRODUCTS

Schirmmützen: Army officer Extramütze

COLORS

On private purchase grade officer caps:

Visor Cap Lining:	Material:
Army: Purplish-gray	Rayon or silk

Emil Wolsdorff makers mark.

Sweatband Colors
On Private Purchase visor caps:
Army: Tan

Visor Reverse Colors:
Army: Tan

GRADES:
None noted

PATENTS or D.R.G.M.:
None noted.

HISTORICAL BACKGROUND:
As a result of postwar political realignments, part of the city of Breslau became Polish after 1945, and was renamed Wroclaw. No records for the Emil Wolsdorff company have yet been found.

Confirmation Sources:
Authentic example

COESFELD

COMPANY: *Hermann Potthoff 'HPC'*

BUSINESS ADDRESS:

WWII period:
Hermann Potthoff
Coesfeld, Kupferstraße 4

OWNER or BUSINESS LEADER:
1. Bernhard Hermann Christian Potthoff, Kappenmacher (cap maker); born 1829
2. Wilhelm Potthoff, born 1869
3. Clemens Fabry, born 1903 (married to the second daughter of Wilhelm)
4. Werner Fabry

CURRENT STATUS:
Closed in 1975 (destroyed in 1945, and rebuilt).

FOUNDED:
March 3, 1856

TRADE NAME If Any:
HPC

SWEATSHIELD SHAPE:
On Extramützen: *Rhomboid*
Note, however, that on caps bearing the HPC mark *together with a distributor's name,* the sweatshield is usually a diamond.

HPC logo on an Army officer's cap. Note the very thin composition sweatband.

This particular HPC mark is highly unusual: The bottom line (partly worn off) appears to read, "...kasse SS". Is the worn off word, in fact, *Kleider...*? No independent SS Kleiderkasse has ever been identified; if there was one, it was probably created well after the war was already underway. Waffen-SS officers generally made use of the Army's 'Offizier Kleiderkasse'. The mystery remains unsolved, and it is possible that this particular makers mark has been doctored (example taken from an Internet offering) – a physical inspection would be required to determine if this were indeed the case.

MAKERS MARKS, WITH LOCATION:

Note: The full name *Hermann Potthoff* never appears on HPC caps.

1. **Fraktur** (Gothic) **font** version: Printed directly on the cap lining, in a variety of ink colors: "**HPC**" in large-sized gothic font. Large *P* positioned between a smaller *H* and *C* within a circle.
First line: "Frischluft" in large-sized script
Second line, within a scrolling rectangle : "D.R.G.M. **1 419 757**"
Third line, in script: "Stirndruckfrei"

2. Roman font-style HPC logo: Found primarily on caps with an accompanying distributor's name. The letter *H*, with the *P* superimposed over the right vertical arm of the *H*; beneath this combo is the letter *C*. These three letters are printed in red ink on the cap lining. The sweat shield is a diamond shape with this logo, not the usual rhomboid. See page 289 for an example.

VERIFIED PRODUCTS:

Schirmmützen:
• Army Officer Extramütze
• Luftwaffe Officer Extramütze
• Waffen-SS Officer Extramütze

COLORS:
On private purchase grade officer/Other Ranks caps:

Visor Cap Lining:	*Material:*
Army/Waffen-SS: Light gold (champagne)	Rayon or silk
Luftwaffe: Pale bluish-green; gold	Rayon or silk

Sweatband colors for Visor caps:	
Army/Waffen-SS : Tan	Leather; thin leather; composition material
Luftwaffe: Tan	

Visor reverse Colors:
Army/Waffen-SS: Reddish tan; tan
Luftwaffe: Green (standard)

GRADES:
No specific grades noted; the "Frischluft" system was simply an option, not a model.

DISTRIBUTOR:
1) Wilhelm Stellrecht (Heilbronn)

PATENTS or D.R.G.M.:
D.R.G.M. No. **1 419 757** Klasse 41b *"Frischluftmütze mit abstehendem Schweissleder und Luftlöchern im Rand"*. Entered in the Patentamt registry on October 20, 1937.
　　The invention covered by this Gebrauchsmuster (termed *Frischluft* by HPC) was a ventilation system that permitted air from outside the cap to enter and circulate about the wearer's head. Unlike the EREL-*Sonderklasse* air vent system (through the cockade center), HPC used a modified oak leaf wreath with rectangular openings along the bottom edge, and multiple holes through the cap band. Inside the cap, a flat leather o

pasteboard plate was affixed behind the sweatband at the forehead, with corresponding holes through which the external air passed into the cap. The plate was equipped with a fold over cover panel, which could be used to completely close off the airflow, if the wearer so desired. The Frischluft system was licensed from HPC by a number of other cap manufacturers, including J.B. Holzinger of Berchtesgaden.

PECULIARITIES OF DESIGN (If Any):
None noted (excluding the Frischluft air hole system).

HISTORICAL BACKGROUND:
On March 3, 1856, a fellow named Bernhard Hermann Christian Potthoff registered his name in the Coesfeld town resident list as a cap maker. Just over a week later, on March 12th, he announced to his fellow citizens that he had settled in the house at Große Viehstraße 14. His business: furrier, cap and umbrella maker. He also produced Russian and balloon caps (what type of headwear is meant by the latter term, is uncertain). Business was good and he later expanded it to include trade in gentlemen's hats. He also acquired a new house at Große Viehstraße 24. By the 1880s, the firm had grown to include 10 to 12 assistants and seamstresses.

In 1894, the house at Große Viehstraße 24 was sold, and the Hermann Potthoff company was moved into a newly acquired house located at Kupferstraße 4. The firm's retail business was dropped as it concentrated on production and wholesale activity. By this time, HPC was already producing "Blaue Mützen" [blue Navy caps] and uniform caps.

Wilhelm Potthoff entered the company in 1896, though it is not clear whether he was the son of [Bernhard] Hermann, or some other family member. Hermann by this time was 67 years old, and it was only a few years later that he passed away at the turn of the century (March 17, 1900), leaving Wilhelm as the sole owner. The company began restructuring its facilities, converting part of the premises to roofed sheds, and installing equipment for the production of packaging materials. Gas and electricity was installed to power the machines. At the same time, construction of a private home (presumably for Wilhelm Potthoff) began on Daruper Straße. The old house was now being used as a storage facility. At this point in time, HPC employed 5 individuals as travelling salesmen and 67 skilled employees in the company's workshops.

In 1909, two new part owners joined the company, which continued as before under the successful leadership of Wilhelm Potthoff. By the start of the First World War in 1914, the company personnel roster had increased to 101 employees. HPC shifted most of its production over to a war footing, and began producing field caps for the Imperial German armed forces. Part of this change included an increase in the number of women employees working for HPC, which was now higher than at any time before in the company's history. With the end of the Great War followed the high inflation period of the 1920s. Business was poor and to survive the company was forced to consolidate its activities and shut down a portion of its production. Starting in 1924, HPC began to intensify its efforts in customer service. It also took part in a number of trade shows in order to increase customer recognition and develop new business. Despite tight finances, the company also invested in new machinery and once again upgraded its facilities – even as the first stages of the worldwide economic depression began to be felt within a German economy that was already reeling under repeated financial crises. HPC nonetheless managed to hold on until the depression finally ran its course and the German economy started on the road to recovery after 1932.

By 1936, with the [military] cap making industry in full bloom, HPC had already shifted much of its production over to uniform (visor) caps for Wehrmacht personnel. The two limited partners from 1909 had left the company by this time, and Wilhelm was, once again, the sole owner. He was now 67 years old.

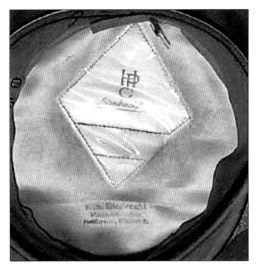

The Roman font-style HPC logo (in red ink) on an Army infantry officer's Extramütze. This cap also bears a distributor's name: Wilhelm Stellrecht, which appears as a simple ink stamp (black ink) beneath the diamond-shaped sweat shield. Stellrecht was also a cap maker, and it is not clear whether caps such as this one were actually manufactured by HPC, or by Stellrecht under license from HPC. This particular piece also has a black velvet (or felt) insert between the sweatband and the cap bottom around the forehead area. *Courtesy of Bill Shea.*

1936 also saw a new partner enter HPC, a fellow named Clemens Fabry, who was married to Wilhelm's second daughter. Wilhelm, however, remained firmly in control.

The company, meanwhile, had certainly not been idle: On October 4, 1937, HPC submitted an application for Gebrauchsmuster protection to the Reichspatentamt. The invention it was seeking to protect with a registered D.R.G.M. was its new air circulation system, officially titled *Frischluft*.

HPC's Frischluft system, in fact, was in direct competition with the cockade vent air system introduced by Robert Lubstein for his *EREL-Sonderklasse* caps (as an optional "extra"). HPC was awarded the desired Gebrauchsmuster, which was assigned the D.R.G.M. number 1 419 757. The Gebrauchsmuster was officially registered at the Patentamt on October 20, 1937.

This particular D.R.G.M. number, in fact, appears on a variety of German visor caps by other makers (not the HPC label), since these firms bought licenses to use the HPC system in their own caps. Regardless of the maker label: if the number 1 419 757 appears somewhere on a cap it means that the manufacturer held a license from HPC.

In 1939, the firm acquired a facility that produced military equipment (what kind of equipment is not certain). During this period, HPC employed as many as 200 workers at times. This acquisition confirmed the company's position as a major player and a prime employer in the Coesfeld area. Through the first two years of the war, HPC continued to produce high quality Extramützen, which it sold under its own label as well as through a number of distributors such as Wilhelm Stellrecht Mützenfabrik (Heilbronn) – a cap manufacturing company in its own right – though it is not clear if Stellrecht (and other distributors) actually manufactured HPC caps at his own factory, or merely distributed them. The HPC label, slightly modified, appeared on these caps together with the distributor's mark; the latter, however, was often only an inked stamp on the cap lining. It is not yet certain whether the company also produced contract visor caps (for Government Issue) and field caps, though this certainly seems likely.

In 1941, the war first struck home when some of HPC's production facilities were destroyed in Allied attacks (probably air raids on the industrial area of Coesfeld). The company was hit again in 1943; these two episodes between them had destroyed approximately 30% of HPC's structures. Wilhelm Potthoff did his best to keep the company running, and in this he succeeded until March 21, 1945 – when the war came knocking at his very door. As Allied forces advanced through the Coesfeld region, HPC's production facilities and business offices were completely destroyed in the fighting: a little more than one month short of the end of the war, the company's fate seemed to have been sealed.

But HPC, existing in name only until near the end of 1945, was not dead. After the dust had settled and the year was drawing to a close, what was left of the company staff gathered together in a small room at a little inn called the "Gastwirtschaft Becker", and went to work once again.

Meanwhile, at the original HPC premises, the ruins were being cleared away. The company was able to relocate its operations into the less cramped quarters of the restored basement of its own former main building from April 1948, where business recommenced. Times were hard and customers limited, but HPC never looked back.

In 1950, the company began manufacturing a type of cap known as the "Chapeaux Claques" (some sort of fold-up headwear). In January of 1951, the firm was at last able to move back above ground into the newly rebuilt upper floors of the HPC main building. It is not known to what extent Wilhelm Potthoff was still involved with the company management by this time. Most of the company's activities were probably directed by Clemens Fabry, though he is not mentioned in available postwar information Wilhelm Potthoff, son of the founder and the driving force behind HPC for over 49 years, passed away on November 25, 1952, at the age of 83.

For the first time in its existence there was no Potthoff at the company helm, bu HPC continued on, eventually closing out its cap manufacture and converting to the production of telecommunications equipment for the well-known Siemens company.

Werner Fabry joined HPC as part owner in 1963, (perhaps the son of Clemens Fabry), and the company moved through the 1960s and into a new decade; this time, however, it would only reach the halfway mark. The dimming lights at HPC were again put out – this time forever – in 1975.

Note: As chance would have it the Heimatverein Coesfeld (a local association which promotes the town's history and area) opened a historical exhibition in 1992 on this particular company – such was HPC's significance for – and impact on – the local area. The exhibition included historical information as well as period photographs and a number of HPC products. Much of the historical information presented here in this brief was originally collected by the Association for use in a paper prepared for the exhibition, and kindly passed on to me by Herr Damberg of the Coesfeld City Archive. My thanks to the members of the Heimatverein Coesfeld and to the personnel of the Stadtarchiv Coesfeld for their friendly assistance in illuminating the interesting history of HPC.

Confirmation Source:
1. Stadtarchiv Coesfeld [city archive], *Herr Norbert Damberg*
2. Heimatverein Coesfeld
3. Bundespatentamt [Federal Patent Office]
4. Authentic examples

DIPPOLDISWALDE

COMPANY: *L.G. Schwind*

BUSINESS ADDRESS:

WWII period:
L.G. Schwind
Dippoldiswalde

OWNER or BUSINESS LEADER:
1. Karl Gotthalf Schwind (from 1897)
2. Karl Gottfeld Rudolf Schwind (from around 1953)

CURRENT STATUS:
Unknown, but presumed out of business. The Handelsregister entry for the company was never officially closed, nor have any other records yet been uncovered.

FOUNDED:
Exact year unknown; first entered in the Dippoldiswalde city Handelsregister on July 11, 1896.

TRADE NAME If Any:
None

SWEATSHIELD SHAPE:
On Extramütze: *Rhomboid*

MAKERS MARKS, WITH LOCATION:
No authentic example of the L.G. Schwind makers mark is currently available.

VERIFIED PRODUCTS:

Schirmmützen (visor caps)

HISTORICAL BACKGROUND:
The company's history extends back beyond the turn of the century, however there is no data currently available that provides an idea of its size, or type of business (i.e. how the company described its business activities). It appears to have been a relatively small firm, though it must have managed its account books and inventory in a form that qualified it for entry in the Handelsregister.

Although the company apparently survived the war (the last Handelsregister entry indicates its existence in 1953), there is no other documentation from the postwar years to shed any light on the company's fortunes during that time. There are no further Handelsregister entries after 1953, nor was the company's entry ever officially closed out (which was the normal action if a company went out of business, or if it could no longer meet the qualification standards for the Handelsregister). Although the fate of the company is thus uncertain, the likelihood of it still being in existence today is extremely slim.

Confirmation Source:
Authentic example

DRESDEN

COMPANY: *Heinz Schmidt*

BUSINESS ADDRESS:

WWII period:
Heinz Schmidt
Dresden, Lauensteinerstr. 14

OWNER or BUSINESS LEADER:
Presumably, Heinz Schmidt

CURRENT STATUS:
Unknown

FOUNDED:
Unknown

TRADE NAME If Any:
None

SWEATSHIELD SHAPE:
On Extramütze: *Rhomboid*

MAKERS MARKS, WITH LOCATION:
No authentic example of the Schmidt makers mark is currently available.

VERIFIED PRODUCTS:

Schirmmützen: Army Officer Extramütze

HISTORICAL BACKGROUND
No historical data has yet been found. This company was probably quite small, and therefore would not have qualified for entry in the city's Handelsregister [commercial registry]. This leaves only the Gewerbeamt's [Business Bureau] company registration records as a means of confirming both the firm's existence and the time frame – but these files were destroyed during the war when the city was firebombed.

Confirmation Source:
1. Authentic example; No *official* confirming documentation for this company has yet been found.
2. Authentic example.

COMPANY: *William Günther Sächsiche Militär-Effekten-Fabrik*

BUSINESS ADDRESS:

WWII period:
William Günther
Dresden, König Georg Allee 15

OWNER or BUSINESS LEADER:
Karl William Günther, merchant

CURRENT STATUS:
No longer exists. First entry in the Dresden Handelsregister was July 1, 1905. The company was probably destroyed during the war, or else commandeered, or closed by the former East German government in the early 1950s. There are no current records for the firm, or for any descendant company.

FOUNDED:
Unknown

TRADE NAME If Any:
None

SWEATSHIELD SHAPE:
On Extramützen: *Rhomboid*

MAKERS MARKS, WITH LOCATION:
No authentic example of the Günther makers mark is currently available.

VERIFIED PRODUCTS

Schirmmützen: Luftwaffe Other Ranks (Extramützen) cap

GRADES:
None

PATENTS or D.R.G.M.:
None

HISTORICAL BACKGROUND:
Aside from a note in the Handelsregister records dated September 22, 1938 (mention of the company entry being carried over from another book), information on William Günther is extremely limited. Frau H. Reim of the Stadtmuseum, Dresden, was able to trace mention of the company to 1944, but besides this and the date of first entry in the Handelsregister (July 1, 1905), there is no further information. This is in part due to the fact that much of Dresden's official records (including those of the Handelsregister and the Gewerbeamt [Business Bureau]) were destroyed when the city was firebombed late in the war; the company as well, may have been destroyed at that time. Much of the Handelsregister files that remain were actually reconstructed after the loss of the originals.

Dresden was part of the former East Germany and during the Communist period most private business enterprises were converted to state-owned businesses – meaning that no Handelsregister records were kept during that time frame.

Once the old Handelsregister records became available (after German Reunification), William Günther's company entry was officially closed (January 27, 1999) – indicating that the Amtsgericht found no evidence of the company's existence. It is relatively certain that the firm was already long gone even if it survived the end of the Second World War.

Confirmation Source:
1. Stadtmuseum Dresden, *Frau H. Reim*
2. Authentic example

ERFURT

COMPANY: *Friedrich Bürger Mützenfabrik*

BUSINESS ADDRESS:

WWII period:
Friedrich Burger Mützenfabrik
Erfurt, Brühler Str. 60

OWNER or BUSINESS LEADER:
Unknown, but presumably Friedrich Bürger.

CURRENT STATUS:
Unknown

FOUNDED:
1905

TRADE NAME If Any:
Unknown

SWEATSHIELD SHAPE:
On Extramützen: *Rhomboid*

MAKERS MARKS, WITH LOCATION:
Line one: "**Friedrich Bürger Mützenfabrik**"
Second line: "Uniformmützen ohne Schirmdruck"*

* *"Uniform caps without forehead pressure"*
No example of the Friedrich Burger makers mark is currently available

VERIFIED PRODUCTS:

Schirmmützen (visor caps)

GRADES:
No specific grades noted.

PATENTS or D.R.G.M.:
Though the Friedrich Bürger advertisement that appears here mentions a Gebrauchs-muster (D.R.G.M.), there is no way to determine – in the absence of the actual D.R.G.M. number – who held the rights. (i.e. under license, or Bürger's own D.R.G.M.)

PECULIARITIES OF DESIGN (If Any):
None noted.

HISTORICAL BACKGROUND:
No historical data yet available.

HISTORICAL BACKGROUND:
No detailed historical information or documentation yet available.

Confirmation Source:
1. *Uniformen-Markt*
2. Authentic example

Advertisement for Friedrich Bürger in the October 15, 1938 issue of *Uniformen-Markt*. The text reads:

Friedrich Burger
Cap Factory Founded 1905
Erfurt, Brühler Str. 60
Specialty: Fine
UNIFORM-CAPS
without forehead pressure D.R.G.M.

COMPANY: *Karl Naubert*

BUSINESS ADDRESS

WWII period:
Karl Naubert
Erfurt, Mainzerhofstraße

OWNER or BUSINESS LEADER:
Tentative – based on city address book entries only.
1. Wilhelm Naubert, Master furier and cap maker
2. Hermann Naubert, cap manufacturer with Karl Naubert, merchant, cap factory
3. Karl Naubert (from 1936)

CURRENT STATUS:
No longer in existence.

FOUNDED:
1900

TRADE NAME (logo) If Any:
HNE

SWEATSHIELD SHAPE:
On Extramütze: *Rhomboid*

Interior of a Luftwaffe Other Ranks Extramütze manufactured by Karl Naubert, with the company's usual, later makers mark (logo) with the initials KNE.

MAKERS MARKS, WITH LOCATION:

Printed in silver on the cap lining, a stylized visor cap facing left and seen from the side, partially superimposed on a disk (full arc of the disk is to the left). Beneath the cap visor, the capital letters *H N E* (Hermann Naubert Erfurt) atop one another, slightly interlocked and staggered. On later caps (post 1940), the *H* changes to *K*.

Below this, the first line, in capitals: "KARL NAUBERT."

Second line, also in capitals: "ERFURT" flanked on each side by dates – the right date being 1925 (the printing on Naubert logos is not commonly very clear).

Third line, in capitals: "LIEFERJAHR" on the left, with the date to the right.

The cap size appears in the same silver ink, below the rest of the logo.

In addition, the company produced caps that bear the visor cap/HNE logo, but without the Karl Naubert name; instead, only the generic "Deutsche Wertarbeit" phrase appears. If a person were unfamiliar with the company identified by the logo, he would be hard pressed to identify the maker with only the three initials to work from. It is probable that this 'semi blank' logo form was the *precursor* to the "Karl Naubert" labeled cap (particularly those later caps with a K letter instead of an H), and which was used while Hermann Naubert was running the company. Once Karl took over, he added his own name beneath the logo, but did not change the logo's HNE to *K*NE until 1937-1938.

VERIFIED PRODUCTS:

Schirmmützen:
• Army Officer Extramütze
• Luftwaffe Other Ranks Extramütze

COLORS:
On private purchase grade Officer/Other Ranks caps:

Visor Cap Lining:	*Material:*
Army (officer): Pale gold (champagne)	Rayon or silk
Luftwaffe: Yellowish-brown	Cotton (linen weave)

Sweatband colors for Visor caps:	
Army (officer): Tan	Material: Thin leather
Luftwaffe OR caps: Tan	Material: Thin leather, or composition material

Visor reverse Colors:
Army: Tan
Luftwaffe: Green (standard)

GRADES:
No specific grades noted

PATENTS or D.R.G.M.:
None

PECULIARITIES OF DESIGN (If Any):
None noted

HISTORICAL BACKGROUND:
The Karl Naubert firm had its beginnings at the turn of the century or earlier, when it was opened in Erfurt by Wilhelm Naubert (no later than 1900). The company's name at

that time is not known, but by the 1920s or early 1930s period it was clearly identified as "Hermann Naubert" – likely Wilhelm's son. The company did well under the expanding economy of the Third Reich, and in 1936 succeeded in gaining registration in the city's Handelsregister. At the time of registration, the firm's recorded name was: "Hermann Naubert", with the owner listed as Karl Naubert. Circumstantial evidence suggests that Karl was a relative or son of Hermann, who passed away some time between 1938 and 1941 (only his widow is listed in the city's address book by that time and the company itself appears under the name of Karl Naubert).

Though the Karl Naubert company survived the war, it was not to last long in postwar Germany. The final entry in the firm's Amtsgericht file is dated August 25, 1947, and states simply that the company was dissolved. In Erfurt's city Address Book of 1948, the company appears, but is identified as: "Mützenfabrik Erfurt vormals Karl Naubert, Landeseigener Betrieb" (*Cap factory Erfurt, formerly Karl Naubert, state-owned business*). This means that after the end of the war, the Karl Naubert company and its assets and property were put under sequester (reason unknown) by the city of Erfurt, and later converted into the property of the [federal] state of Thüringen – at which time it was probably liquidated. The company is absent from any subsequent address books.

Confirmation Source:
1. Stadtarchiv Erfurt, *Stadt Chronistin [City Chronicler] Frau Astrid Rose*
2. Registergericht Erfurt, *Herr Keller*
3. Authentic examples

ERLANGEN

COMPANY: *Michael Keck Militäreffekten*

BUSINESS ADDRESS:

WWII period:
Michael Keck Militäreffekten
Erlangen, Luitpoldstraße 50

OWNER or BUSINESS LEADER:
Unknown

CURRENT STATUS:
Unknown. No confirmation documentation for this company has yet been received from the responsible authorities.

FOUNDED:
Unknown

TRADE NAME If Any:
None

SWEATSHIELD SHAPE:
On Extramütze: *Rhomboid*

MAKERS MARKS, WITH LOCATION:
In a large-size font, in black ink printed directly on the lining material, the name "Michael Keck."

The makers mark for Michael Keck. *Courtesy of Bill Shea.*

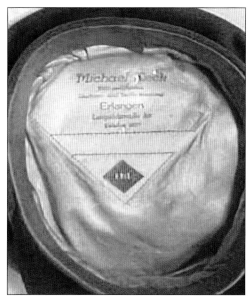

Full view of a Michael Keck cap interior. Note that the visor reverse is black rather than the much more common crème, or tan. *Courtesy of Bill Shea.*

Second line (small font): "Militär-Effekten" [Military Goods]
Third line (small font): "Uniform- und Reitbekleidung" [Uniforms and riding clothes]
Fourth line (larger font): "Erlangen"
Fifth line (small font): "Luitpoldstraße 50"
Sixth line (Small font): "Telefon 2671"

Below the maker information (within the lower point of the sweatshield) appears a red diamond, with the letters BMI (meaning unknown)

VERIFIED PRODUCTS:

Schirmmützen:
Army Extramütze

COLORS:
On private purchase grade officer caps:

Visor Cap Lining:	*Material:*
Army: Light gold (champagne)	Rayon or silk

Sweatband colors for Visor caps:	
Army: Dark tan; brown	Leather, composition material

Visor reverse Colors:
Black

GRADES:
No specific grades noted

PATENTS or D.R.G.M.:
None

PECULIARITIES OF DESIGN (If Any):
None noted

HISTORICAL BACKGROUND:
Historical documentation on this company has not yet been located. The firm may have been a seller only (selling caps made for it by an unnamed maker, under the Michael Keck mark) with no manufacturing, but this can not be confirmed one way or another from the minimal data as yet available, Research efforts continue.

Confirmation Source:
Authentic examples

FELLBACH

COMPANY: *Sport Metzger*

BUSINESS ADDRESS:

WWII period:
Sport Metzger
Fellbach, Uannstatterstr. 119

OWNER or BUSINESS LEADER:
Max Metzger

CURRENT STATUS:
Unknown, but presumed out of business.

FOUNDED:
Unknown

TRADE NAME If Any:
None

SWEATSHIELD SHAPE:
On Extramütze: *Rhomboid*

MAKERS MARKS, WITH LOCATION:
Top line: "Die Deutsche Qualitätsmütze"
Second line: "**SPORT METZGER**" in capitals.
Third line: "Inh. Max Metzger"*
Third line: Fellbach-Stuttgart
Fourth line: "Uannstatterstr. 119"

No authentic example of the Sport Metzger makers mark is currently available.

* *Inh. Is the abbreviation for Inhaber [owner]*

VERIFIED PRODUCTS:

Schirmmützen (visor caps)

HISTORICAL BACKGROUND:
No historical documentation has yet been located.

Confirmation Source:
Authentic example

FLENSBURG

COMPANY: *Friedrich Methmann Nordmark Mütze*

BUSINESS ADDRESS:

WWII period:
Friedrich Methmann
Flensburg

Postwar:

OWNER or BUSINESS LEADER:
Friedrich Methmann

CURRENT STATUS:
Unknown

Methmann makers mark on a Luftwaffe Other Ranks summer cap (white top). *Courtesy of Bill Shea.*

Complete interior (including the visor reverse) of a Methmann cap. Luftwaffe white top Sommermütze for FLAK artillery. *Courtesy of Bill Shea.*

FOUNDED:
Unknown

TRADE NAME If Any:
Nordmark Mütze

SWEATSHIELD SHAPE:
On Extramütze: *Rhomboid*

MAKERS MARKS, WITH LOCATION:
In silver ink, printed on the sweatshield reverse:
Top left corner, angling upward to the right "**Methmanns**" in cursive script and with the tail of the *S* as an underscore; to the lower right of the name is the shape of a cook's hat, with the words "**Nordmark Mütze**" in Fraktur font within the hat shape.
To the right of this, also angling upward from left to right, is the phrase "bürgt für Qualität"* in cursive script.

* [*stands*] *for quality*

VERIFIED PRODUCTS:

Schirmmützen:
Luftwaffe Other Ranks white top (Sommer) Extramütze

COLORS:
On private purchase grade Other Ranks caps:

Visor Cap Lining:	*Material:*
Luftwaffe: Deep gold	Rayon (ribbed)

Sweatband colors for Visor caps:	
Luftwaffe: Dark tan	Leather

Visor reverse Colors:
Luftwaffe: Green

GRADES:
None

PATENTS or D.R.G.M.:
None noted.

PECULIARITIES OF DESIGN (If Any):
None noted.

HISTORICAL BACKGROUND:
No historical data yet available

Confirmation Source:
Authentic example

FRANKFURT

COMPANY: *F. Wendl & Sohn*

BUSINESS ADDRESS:

WWII period:
F. Wendl & Sohn
Frankfurt (street address unknown)

OWNER or BUSINESS LEADER:
F. Wendl

CURRENT STATUS:
Unknown, but presumed out of business.

FOUNDED:
Unknown

TRADE NAME If Any:
None

SWEATSHIELD SHAPE:
Unknown

MAKERS MARKS, WITH LOCATION:
No authentic example of the Wendl makers mark is currently available.

VERIFIED PRODUCTS:

Schirmmützen (visor caps)

HISTORICAL BACKGROUND:
No information yet available

Confirmation Source:
Authentic example

FREIBURG-SCHLESIEN

COMPANY: *Ostland*

BUSINESS ADDRESS:

WWII period:
Ostland
Freiburg-Schlesien

OWNER or BUSINESS LEADER:
Unknown

CURRENT STATUS:
Unknown; presumed no longer in existence

The Ostland makers mark. This cap is without a sweatshield.

FOUNDED:
Unknown

TRADE NAME If Any:
None

SWEATSHIELD SHAPE:
On Extramützen: *Rhomboid*

MAKERS MARKS, WITH LOCATION:
Printed directly on the cap lining in black ink, "**Ostland**" in a large font, forming an arc. Beneath this a wide, bordered bar formed from a set of black-colored upper and lower lines (with space between them).
In the space between the lines, in small print: *"Mützen u. Bekleidungsindustrie"*
Third line (larger font size): "Freiburg-Schlesien"
Fourth line: (Year)
Fifth line (separate from the rest): Cap size

VERIFIED PRODUCTS:

Schirmmützen

HISTORICAL BACKGROUND:
No information available, however this company was located in the East and most likely did not survive the war.

Confirmation Source:
Authentic example

GLAUCHAU

COMPANY: *Felix Weissbach* Mützenfabrik*
*(also spelled: Weißbach)

BUSINESS ADDRESS:

WWII period:
Felix Weissbach Mützenfabrik
Glauchau/Sa. (street address unknown)

OWNER or BUSINESS LEADER:
Rudolf Felix Weissbach, Manufacturer
Felix Weissbach (Junior) Merchant, and
Karl William Johannes Weissbach, Merchant

CURRENT STATUS:
No longer in existence. The company was technically listed as 'closed' by 1939.

FOUNDED:
January 29, 1893

TRADE NAME If Any:
None

SWEATSHIELD SHAPE:
On Extramützen visor caps: *Rhomboid*

MAKERS MARKS, WITH LOCATION:
In gold ink printed on the cap lining:
Within a circle, a stylized eagle (outline), vertically oriented. Within the eagle, on two lines, the words '**Deutsche Wertarbeit**' [German Value Craftsmanship] in Fraktur font (German gothic).
First line below the circle/eagle, the name "*Felix Weissbach*" in cursive script.
Second line: "Glauchau" | Sa."
Third line: "Lieferjahr" with the year
On some caps, the word "Frischluft" may also appear. This may indicate the cap was manufactured using the HPC Frischluft system (under license).

VERIFIED PRODUCTS:

1) *Schirmmützen:*
Army Officer/Other Ranks Extramütze
Luftwaffe Other Ranks Extramütze

2) *Sommermützen:* Luftwaffe white top summer caps

The Felix Weissbach makers mark. The "Lieferjahr" in this case, is 1938 (see the historical background details below).

COLORS:
On private purchase grade officer/Other Ranks caps:

Visor Cap Lining: *Material:*
Army: Reddish tan (champagne)   Other Ranks: Chintz (waterproof)
Luftwaffe: Orange-tan (fliegerblau cap)

Sweatband colors for Visor caps:
Army: Tan   Leather; composition material
Luftwaffe: Tan

Visor reverse Colors:
Army: Reddish tan
Luftwaffe: Green (standard)

GRADES
No specific grades noted

PATENTS or D.R.G.M.:
D.R.G.M. No. **1 457425** Klasse 41b. "*Stütze für Uniformmützen*". Weissbach's D.R.G.M. covered the company's version of a uniform cap *Stütze* (internal crown support post) for use on Luftwaffe white summer caps (removable top). The Gebrauchsmuster was issued in 1939.

PECULIARITIES OF DESIGN (If Any):
None noted.

HISTORICAL BACKGROUND:
Rudolf Felix Weissbach opened his business prior to 1900; records indicate the year was 1893. The firm's business activity was identified as 'The manufacture and sale of patented fold-up caps, silk piping; ventures in associated undertakings." Cap-making was part of Weissbach's original business, but the company apparently involved itself in a number of associations with other businesses in related products.

Uniformen-Markt Advertisement for Felix Weissbach from a
1935 issue. The text reads:

[*Within the eagle logo:*] German Value Craftsmanship
Uniform caps
RW-Extramützen *
Party caps
Felix Weissbach, Mützenfabrik, Glauchau/Sa.

*RW= Reichswehr

Weissbach's company either started as, or was converted to an *Aktiengesellschaft* (stock company) early in its history and appears to have remained one throughout its active life. It did well for a time with Rudolf Felix running the firm together with his son Felix (Jr.), and another Weissbach family member (exact relationship unknown).

Surviving the economic hardships of the post World War I period and the Weimar Republic, the company was entered in the Glauchau Handelsregister (commercial registry) on November 22, 1928. Its actual financial condition, however, seems to have been sailing through rough waters even earlier: At least two times, the firm had already been forced to reduce its operating capital by a considerable amount – perhaps to buy back shares; it also revalued its stock. By 1933, the situation had reached such a critical point that a general meeting of shareholders voted to dissolve the company's management system and charter – and with that, eventually the company itself. This vote and decision was noted in the Handelsregister on July 5, 1933. Felix Weißbach Jr. was assigned as one of two *Abwickler* [liquidators]; this position apparently entailed either selling off the firm's physical assets, or divesting it of loss-incurring business departments/affiliations (or both); the records do not make clear which was the case. The liquidation proceeded quite slowly, however, and was even halted for a time after the death of founder Rudolf Felix Weissbach on February 27, 1935.

Throughout the liquidation period and the break following Weissbach's death, the cap-making part of the company at least seems to have remained fully active. Perhaps it was independent; the cap-making industry was certainly booming for everyone at this time, and it seems illogical that this part of the business would be losing any money. A Weissbach cap advertisement appeared in the September 1935 issue of *Uniformen-Markt*, which gives no hint of any problems.

The Handelsregister file records the official dissolution of the firm Felix Weissbach – at least as a stock company – on March 31, 1936, and also notes the resumption of liquidation activities on July 26, 1937. Yet again, the cap-making business was still operating as late as 1939, which another ad in the *Uniformen-Markt* paper from 1939 clearly shows (exact same format as the 1935 ad). In addition, the company's Gebrauchsmuster application for a "crown support" system used for Luftwaffe removable white top summer visor caps was finally approved and issued by the Reichspatentamt in 1939. A bit too late it turns out – since the production of these caps was suspended by order of the military for the duration of hostilities which, of course, began in 1939.

No further entries occur in the Handelsregister since the company's file was officially terminated with its dissolution in 1936. No Gewerbeamt [Business Bureau] records for the firm have yet been found; the fate of the cap-making factory remains unknown. It is not likely to have survived the war, but if it did, its location in postwar communist East Germany would have spelled its eventual doom in any case (prior to 1955).

Confirmation Source:
1. *Uniformen-Markt*
2. Bundespatentamt
3. Staatsarchiv Dresden (Glauchau) [State Archive]
4. Authentic examples

HAMBURG

COMPANY: *Gustav Oelkers Mützenfabrik*

BUSINESS ADDRESS:

WWII period:
Gustav Oelkers
Hamburg 11, Rödingsmarkt 19/20

OWNER or BUSINESS LEADER:
Unknown; Presumably Gustav Oelkers himself at founding. Available records show only Herbert Erich Sylvester Oelkers as *Einzelprokura* [head clerk or deputy] as of 1940.

CURRENT STATUS:
Uncertain, but presumably no longer in existence.

FOUNDED:
Unknown

TRADE NAME If Any:
None

SWEATSHIELD SHAPE:
Unknown

MAKERS MARKS, WITH LOCATION:
Unknown, no authentic example available.

VERIFIED PRODUCTS:

Schirmmützen:
• for the Armed Forces (assume all services), in particular caps for the Kriegsmarine
• Party and organization caps
• Other government organizations (non-military)

PATENTS or D.R.G.M.:
None

HISTORICAL BACKGROUND:
Very little information is available for the Gustav Oelkers company beyond occasional advertisements appearing in the *Uniformen-Markt* industry (trade) paper. No authentic example of a Gustav Oelkers cap has been reported to date.

   The company was probably not very large, nor does it ever seem to have qualified for entry in the Hamburg Handelsregister (no records have been found thus far). The local Gewerbeamt [Business Bureau] records for the pre-war period no longer exist. In any case, there is no current listing in Hamburg for any company under this name.

Confirmation Source:
1. Amtsgericht Hamburg
2. Stadtarchiv Hamburg
3. *Uniformen-Markt*

Gustav Oelkers advertisement from the October 1, 1936, issue of *Uniformen-Markt*. The text reads:

Uniform Caps
for the Wehrmacht,
NSDAP and all National Socialist
associations.
Furthermore,
a large selection of
blue caps.

Note: "Blue caps" refers to the blue-top Kriegsmarine visor cap.

COMPANY: *Opolka & Müller Mützenfabrik*

BUSINESS ADDRESS:

Pre-war and WWII period:
Opolka & Müller Mützenfabrik
Hamburg 15, Haus Hammaburg

(by 1941):
Opolka & Müller Mützenfabrik
Hamburg 1, Spaldingstr. 210/212

OWNER or BUSINESS LEADER:
L. and Heinrich Leonhard Winter, merchants (wartime, with Heinrich continuing into the postwar period)

Subsequent: (from 1966)
Jürgen Hische, merchant (who apparently first started with the company as the *Prokura* [head clerk/deputy])

CURRENT STATUS:
Presumed out of business. No current records for the firm, or any descendant company.

FOUNDED:
August 16, 1901 (according to *Uniformen-Markt*, which records the company's 40 year anniversary as being August 16,1941). The company itself listed 1903 as its founding date.

TRADE NAME If Any:
None

SWEATSHIELD SHAPE:
Unknown

MAKERS MARKS, WITH LOCATION:
To date, no authentic example of an Opolka & Müller cap is available.

VERIFIED PRODUCTS:

Schirmmützen. Army Other Ranks Extramütze

GRADES:
No specific grades noted

PATENTS or D.R.G.M.:
Unknown

CERTIFICATIONS:
R.Z.M. [License] No. **60**

PECULIARITIES OF DESIGN (If Any):
None noted

HISTORICAL BACKGROUND:

This Hamburg company was founded in 1903 (according to the company). The origin of the name is not clear; the only available records of the company present only the two Winters gentlemen L. and H. Winters, of which the person identified as *H.* was presumably Heinrich Leonhard Winters.

The firm achieved entry in the Handelsregister for the first time on January 5, 1904, and apparently did reasonably well for itself, despite the difficulties all companies faced after the end of The Great War (World War I) and the ensuing economic turmoil of the 1920s and early 1930s. Under the Third Reich, Opolka & Müller produced military caps for the revitalized Armed Forces. This continued throughout the war, with the firm somehow managing to remain intact despite the punishing bombing raids made on Hamburg. The company relocated within the city sometime after 1935 (but before 1941) from it original business premises at *Hamburg 15*, Haus Hammaburg, to *Hamburg 1*, Spaldingstr. 210/212 (perhaps this was expansion into a larger facility). The war finally came to a close, with Opolka & Müller still operational – or relatively so.

In 1946, the company's business activity began afresh, though it is not certain whether it continued on as a *Mützenfabrik* or switched to a different – but related – business area within the clothing industry.

In either late 1965 or early 1966, Heinrich Leonhard Winter passed away and direction of the company was taken up by Jürgen Hische, who had begun as the firm's head clerk (Prokurist). It was not possible to hold both positions simultaneously, and Hische therefore resigned as clerk to take over as either the company owner, or the personally liable (financially, that is) chief officer. Hische remained in place until the company finally closed.

It was officially stricken from the active Handelsregister listing on November 28, 1966, though it may or may not have continued in business for a few years more. No current records for any contemporary firm exist, listed under this name. It seems likely that Opolka & Müller closed its doors for the last time at a date not much later than its 1966 departure from the Handelsregister.

Confirmation Source:
1. Amtsgericht Hamburg
2. Stadtarchiv Hamburg
3. *Uniformen-Markt*

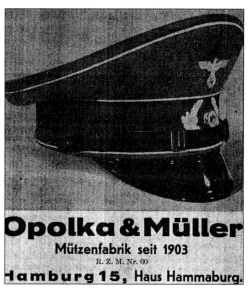

Advertisement for Opolka & Müller from the August 1, 1935, issue of *Uniformen-Markt.* The text reads:

Opolka & Müller
Cap Factory since 1903
R.Z.M. No. 60
Hamburg 15, Haus Hammaburg

COMPANY: *Steinmetz & Hehl Mützenfabrikation*

BUSINESS ADDRESS:

Pre-and WWII period:
Steinmetz & Hehl Mützenfabrik
Hamburg, Stubbenhuk 25

Postwar:
Steinmetz & Hehl GmbH & Co.
Marine-Uniformen Tropen- u. Wassersportbekleidung
Rödingsmarkt 20
D-20459 Hamburg

OWNER or BUSINESS LEADER:

Prewar:

1. Hinrich Wilhelm Martin Hehl and Claus August Gustav Hehl, merchants in Hamburg
2. Maria Hehl (neé Schulten) – widow upon Gustav's death in late 1944

Postwar (1945 -):
1. Martin Hinrich Karl Uwe Hehl and Jürgen Klaus Heinrich Hehl, merchants, (the latter from April 26, 1967) – probably the older and younger sons of either Hinrich or Claus
2. Jürgen Klaus Heinrich Hehl (after Martin's death in 1988) and Felicitas-Eike Hehl (neé Hellwig) as Kommanditistin (see historical background info below)
3. Hehl Verwaltung GmbH [a management company, Hehl subsidiary] (1991) with degreed Engineer Karl-Heinz Erren and Edith Erren (neé Kremer) in the position of Kommanditisten [limited partners] from 1996 and Dagmar Erren in place of Karl-Heinz in the year 2000 (see historical background below).

CURRENT STATUS:
Insolvent, business activities restricted, as of 2000. Now closed.

FOUNDED:
1873

TRADE NAME, If Any:
None

SWEATSHIELD SHAPE:
On Extramützen visor caps (Kriegsmarine): *Diamond, horizontally positioned*

Typical Steinmetz & Hehl makers mark, this one in a Kriegsmarine admiral's visor cap. *Courtesy of Bill Shea.*

MAKERS MARKS, WITH LOCATION:
Large, thick, highly visible upward arcing letters in white ink on the cap lining: "Steinmetz & Hehl". The font is unusual – blocky with curls and a slight gothic look. Second line: "Hamburg"
Third line: "Stubbenhuk 25"

VERIFIED PRODUCTS:

Schirmmützen: Kriegsmarine (Navy) Officer and Other Ranks Extramütze

COLORS
On private purchase grade officer caps:

Visor Cap Lining:	*Material:*
Kriegsmarine: Dark blue	Rayon or silk in a flat weave.
Sweatband colors for Visor caps:	
Kriegsmarine: Tan	Leather
Visor reverse Colors:	
Kriegsmarine: Black	

GRADES:
No specific grades noted

PATENTS or D.R.G.M.:
None

PECULIARITIES OF DESIGN (If Any):
None noted

HISTORICAL BACKGROUND:
Steinmetz & Hehl opened in Hamburg in 1873. The original founders must have included a Steinmetz, but the identity of that individual is lost; the business eventually came into complete possession of the Hehl side of the arrangement, and remained associated with that family in some fashion. The company's original business activity is also nowhere identified, but cap-making played a large part in it, most likely as the root activity. The company was successful enough to achieve registration in the city's Handelsregister by February 18, 1875 – a scant two years after its founding.

Circumstantial evidence suggests that Steinmetz & Hehl was heavily oriented toward Hamburg's maritime industry from early on – not surprising in a city with an active port and considerable shipping activity. The company produced caps for naval officers and sailors in particular. The First World War did not have any major effect on the firm and it also successfully weathered the twenty one years leading up to World War II. With the rise of the Third Reich, Steinmetz & Hehl's business increased and it manufactured caps for other military services – though it remains most noted for its Kriegsmarine "blaue Mützen" ([non-removable] blue-top visor caps) for officers and Navy NCO ranks. Examples of Steinmetz & Hehl caps for other services (that is, other than Navy caps) rarely ever appear on the militaria market.

The company survived the destruction of World War II, and soon took up business again. The firm's Handelsregister entry remained open. Martin Hehl led the company through the difficult years of the late 1940s and early 1950s as the company – and the country – struggled through the aftermath of the war. Its business in the immediate postwar years is uncertain, but perhaps it supported itself making civilian caps as well as caps and uniforms for maritime marine personnel. At some point the company also turned to tropical clothing and marine sportswear, a line which may have kept it afloat while its remaining local peers (except for Willy Sprengpfeil) were slowly sinking.

With the death of Martin Hehl some time around 1988, Jürgen Hehl – most likely Martin's son – took charge of the company as the business owner with financial responsibility for the firm. He was joined by Felicitas-Eike Hehl (probably his wife), who took the position of Kommanditist [limited partner] and with this, the company's business form was changed to a **K**ommanditg**e**sellschaft (a limited partnership company), and the firm renamed Steinmetz & Hehl *KG*. The company moved forward into the 1990s, and in 1991, financial responsibility for the company was assumed by an offspring firm, Hehl Verwaltungs GmbH [Hehl Management L.L.C.]. Jürgen resigned as the financially responsible owner and took a new position as Kommanditist. These last two original Hehls departed the company in January of 1996 – perhaps to retire. In their stead came Karl–Heinz and Edith Erren who served as limited partners until Karl-Heinz Erren's death in late 1999 or January of the year 2000. Daughter Dagmar Erren (born 1963) took his place; this is the latest entry for the company logged in the Handelsregister (February 2, 2000). There was no longer a Hehl at the helm. Sometime during the year 2000, an insolvency investigation of the company was begun. As a result, its business activities were restricted, and the company eventually closed.

Confirmation Source:
1. Amtsgericht Hamburg [Handelsregister], *Justizangestellter Peter Krause*
2. Staatsarchiv [State Archive], *Herr Ulf Bollman*
3. Authentic examples

Willy Sprengpfeil maker stamp on the lining of a field gray Army M34 cap. *Courtesy of Bill Shea.*

COMPANY: *Willy Sprengpfeil Mützenfabrik*

BUSINESS ADDRESS:

WWII period:
1. **Willy Sprengpfeil Mützenfabrik**
Hamburg 19, Eduardstr. 46-48

Postwar: (from 10/28/83)
Willy Sprengpfeil Hut- und Mützenfabrik
(Post code unknown) Hamburg, Conventstraße

OWNER or BUSINESS LEADER (financially liable):
1. Richard Friedrich "Willy" Sprengpfeil, manufacturer, and Paul Alfred Otto Sprengpfeil, merchant
2. Richard Friedrich Willy Sprengpfeil (from 1921) sole owner
3. Richard Friedrich Willy Sprengpfeil and Hans Gloy, merchant (the latter from 1939)
4. Richard Friedrich Willy Sprengpfeil (sole owner again)
5. Minna Caroline Louise Sprengpfeil [neé Bohlmann] (from 1956)
6. Dr. Werner Oetker, merchant
7. Minna Caroline Louise Sprengpfeil and Karl Klan, merchant [from Stuttgart] after the death of Werner Oetker in 1958.
8. Dr. Günter Betthäuser, merchant (from 1961, after Minna Sprengpfeil's death and Karl Klan's departure)
9. Dr. Betthäuser Beteiligungsgesellschaft* with limited liability
10. Others from the Betthäuser family

[*Holding Company]

CURRENT STATUS:
Unknown. The company appears in Handelsregister records as late as 1996, and the company's registry file was never officially closed, however the firm does not appear in either current Hamburg business listings or phone books, and is therefore presumed to be out of business.

FOUNDED:
1911 in Hamburg. With its entry in the Hamburg Handelsregister on March 12, 1919, the company changed its business form to a general partnership company *Offene Handelsgesellschaft* (officially from March 1, 1919).

TRADE NAME (logo) If Any:
Logo (see below)

SWEATSHIELD SHAPE:
Unknown

MAKERS MARKS, WITH LOCATION:
Though Sprengpfeil makers mark stamps in ink appear from time to time on field caps and the company's ads certainly indicate that it produced a full range of visor caps for the military services and non-military organizations, no authentic visor cap example (whether an Extramütze, or contract cap) made by Willy Sprengfeil has yet been encountered.

VERIFIED PRODUCTS:

1) *S**chirmmützen:***
- Army Officer/Other Ranks Extramütze; Contract caps (Government Issue)
- Luftwaffe officer /Other Ranks Extramütze
- Kriegsmarine Extramütze
- Party caps (non-military) [under RZM auspices]

2) ***Einheitsmütze M34*** *(government contract)*

COLORS:

Visor Cap Lining: *Material:*
Unknown

Field Cap Linings:
Army M34: Gray   Cotton twill (herringbone)

GRADES:
No specific grade names noted

PATENTS or D.R.G.M.:
None

CERTIFICATIONS:
R.Z.M. License **A 1/84**

PECULIARITIES OF DESIGN (If Any):
None noted

HISTORICAL BACKGROUND:
First opened in 1911 under the original name of *Gebrüder Sprengpfeil* (the Brothers Sprengpfeil), the firm did well in business and succeeded in being accepted for entry in the Hamburg Handelsregister eight years later in 1919. The company weathered the initial post World War I years and the following difficult decade with no major crisis. By 1921, Richard F. "Willy" Sprengpfeil had assumed sole control over the company (there is no indication of what happened to Paul Alfred Otto), and it was he who carefully steered the firm through the difficult economic period of the 1920s and into the Third Reich era. German remilitarization brought with it a return to a very brisk business for the company; no doubt Herr Sprengpfeil had his hands full.

Particularly interesting in this ad (shown at right) is the ADEFA mark. Rarely seen in advertisements, this mark indicated that the goods bearing it were certified as having been produced by "Aryan" workers – therefore, presumably an "Aryan" company. Exactly what agency controlled the ADEFA certification is not known.

In 1939, Willy Sprengpfeil welcomed another Hamburg merchant named Hans Gloy as either a partner, or as a senior company officer sharing in the firm's financial liability (in this system, were it to go under, the financially liable owner/partner(s) would be held fully accountable for debts). Gloy apparently died in late 1942 or early 1943, however, and Willy Sprengpfeil was once again solo, leading the company through the end of the Second World War and on into the postwar years. A devastated early postwar economy notwithstanding, Sprengpfeil started rebuilding the company while Hamburg began clearing away the war rubble and began its own process of rebuilding. It is not known who Sprengfeil's customers were during this time frame, only that he chose not to cease cap production or shift to some other, related industry. Since a na-

Sprengpfeil advertisement from the October 15, 1938 issue of *Uniformen-Markt.* Note the detailed "Fabrik Marke" [Factory trademark] symbol. The text reads:

WILLY SPRENGPFEIL, Mützenfabrik
Hamburg 19, Eduardstrasse 46/48
Army Extramützen
Airman Extramützen *[i.e. Luftwaffe]*
Navy Extramützen
Party Extramützen
and all other uniform caps
[Sprengpfeil logo 'Fabrik Marke']
[ADEFA "The sign for goods made by Aryan hand"]
R.- Z.-.M. – License
A 1/84
of the best quality and workmanship

*M*einen geschätzten Abnehmern die besten Wünsche für ein glückliches und erfolgreiches neues Jahr!
Willy Sprengpfeil
Hamburg 19 - Eduardstraße 46—48

tional military force no longer existed, he likely moved to civilian production while also servicing the merchant marine and the needs of local government agencies (police, fire departments, etc.).

With Willy Sprengpfeil's death sometime prior to 1956, Minna Caroline Louise Sprengpfeil stepped into the leading role. Although the term "Witwe" [widow] is not specifically noted in the 1956 Handelsregister entry recording this change of ownership, it seems logical to assume that she was his wife. From this point on, the company went through a number of other owners and directors, changing also to a Kommanditgesellschaft [limited partnership company] (KG) in 1957. Minna left the company with her death around 1961. With her passing the Sprengpfeil reign ended; there was never to be another Sprengpfeil at the helm of the business – but the company did move on.

In fact, the firm was apparently doing quite well: In 1964, it expanded with a branch in Berlin, and in 1969 with a second branch in Lindberg/Allgäu. Over the long term, however, these branches were apparently not sustainable. The Berlin branch was closed first, in 1972 (after only eight years in operation), followed by the Lindenberg/Allgäu branch a decade later, in 1981. No reasons for the closings are given in the records; one can only assume that the branches had become financial liabilities rather than assets. The main company in Hamburg survived the closings – it apparently shed these branches in time to avoid any serious financial crisis.

In a 1984 Handelsregister notation of the company's restructuring into a GmbH company [limited liability company – i.e. L.L.C.], the firm's business activity is identified as:

a.) Hat and cap factory Willy Sprengpfeil GmbH.,
d.) Hamburg
e.) (1) The manufacture of hats and caps and all forms of business associated with this.
(2) The company may take part in other business of the same or similar type...and may also set up branch offices.

The last official entry in the Handelsregister file for Willy Sprengpfeil Hut- und Mützenfabrik GmbH is dated 1996. Although the file was never officially closed out and terminated by the responsible Amtsgericht, the company disappears from other Hamburg public records not long after the date of this entry and is no longer listed in either the city's business records, directories, Address Book or phone books. Its sudden disappearance and apparent demise remain unclarified.

Confirmation Source:
1. Amtsgericht Hamburg
2. Staatsarchiv [State Archive], *Herr Ulf Bollman*
3. *Uniformen-Markt*
4. Authentic examples (field caps only)

HANAU

COMPANY: *F. Eckhard*

BUSINESS ADDRESS:

WWII period:
F. Eckhard
Hanau (street address unknown)

OWNER or BUSINESS LEADER:
Unknown

CURRENT STATUS:
Unknown

FOUNDED:
Unknown

MAKERS MARKS, WITH LOCATION:
Makers mark on the cap lining:

First line: "**F. ECKHARD HANAU**"
Second line: "Stirndruckfrei"
Third line: "Deutsches Reichspatent"

VERIFIED PRODUCTS:

Schirmmützen (Private Purchase and government contract [issue]):
• Army Other Ranks Extramütze

GRADES:
Unkown

PATENTS or D.R.G.M.:
Unknown

HISTORICAL BACKGROUND:
None yet available.

Confirmation Source:
Authentic example

HANNOVER

COMPANY: *C. Louis Weber Uniformen*

BUSINESS ADDRESS:

WWII period:
C. Louis Weber Uniformen
Hannover M, (street address unknown)

Standard C. Louis Weber makers mark on an Army officer's visor cap (Extramütze). *Courtesy of Gerard Stezelberger of Relic Hunter.*

Postwar:

C. Louis Weber
Hannover

OWNER or BUSINESS LEADER:
Unknown

CURRENT STATUS:
Unknown

FOUNDED:
1862

TRADE NAME (logo) If Any:
C and *L* to the left and right of the central arm of a larger, rounded *W*, all within a silver disk.

SWEATSHIELD SHAPE:
On Extramützen: *Rhomboid*

MAKERS MARKS, WITH LOCATION:
In silver ink printed on the sweatshield reverse, the circular C.L. Weber logo; beneath this:
First line: "C. Louis Weber" in cursive script, with "Weber" underlined;
Second line: "UNIFORM-FABRIK" printed in capital letters;
Third line: "HANNOVER" printed in capital letters.

VERIFIED PRODUCTS:

Schirmmützen*: Army officer Extramützen*

COLORS:
On private purchase grade officer caps:

Visor cap lining:	*Material:*
Army: Pale gold	Rayon or silk

Sweatband	
Army: Tan	Leather

Visor reverse Colors:
Army: Tan

GRADES:
No specific grade names noted.

PATENTS or D.R.G.M.:
None

PECULIARITIES OF DESIGN (If Any):
None noted

HISTORICAL BACKGROUND:
The C.L. Weber company was a successful uniform tailoring business that was founded in 1862, and it was a natural progression for the firm to include uniform caps in its

inventory. Many tailors catering to the needs of military personnel probably hired a professional, experienced cap maker and a seamstress or two, who could produce high quality "Extra"-grade uniform caps for their customers. These businesses were not usually large enough (unless extremely successful) to get involved with mass production operations required for contract cap manufacturing. Judging from C.L. Weber's advertisements, this company may have been one of the exceptions, however: large enough to maintain an "on-hand" stock of caps (as opposed to *made to order*), and perhaps even large enough to fill small-sized government contracts as a member of a LAGO, or AGO association. In any case, the company surely did a good business in high-quality Extramützen.

C. Louis Weber's main business premises were located in Hannover, but the company also had branch shops in Hamburg and Kassel. Unfortunately, no other information about the company or the branches is available: There has not yet been any response to several requests for information made to the appropriate government offices in Hannover. Whether any records for the company exist, remains unknown. Is the company still in business today? Though it is likely that it survived the end of the war and its aftermath (a tailoring business, after all, can certainly produce civilian clothing just as easily as uniforms), the question of whether it still exists today remains unanswered.

Confirmation Source:
1. *Uniformen-Markt*
2. Authentic examples

Advertisement for the C. Louis Weber company in the *Uniformen-Markt* trade paper Issue 1, January 1, 1937 noting the company's 75th anniversary (1862-1937).

COMPANY: Georg Grote

BUSINESS ADDRESS:

WWII period:
Georg Grote
Hannover 1, (street address unknown)

OWNER or BUSINESS LEADER:
Unknown

CURRENT STATUS:
Unknown

FOUNDED:
Unknown

TRADE NAME (logo) If Any:
None

SWEATSHIELD SHAPE:
Unknown

The Kurt Dallüge makers mark (the cap in this example is non-military).

MAKERS MARKS, WITH LOCATION:
No example of a Georg Grote makers mark is currently available.

VERIFIED PRODUCTS:

Schirmmützen: Extramützen and government contract [issue] visor caps

GRADES:
No specific grade names noted

PATENTS or D.R.G.M.:
None

PECULIARITIES OF DESIGN (If Any):
None noted

HISTORICAL BACKGROUND:
Information on this company is not yet available.

Confirmation Source:
1. *Uniformen-Markt*
2. Authentic examples

COMPANY: *Kurt Dallüge*

BUSINESS ADDRESS:

WWII period:
Kurt Dallüge
Osterstr. 2 – 3/Schmiedstr. 38 39

OWNER or BUSINESS LEADER:
Unknown

CURRENT STATUS:
Unknown

FOUNDED:
Unknown

TRADE NAME (logo) If Any:
None

SWEATSHIELD SHAPE:
On Extramützen: *Rhomboid*

MAKERS MARKS, WITH LOCATION:
Printed in black ink directly on the cap lining, a twelve-pointed star surrounding a wide ring with a disk at its center (the disk in the lining color). Within the disk, a circle enclosing the letters *ED*, topped by a symbol. Between the inner and outer edge of the ring are the words '**MARKE DALLÜGE**' in capitals, with 'Marke' filling the ring's top arc, 'Dallüge' the bottom arc.

Beneath the star symbol, first line: "**DALLÜGE**" in large font size
Second line: "Hannover"
Third line: "Oesterstr. 2-3"
Fourth line: "Schmiedstr. 38 39"

VERIFIED PRODUCTS:

Schirmmützen:
• Army Officer Extramütze
• Waffen-SS Officer Extramütze
• Organizational (non-military) caps, e.g. Hitler Jugend Officer Extramütze

GRADES:
No specific grade names noted

PATENTS or D.R.G.M.:
None

PECULIARITIES OF DESIGN (If Any):
None noted

HISTORICAL BACKGROUND:
Information on this company is not yet available.

Confirmation Source:
Authentic example

HEILBRONN

COMPANY: *Wilh. Stellrecht Mützenfabrikation*

BUSINESS ADDRESS:

WWII period:
Wilh. Stellrecht Mützenfabrikation
Heilbronn, Wilhelmstr. 30

OWNER or BUSINESS LEADER:
Unknown; presumably Wilhelm Stellrecht at startup.

CURRENT STATUS:
Unknown

FOUNDED:
Unknown

TRADE NAME (logo) If Any:
None

SWEATSHIELD SHAPE:
On Extramütze caps: *Rhomboid*

Stellrecht ink stamp makers mark on a visor cap lining.

MAKERS MARKS, WITH LOCATION:
Ink stamped on the cap lining:

First line: "**Wilh. Stellrecht**"
Second line: "Mützenfabrikation"
Third line: "Stirndruckfrei"
Fourth line: "Heilbronn, Wilhelmstr.30"
Stellrecht makers mark stamps appear in black ink on the lining of caps made under another maker's name (i.e. under license). No authentic example of Wilhelm Stellrecht's *own* makers mark (i.e. a Stellrecht cap not bearing another firm's label) is currently available.

VERIFIED PRODUCTS

Schirmmützen: Extramützen and government contract [issue] visor caps

GRADES:
No specific grade names noted

PATENTS or D.R.G.M.:
None; Used the HPC "Frischluft" system under license from HPC (Hermann Potthoff, Coesfeld)

PECULIARITIES OF DESIGN (If Any):
None noted

DISTRIBUTORS:
None; Stellrecht, however, appears to have served as a distributor/licensee for **HPC**.

HISTORICAL BACKGROUND:
No historical information on Wilhelm Stellerecht Mützenfabrikation is yet available.

Confirmation Source:
Uniformen-Markt

INNSBRUCK (AUSTRIA)

COMPANY: *Hermann Gollhofer Kappenerzeugung*
[cap production]

BUSINESS ADDRESS:

WWII period:
Hermann Gollhofer
Innsbruck, Anichstr. 5
(Österreich) Austria

OWNER or BUSINESS LEADER:
Unknown, presumably Hermann Gollhofer

CURRENT STATUS:
Unknown

SWEATSHIELD SHAPE:
Unknown

MAKERS MARK, WITH LOCATION
In silver ink printed on the sweatshield reverse;
Line one: "**Herm. Gollhofer**"
Line two: "Kappenerzeugung"
Line three: "Innsbruck"
Line four: "Anichstr. 5"

VERIFIED PRODUCTS:

Schirmmützen:
• Army Other Ranks Extramütze caps

COLORS:
On private purchase grade caps:

The only known example of a Hermann Gollhofer makers mark.

Lining:	*Material:*
Army: Purplish gray	Cotton (linen weave)

Sweatband:	
Army: Reddish tan	Leather (*V*-style attachment)

Visor reverse Colors:
Army: Yellowish tan

PATENTS or D.R.G.M.:
None

HISTORICAL BACKGROUND:
No information yet available from Innsbruck.

Confirmation Source:
Authentic example

KASSEL

COMPANY: *Kurt Triebel Erste Kasseler Mützenfabrik*
(First Kassel Cap Factory)

BUSINESS ADDRESS:

WWII period:
Kurt Triebel Uniformmützenfabrik
Kassel, Hohenzollerstr. 87

OWNER or BUSINESS LEADER:
1. Wilhelm Reitz (as a *Kürschnerei* [furrier] business)
2. Kurt Triebel (born November 28, 1903)

CURRENT STATUS:
Destroyed in an Allied bombing attack in late 1943 or early 1944.

FOUNDED:
1906

TRADE NAME if any:
None

SWEATSHIELD SHAPE:
On Extramützen: *Rhomboid*

MAKERS MARK, WITH LOCATION:
No example available

VERIFIED PRODUCTS:

Schirmmützen:
• Army Other Ranks contract (Government Issue) caps

PATENTS or D.R.G.M.:
None

PECULIARITIES OF DESIGN (If Any):
None noted.

COMPANY DATA
No specific data available.

HISTORICAL BACKGROUND:
The history of Kurt Triebel Uniformmützenfabrik is particularly interesting as this company is one of only a few cases in which destruction through Allied combat actions during the Second World War can actually be confirmed.

Although a Mützenmacher [cap maker] by trade, Mr. Wilhelm Reitz established his company instead as a furrier business in 1906 at Königstraße 72 in Kassel. This was quite a common practice, since both furrier activities and cap-making were related handcrafts.

The city's official Address Book records the company for the first time in this year. In 1913, Herr Reitz moved the firm's business premises to a new location at Hohenzollerstraße 113, where it remained for the next twenty-five years.

At some point during the 1920s, a young man named Kurt Triebel made acquaintance with the aging Reitz. A trained furrier in his own right, Triebel (in his mid-twenties) likely took a job with the company in some capacity. It seems he also took a liking to Reitz' daughter Marie (or perhaps it was through Marie that he first met Mr. Reitz), and the two were married on December 19, 1928. Wilhelm Reitz turned the reins of the company over to his son-in-law in 1929 and passed away the following year. The first entry for the company (still as a furrier business) under the Kurt Triebel name appears in the 1930 edition of Kassel's city Address Book. Kurt wasted no time implementing his own plan for the company, and immediately converted it from a straight fur business to cap production in 1931. From 1933 through 1937 the company listed its business activity as "Cap manufacture and Furrier". In 1938, Triebel moved the business to a new facility at Hohenzollerstraße 87, and from this point on he identified the business (officially) as a "Uniformmützenfabrik" (uniform cap factory). In addition, somewhere along the line he added the title of *Erste Kasseler Mützenfabrik,* The First Kassel Cap Factory; there never seems to have been a second.

The early war years were productive for the company, which secured (or was assigned) at least some government contracts – the Triebel maker stamp on a Government Issue (Other Ranks) cap identifies the firm as a *Lieferant* – contractor (or supplier). Unfortunately, no records pertaining to the period from 1940 through the end of the war are available for the company. This is due in part to the fact that no city Address Books were kept from 1940 until 1949, and also due to the destruction of the city's Handelsregister [commercial register] on October 22,1943 (presumably through fires caused by an air raid).

Nonetheless, the Kurt Triebel Uniform Cap Factory seems to have done well for itself until late 1943, when the war landed directly in Triebel's lap. Though no direct records exist, detailed cross-referencing by Kassel's dedicated city archivist, Mr. Klaube, turned up a map of damaged buildings and houses in the city's old, war-period record files. This map identifies buildings that were damaged in air raids which took place in late 1943 and early 1944. The building at Hohenzollerstraße 87 is identified as having been "heavily damaged", and the map notes specifically mention a workshop located in a side wing that was completely destroyed – most certainly the Kurt Triebel factory.

In addition, a check of Kassel city resident registration card files for 1943 and 1944 produced a record of the Triebel family's having departed Kassel in February of 1944 – as homeless refugees, perhaps, looking for a safer haven in which to wait out the end of the war?

These two mutually supporting records are convincing evidence that Triebel's factory at Hohenzollerstraße 87 was a total loss after being hit by bombs during an Allied air raid on the city at the end of 1943 or in January of 1944. Whether Kurt Triebel merely lost all of his assets, or also his life along with the factory, is not known. The Triebel family at least, left the city behind and with it, the history of Kassel's First (and last) Uniform Cap Factory.

Special thanks to Herr Klaube of the Kassel City archive.

Confirmation Source:
1. Kassel City Archive
2. Kassel Einwohnerbuch [Resident Address Book] 1939 and 1940

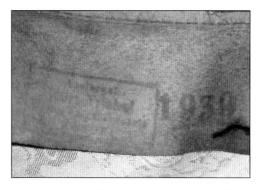

No example of a Kurt Triebel-made Extramütze has yet been encountered. This faint ink stamp on the sweatband reverse of an Other Ranks Government Issue visor cap is one of very few Triebel marks that have been seen. The mark reads: "Lieferant' [contractor] Kurt Triebel Uniformmützenfabrik KASSEL".

KIEL

COMPANY: *August Geiger Mützenfabrikation*

BUSINESS ADDRESS:

WWII period:
August Geiger Mützenfabrikation
Kiel, Dänische Str. 24

OWNER or BUSINESS LEADER:
Unknown; original owner presumably August Geiger

CURRENT STATUS:
Unknown

FOUNDED:
1872

MAKERS MARKS, WITH LOCATION:
Unknown; No authentic example available.

Geiger advertisement from the October 1, 1935 issue of *Uniformen-Markt.* The text reads:

Aug. Geiger
Kiel
Dänische Str. 24, founded 1872.
Uniform Caps
of All Types,
Quick and moderately priced.

VERIFIED PRODUCTS:
Schirmmützen: Visor caps

HISTORICAL BACKGROUND:
No information as yet on August Geiger has been received from the city of Kiel.

Confirmation Source:
Uniformen-Markt

KOBLENZ a/Rh.

COMPANY: *Almi Uniformen-Mützen*

BUSINESS ADDRESS:

WWII Period
Almi Uniform- und Mützenfabrik
Koblenz, Schützenstr. 20

OWNER or BUSINESS LEADER:
1. Alfred Mieß
2. Elisabeth Josefine Mieß (née Skupnik) – widow and heiress

FOUNDED:
Unknown

CURRENT STATUS:
Does not appear to have survived the end of the war

TRADE NAME (logo) if any:
Almi

MARKS, WITH LOCATION:

First line: "**ALMI UNIFORMEN MÜTZEN**"
Second line: "**KOBLENZ**"
No example of an authentic **Almi** makers mark is currently available.

VERIFIED PRODUCTS

Schirmmützen: Visor caps

HISTORICAL BACKGROUND
The company owner, Alfred Mieß, followed the German penchant for acronyms and used the *Al* from Alfred and the *Mi* from Mieß to create his company trade name. Little information is available for this firm, unfortunately. Its founding date does not appear in the files of the Koblenz Amtsgericht, nor are there any documents that were submitted by the company itself. In accordance with the time validity statute used by Koblenz for the maintenance of old records, inactive files older than a certain number of years (including those from pre-war and wartime periods) have already been destroyed. The last entry for Almi in the Koblenz city Address Books is for the year 1937/1938.

Sweatband Colors:
for visor caps
Army: Light gray   Leather

Sweatband reverse Colors:
Army: Gray

HISTORICAL BACKGROUND:
Qualifying for registration in the Handelsregister in 1938 under its original founder Wilhelm Schreiber, it appears that a new owner or partner, Anton Bippi, took control of the firm exactly one year later on March 31 1939. Unfortunately, no detailed documentation on the company exists. The firm did, however, survive the war. On June 14, 1948, the company's title was changed to 'Bavarian Business Clothing and Cap Factory Wilhelm Schreiber' and it continued in business under the leadership of Heinz Bippi (presumably Anton's son). By the mid-1950s, the company failed to qualify for continued listing in the Handelsregister and the Amtsgericht officially closed the company file on March 22, 1955. It is not certain whether the firm continued on for a time under another business structure (such as a G.m.b.H. – a limited liability corporation) after this point, or for how long; it never again qualified for the Handelsregister, in any case. No contemporary records for the company exist.

Confirmation Source:
1. *Uniformen-Markt*
2. Landeshauptstadt München, Stadtarchiv (State Capitol Munich, City Archive)
3. Amtsgericht München [Registergericht]

The makers mark for Bayerische Mützenfabrik *(BAMÜFA)* owned by Wilhelm Schreiber, on an Army infantry officer's Extramütze. *Courtesy of Gerard Stezelberger of Relic Hunter.*

COMPANY: Johann Frey

BUSINESS ADDRESS:

WWII period:
Johann Frey
München, Bahnhofplatz 5, Dachauerstr.

Postwar:
Johann Frey
Bahnhofplatz 5, Dachauerstr.
München

OWNER or BUSINESS LEADER:
1. Johann Frey
2. Ther. Ebert

CURRENT STATUS:
Unknown

FOUNDED:
Frey's founding date is unknown, nor can the firm's first entry in the city's Handelsregister be determined.

TRADE NAME (logo) If Any:
None

Frey logo on an Extramütze (exterior view below).

Frey-made Luftwaffe Other Ranks Extramütze. The Waffen-farbe is for Administration.

SWEATSHIELD SHAPE:
On Extramützen: *Diamond*

MARKS, WITH LOCATION:
In silver ink printed on the lining, "Johann Frey" in vertical script (slightly gothic). Second line, in capitals: "MÜNCHEN"

VERIFIED PRODUCTS:

Schirmmützen:
• Army Officer Extramütze (private purchase cap)
• Luftwaffe Other Ranks Extramütze

COLORS:
On private purchase grade officer /Other Ranks caps :

Visor Cap Lining:	*Material:*
Luftwaffe: Gray	Cotton or rayon
Sweatband colors	
Luftwaffe: Reddish Tan	Thin leather
Visor reverse Colors:	
Luftwaffe: Green	

PATENTS OR D.R.G.M.:
None

HISTORICAL BACKGROUND
The beginnings of the Johann Frey company have been lost to history. The company appears in city address books dated 1950, under the entry: "Johann Frey (owner Ther. Ebert), Hats, Caps, Men's goods, Bahnhofplatz 5, entrance to Dachauerstr." Munich's Handelsregister records the company in 1970, but also notes that the "date of original entry can no longer be determined." Since authentic war-vintage caps with the Frey makers mark have been confirmed, it seems safe to assume that the company existed much earlier than the currently available records indicate.

The Handelsregister file does not include any other information, and the firm's registration was officially closed out in any case on May 19, 1980. This is also the last year that the firm Johann Frey appears in the Munich city Address Book (1980 edition), and likely represents the end of the company (or else it relocated to another city).

Confirmation Source:
1. Landeshauptstadt München, Stadtarchiv [State Capitol Munich, City Archive], *Herr Löffelmeier, Archivamtmann [Archival steward]*
2. Amtsgericht München (Registergericht)
3. Authentic example

COMPANY: *M. Drecksler Mützenfabrik*

BUSINESS ADDRESS:

WWII period:

M. Drecksler Mützenfabrik
München, Dachauerstr. 26

OWNER:
Unknown; initially, presumably M. Drecksler.

CURRENT STATUS:
Unknown

FOUNDED:
Unknown

TRADE NAME (logo) If Any:
None

SWEATSHIELD SHAPE:
On Extramützen: *Rhomboid*

MARKS, WITH LOCATION:
Printed in silver ink on the reverse of the rhomboid-shaped sweatshield: "M. Drecksler"
in cursive-style lettering.
Second line: "Mützenfabrik"
Third line: "München"
Fourth line (small font): "Dachauerstr. 26"

The M. Drecksler makers mark. *Courtesy of Bill Shea.*

VERIFIED PRODUCTS:

Schirmmützen:
• Army officer/Other Ranks Extramütze
• Non-military organization caps (e.g. RLB)

COLORS
On private purchase grade officer/Other Ranks caps:

Visor Cap Lining: *Material:*
Army: Pale gold (champagne)   Rayon or silk

Sweatband colors
Army: Tan   Leather
Note: Drecksler sweatband ends often overlap (join) on the left side, rather than at the
rear of the cap.

Visor reverse Colors:
Army: Tan

PATENTS OR D.R.G.M.:
None

HISTORICAL BACKGROUND:
Historical documentation for the M. Drecksler Mützenfabrik has not yet been located
in Munich. Efforts continue.

Confirmation Source:
Authentic examples

MUNSTER

COMPANY: *Anton Freiting*

BUSINESS ADDRESS:

WWII period:
Anton Freiting
Munster (street address unknown, possibly Adolf Hitlerstr.)

OWNER or BUSINESS LEADER:
Anton Freiting

CURRENT STATUS:
Unknown

FOUNDED:
Unknown

TRADE NAME (logo) If Any:
None

SWEATSHIELD SHAPE:
On Extramützen: *Diamond*

MARKS, WITH LOCATION:
Freiting's makers mark consisted of:

First line: "**Anton Freiting**"
Second line: "Munster"
Third line: "Stirndruckfrei"
Fourth line: "Deutsches Reichspatent"

VERIFIED PRODUCTS:

Schirmmützen:
• Army officer/Other Ranks Extramütze

PATENTS or D.R.G.M.:
Since no patent number accompanies the phrase *Deutsches Reichspatent*, it is not clear whether Freiting held his own patent, or used another company's patented system under license.

HISTORICAL BACKGROUND:
No documentation yet available.

Confirmation Source:
Authentic example

NEISSE

COMPANY: *Paul Kap*

BUSINESS ADDRESS:

WWII period:
Paul Kap
(street address unknown)
Neisse

OWNER or BUSINESS LEADER:
Paul Kap

CURRENT STATUS:
No longer exists

FOUNDED:
Unknown

TRADE NAME (logo) If Any:
None

SWEATSHIELD SHAPE:
On Extramützen: *Diamond*

MARKS, WITH LOCATION:
Vertically oriented double oval topped by a crown; in the space between the two oval borders: "PAUL KAP NEISSE". Centered within the inner oval, a stylized visor cap facing right (to the left as viewed); to the bottom right of this, in fine print: "FABRIK MARKE" [factory trade mark]. No cap available for photography.

VERIFIED PRODUCTS:

Schirmmützen:
• Army general /officer/Other Ranks Extramütze

HISTORICAL BACKGROUND:
Little documentation is available for the company owned by Paul Kap, primarily due to the fact that the town of Neisse lay East of the Oder-Neisse [river] line, which territory by political accord became part of Poland after the war. An industry news note in the *Chronicle* section of the July 1,1941 issue of *Uniformen-Markt* announced Paul Kap's death on 11 June at the age of 73 years; unfortunately, no mention is made as to whether the company would continue in operation, and if so, under whose management. Given the firm's location in the province of Silesia, it is unlikely that it survived the Soviet advance through the area in late 1944/early 1945.

Confirmation Source:
1. *Uniformen-Markt*
2. Authentic example

OFFENBACH am MAIN

COMPANY: *Herman Schellhorn Feuerwehrausrüstungen*
[Fire Department Equipment]

BUSINESS ADDRESS:

WWII period:
Hermann Schellhorn
Offenbach am Main, Spießstr. 30

Postwar:
Hermann Schellhorn
D-6050• Offenbach am Main, Spießstr. 30
(•old-style postal code)

OWNER:
Hermann Schellhorn

CURRENT STATUS:
Officially closed by Mrs. Schellhorn on June 30, 1980.

FOUNDED:
1927

TRADE NAME (logo) If Any:
None

SWEATSHIELD SHAPE:
On Extramützen: *Diamond*

An example of a non-military Schellhorn cap that bears the Peküro (Peter Küpper) diamond logo along with Schellhorn's own makers mark.

MARKS, WITH LOCATION:
In silver ink on the cap lining, "Herm. Schellhorn", and beneath this, "Offenbach" in a loose, cursive script style.

VERIFIED PRODUCTS:

Schirmmützen:
• Army officer/Other Ranks Extramütze as a distributor for Schellenberg
• Extramützen for non-military governmental agencies (also under the 'Peküro' logo).
• Non-military visor caps (Fire police, etc.)

DISTRIBUTORS:
None; however Schellhorn either produced (under license) or distributed caps for:
1. August Schellenberg (Berlin)
2. Peter Küpper (Wuppertal-Ronsdorf) for non-military caps (Feuerwehr, etc.)

HISTORICAL BACKGROUND:
Hermann Schellhorn opened his company in Offenbach on July 21,1927 – or at least, that is the date he first registered the business with the city's Gewerbeamt [Business Bureau]. His company achieved entry into the Offenbach am Main Handelsregister eight years later, on October 21, 1935. The company's business premises was located at Spießstraße 30, and he later moved his residence into the same building. Schellhorn remained at this location until his death in 1980.

The company's products were geared specifically toward the market for equipment and uniforms used by firefighters, and this remained the case throughout the Second World War *and* on into the postwar era.

Though there is neither pre-war, nor wartime data available that offers figures on Schellhorn's production output during these periods, the company seems to have done a small but brisk business. With Germany's remilitarization under the Third Reich, the rapid expansion of the military cap market offered much greater opportunities – none of which was lost on Mr. Schellhorn. To take advantage of this, he began offering military caps as well. These seem to bear not only his own makers mark, but also those of two major makers: August Schellenberg of Berlin, and, (on some of his *non*-military caps) Peter Küpper's 'Peküro' diamond logo. Schellhorn's relationship with these two firms is not entirely clear. Did he produce caps under license? Or, did they produce the caps for him, with his Schellhorn mark added (which he then sold)? Unfortunately, there will never be an answer to these questions.

At the war's end, Schellhorn's company remained intact and he returned to his original product line of fire department equipment and uniforms. His workshop never moved from the original address until it was closed (and the space presumably sold) with Schellhorn's death in 1980. His widow Maria submitted a form to the Offenbach Amtsgericht and IHK on July 15, 1980, noting the firm's closure fifteen days earlier, on June 30, 1980.

Confirmation Source:
1. Amtsgericht Offenbach am Main, *Frau Kiehle*
2. IHK [Chamber of Commerce and Industry] Offenbach, *Frau H. Schlegl*

PFUNGSTADT

COMPANY: *Gebr. Statter*

BUSINESS ADDRESS:

WWII period:
Gebr. Statter
Pfungstadt

OWNER or BUSINESS LEADER:
Unknown

CURRENT STATUS:
Unknown

FOUNDED:
Unknown

TRADE NAME if any:
None

MAKERS MARKS, WITH LOCATION:
Ink stamp on the cap lining: "**GEBR. STATTER**"
Second line: "**PFUNGSTADT**" and date.

VERIFIED PRODUCTS:

Bordmütze:
• Kriegsmarine Other Ranks Bordmütze (garrison cap)

HISTORICAL BACKGROUND:
No information has yet been received from Pfungstadt government authorities. Research efforts continue.

Confirmation Source:
Authentic example

POTSDAM

COMPANY: *H. Ahlers*

BUSINESS ADDRESS:

WWII period:
H. Ahlers
Potsdam, Brandenburger Str. 55

Postwar:
H. Ahlers
Potsdam, Brandenburger Str. 55

OWNER or BUSINESS LEADER:
Gerhard Ahlers

The colorful H. Ahlers makers mark in a fine Army officer's Extramütze. *Courtesy of Bill Shea.*

Exterior of an Ahlers cap. Superb design, with top-quality doeskin wool. Thick crown piping, and a clean saddle shape. *Courtesy of Bill Shea.*

CURRENT STATUS:
Unknown

FOUNDED:
Unknown

TRADE NAME (logo) If Any:
None

SWEATSHIELD SHAPE:
On Extramützen: *Diamond*

MARKS, WITH LOCATION:
The H. Ahlers company used an unusual three-colored makers mark in black, red and white. (Most manufacturers used only one ink color for their marks.) Within a large white circle, double-outlined in black, appears the word *Tradition* in large-sized, red letters angling upward from the lower left to the upper right. Beneath this, in the lower portion of the circle in black letters: "H.AHLERS Potsdam". A crest (shield) overlaps (bottom half) the top of the circle. Within the shield is a black eagle surmounted by the stone wall of a medieval town (also black).

VERIFIED PRODUCTS:

Schirmmützen:
• Army officer/Other Ranks Extramütze

COLORS:
On private purchase grade Officer/Other Ranks caps:

Visor Cap Lining:	*Material:*
Army: Pale gold (champagne)	Rayon or silk

Sweatband colors	
Army: Gray; dark brown	Leather

Visor reverse Colors:
Army: Oxide green

Complete interior view. Ahlers often used the less common oxide green color for its visor reverse paint. *Courtesy of Bill Shea.*

HISTORICAL BACKGROUND:
The founding date for the H. Ahlers company is not known. The original founder was presumably H. Ahlers himself, followed by his son Gerhard Ahlers. The September 13, 1938 edition of the Potsdam city address book lists Gerhard Ahlers as the owner in an entry under the topic: *Herrenartikel* (Gentlemen's goods). The company appears to have remained a small affair throughout the war, and probably produced only top quality Extramützen (no contract caps). Judging from existing examples, Ahlers' products were very well made using first class materials, with an excellent appearance.

Gerhard Ahlers' company made it through the war intact. He continued producing caps and hats, though no doubt these were now of the civilian variety, perhaps also some for government agencies such as the Polizei or Fire Department. As a small business, his company did not pose a significant threat to larger competitors in nearby Berlin. In the Potsdam address book for 1947, the firm is listed under the heading "Hats", as "Gerhard Ahlers *Hutgeschäft, Mützen eigener Herstellung*" [Gerhard Ahlers Hat business, caps of own manufacture]. The company was never large enough or organized in a manner that qualified for entry in the Handelsregister (commercial register); no documents are therefore available from that source. Whether the firm remains in business today, or passed into history long ago remains an unanswered question. Efforts to track down more information continue.

Confirmation Source:
1) Ordnungsamt Potsdam [Handelsregister], *Frau Lampe*
2) Potsdam City Address Books

RASTATT

COMPANY: *Biehler-Mütze*

BUSINESS ADDRESS:

WWII period:
Biehler-Mütze
Rastatt, Kaiserstr. 4

Postwar:
Huthaus-Otto Biehler/Hutgeschäft Biehler
Rastatt, Kaiserstr. 4

OWNER or BUSINESS LEADER:
1911-1921 P.M. Gräfinger
1927-1933 Mrs. Gräfinger (widow of previous owner)
1934 Otto Späth, Schneidermeister [Master (certified) tailor]
1936 Otto Biehler, Mützenmacher [cap maker]

CURRENT STATUS:
Closed in 1966

FOUNDED:
1911

TRADE NAME if any:
None

SWEATSHIELD SHAPE:
On Extramützen: *Rhomboid*

MAKERS MARKS, WITH LOCATION:
Printed directly on the cap lining, "**Biehler-Mütze**" in cursive script canting upward on the right.
Second line: 'Rastatt'

VERIFIED PRODUCTS:

Schirmmützen:
• Army Officer Extramütze

COLORS:
On private purchase grade officer's caps:

Biehler-Mütze makers mark on an Army officer's Extramütze.

Visor Cap Lining:
Army: Pale gold (Champagne)*

* *Sides only:* Reddish tan

Sweatband colors for Visor caps:
Army: Tan

Visor reverse Colors:
Army: Creme

Material:
Rayon or silk

Thin, suede-like material

GRADES:
No specific grade names

PATENTS or D.R.G.M.:
None

PECULIARITIES OF DESIGN (If Any):
The lining on some visor caps made by this company only cover the top of the cap – that is, the sides are *not* lined with the usual rayon in the normal manner. The lining in the top of the cap is very taught, resting directly against the wool cover material with no sag whatsoever. Along the top seam, the lining forms a rounded ridge. The sides are in fact lined – but with a very fine, very thin reddish brown suede-like material, which lays tight against the inner side of the cap wool. It extends only to the inner cap band – which is fully exposed behind the sweatband. Where the suede-like lining reaches the cap's top seam, it is folded over, creating a narrow strip that runs along the seam just below the ridge formed by the upper lining. The fact that the cap does not make use of the standard, full lining (which normally covers the cap sides), means that the front crown support (Stütze) is completely visible. This is a wide, linen-covered strip of very stiff material which has a vertical metal pin wrapped into the front side. Such an interior set up is not normally seen, and affords collectors and historians a unique opportunity to see what one form of crown support looks like – without needing to damage an authentic cap lining in order to do so. No other officer caps (than Biehler's) with this construction style have been seen by the author to date. For a photograph of the Biehler-Mütze interior, see Chapter 2.

HISTORICAL BACKGROUND:
The "Biehler-Mütze" cap originated from a small, local shop in Rastatt which served civilian residents prior to the Third Reich era. There is no documentation that indicates whether the company ever produced any military caps during the World War I period. Very little historical or business data is available on the company, as the Gewerbeamt records for the period are no longer complete and the firm was never big enough to qualify for the local Handelsregister.

The business appears to have been founded in 1911 by a gentleman named P.M. Gräfinger, though the name of the shop at that time is unknown. By 1927, it bore the name "Hutgeschäft Hauser" – though Hauser does not appear as the owner until 1927, after Gräfinger's death. The owner from 1930 to 1932 is recorded as Herr Gräfinger's widow, and during this time there is no entry in the city address books for a firm identified as either Hauser, or Biehler. Herr Otto Späth, whose occupation was Master tailor, is listed from 1933. At this time the shop is first recorded under the name "Hutgeschäft Otto Biehler", which was later changed to "Hut-Haus Biehler". Otto Biehler was a cap maker by trade, and he must at least have worked at the shop, if not been part owner. Whether Biehler actually came to own the business himself – or why it bore his name – is not clear from the available records. At this time, the small company was identified as a *Mützengeschäft* (cap shop/business).

View of the street where the shop was located. The photo is from the 1940s, though which year is not certain. Perhaps, if the car models could be identified, an earliest date might then be established. The "Hut-Haus Biehler" store is located just forward and to the right of the man walking the dog. Note the two large side by side roof windows in the foreground (barely visible next to the wall of the neighboring building's third floor). *Courtesy of Karl-Josef Fritz, Historischer Verein Rastatt e.V.*

Close up of the store; Though only the *t* of "Hut" is visible, the *Haus Biehler* is quite clear in this picture. *Courtesy of Karl-Josef Fritz, Historischer Verein Rastatt e.V.*

This photograph, taken in May 2000, shows the former Biehler building façade. Note once again the two large, side by side roof windows. The former *Hut-Haus Biehler* is now a drugstore ('Apotheke'). *Courtesy of Karl-Josef Fritz, Historischer Verein Rastatt e.V.*

Whether or not Otto Biehler actually made any caps at the shop is not known; there are no records to either confirm or deny that any manufacturing took place there. Given the fact that he was a cap maker by trade, however, it seems logical to assume that he did, in fact, make caps "to order" – perhaps in the back of the shop. There is no record of any other facility associated with Hutgeschäft Otto Biehler, in which caps might have been produced.

As already mentioned, the interior construction of caps bearing the Biehler mark differs significantly from that of other makers. It may have been Otto Biehler's own, unique design – but if so, he does not appear to have applied for a patent or Gebrauchsmuster for it: no patent or D.R.G.M. notice of any kind appears in Biehler caps. As a small business, it is likely that Biehler made caps for local boys in the military – who showed some native loyalty by buying from a 'home-town' shop. They were not sacrificing: Biehler caps have a sharp appearance and are very well made. The relatively small number of customers means that the appearance of Biehler caps on today's militaria market is very limited. It is unlikely that a large quantity were produced, and of those that were, the number that survived both the end of the war and the fifty-six intervening years until the present time is undoubtedly quite small.

The quiet town of Rastatt also survived the end of World War II without any major bomb damage. The shop continued its business, probably retailing both hand-made and mass-produced caps to the local civilian populace. It is not known whether any of the caps sold at the renamed Hut-Haus Biehler in the postwar years were actually produced on the premises; in any case, the business managed to stay afloat through the early years of the postwar economy and on into the 1960s. There are no records indicating whether Herr Otto Biehler himself – or Herr Späth – were still alive at this time, or still running the business.

For reasons unknown (perhaps Otto – if still alive – retired, or the store could not do enough business to survive), the doors closed on the former Hut-Haus Biehler for the final time in 1966. The building that once housed this little cap shop still exists today, but the shop space now hosts an Apotheke (pharmacy).

Confirmation Source:
1. Historischer Verein Rastatt e.V. [Rastatt (registered) Historical Association], *Mr. Karl-Josef Fritz, Director* (with many thanks)
2. Authentic examples

REGENSBURG

COMPANY: *J. Sperb Mützenfabrikation*

BUSINESS ADDRESS:

WWII period:
J. Sperb Mützenfabrikation
Regensburg, Maximilianstraße 25

Postwar:

J. Sperb GmbH & Co. Mode KG
Maximilianstraße 24
D-93047 Regensburg

OWNER or BUSINESS LEADER:
Original owner, Jakob Sperb.
As of the year 2000, the J. Sperb GmbH & Co. Mode KG is represented by Herr Gerhard Sperb.

CURRENT STATUS:
Still an active company, however the firm's primary product line is textiles and fashion clothing, having closed its cap manufacturing operation many years ago.

FOUNDED:
1914

TRADE NAME if any:
None

SWEATSHIELD SHAPE:
On Extramützen:
1. *Rhomboid (with the standard makers mark)*
2. *Diamond (with the Anniversary-style makers mark)*

MAKERS MARKS, WITH LOCATION:
The Sperb Adler (eagle) symbol and name: In two forms, both printed directly on the lining material in black or bluish black ink.

1. Standard version: An eagle's head within a black oval at the top of a larger, black-colored circle. The head faces from the viewer's right to left, with beak open. Text in the larger circle below the eagle reads: "J. SPERB HÜTE-MÜTZEN REGENSBURG, MAXSTR."

2. Anniversary version: Since the company was founded in 1914, its 25 year anniversary occurred in 1939. With this mark, the usual, round form has a new portion added

Standard (round) Sperb makers mark on an Army officer's visor cap.

This anniversary version comes with a diamond-shaped sweatshield, rather than a rhomboid.

J. Sperb cap exterior. A well made, solid example of high-quality workmanship. The cap cover is a plush, fine grade [Chinchilla] Eskimo wool.

above the eagle-bearing oval. The upper edge of this extension arcs slightly upward in the middle – in fact, the new mark looks like a medieval shield. The additional space at the top provides room for a new line of text: *seit 25 Jahren* [for 25 years]. The remaining text is unchanged from the standard version. With this anniversary mark, the sweatshield shape is a [*vertical*] diamond – not a rhomboid.

VERIFIED PRODUCTS:

Schirmmützen:
• Army Officer Extramütze

COLORS:
On private purchase grade officer caps:

Visor Cap Lining:	*Material:*
Army: Pale gold (Champagne)	Rayon or silk

Sweatband colors for Visor caps:
Army: Dark brown; tan

Visor reverse Colors:
Army: Creme

GRADES:
No specific grade names

PATENTS or D.R.G.M.:
None

PECULIARITIES OF DESIGN (If Any):
The lining on visor caps made by J. Sperb is nearly always *machine-sewn* to the cap interior around the level of the cap band top. Normally, the bottom edge of the cap lining is tacked to the cap material behind the sweatband using large, sawtooth stitches. On caps made by Sperb however, the excess lining below the machine-stitch line is *not* secured at all and simply hangs down (loose) behind the cap band.

COMPANY DATA
No specific data available. The company itself no longer maintains old records, and the Regensburg IHK no longer has any records dating earlier than 1946.

HISTORICAL BACKGROUND
J. Sperb Mützenfabrikation was founded by Herr Jakob Sperb in 1914 as a company that manufactured caps for grade school boys and students. Although the company opened in the same year that saw the start of the First World War, there is no indication that the firm manufactured any military caps for Imperial Germany's armed forces – though it is possible that it may have done so. The years following the end of the Great War were very difficult for Sperb, as they were for all German companies, yet there was always a requirement for student caps and the company was by now well established in this niche. It survived into the Third Reich era.

With the improvement in the German economy and remilitarization under the Third Reich during the 1930s, the J. Sperb company decided that it could not pass up the extraordinary business opportunity offered by a resurgent military headgear market. Sperb too, shifted much of its production over to military visor caps and caps for political leaders. The company's military caps were all apparently intended for private purchase since there is no indication that the firm ever mass-produced any caps for government contracts.

Meanwhile, Sperb continued to offer at least part of its original civilian product line, as well. The company officially dates the start of its military cap production to 1939, though it was probably earlier. It produced military caps throughout the war until 1946. Unfortunately, there are no longer any wartime records at the Regensburg IHK (Industrie und Handelskammer – Chamber of Commerce and Industry), and none have yet been received from the Regensburg Amtsgericht; it is therefore not certain whether the company had an active Handelsregister listing during that period, or not.

With the end of hostilities in 1945, the J. Sperb company found itself in relatively fair shape – but with a very small market in the postwar economy. The management deliberated what to do in this dilemma, and decided to target the newly emerging cap market for one of the equally new West German government agencies: the *Bundespost* (the German Federal Postal Service). Sperb thus survived the very lean and difficult postwar years by manufacturing caps for postal personnel around the new West German nation. Cap-making, however, was slowly but surely proving to be a field with little future. Competition was ferocious: a number of well-established competitors and fellow war survivors were also [still] producing caps for a limited market. The Sperb management again wisely concluded that it needed to make major changes while the going was still good and it still had some freedom to maneuver. It closed down its cap production and refocused its efforts into a new (though related) product line: textiles and fashions.

In 1987, the company changed its business structure and registered with the Regensburg IHK. It is still in business today, though it has not manufactured caps for many years. The historical dates presented in this company history were all provided by the J. Sperb company, and one of the firm's senior management personnel was kind enough to rummage through the few remaining old files and dig up a copy of the old J. Sperb company's original wartime letterhead (also giving permission for its reproduction here).

Confirmation Source:
1. J. Sperb GmbH & Co. Mode K.G. *(with many thanks to the company management)*
2. IHK Regensburg, *Frau Evelyn Bachfisch (with many thanks for her dedicated efforts)*
3. Authentic examples

Pre-1945 period J.Sperb company letterhead. The original was kindly provided courtesy of Sperb GmbH & Co. Mode KG.

ROTHENBURG

COMPANY: *Kurt Kläber*

BUSINESS ADDRESS: (primary)

WWII period:
Kurt Kläber
Rothenburg (street address unknown)

OWNER or BUSINESS LEADER:
Unknown, but the original owner was presumably Kurt Kläber.

CURRENT STATUS:
No longer exists

FOUNDED:
Unknown

TRADE NAME if any:
None

SWEATSHIELD SHAPE:
On Extramützen: *Rhomboid*

MAKERS MARKS, WITH LOCATION:
No authentic example of the Kurt Kläber makers mark is currently available.

VERIFIED PRODUCTS:

Schirmmützen:
• Luftwaffe Officer White-top Sommermütze

COLORS:

Visor Cap Lining:
Luftwaffe: Pale gold (Champagne)

Material:
Rayon or silk

Sweatband colors for Visor caps:
Luftwaffe: Tan; brown

Visor reverse Colors:
Luftwaffe: Green (standard)

GRADES:
No specific grade names

PATENTS or D.R.G.M.:
None

PECULIARITIES OF DESIGN (If Any):
None noted

HISTORICAL BACKGROUND:
No historical information/documentation yet available.

Confirmation Source:
Authentic example

ROTHENBURG-ODER
[City not confirmed]

COMPANY: *ADHERO*

BUSINESS ADDRESS:

WWII period:
Unknown

OWNER:
Unknown

CURRENT STATUS:
Unknown

TRADE NAME if any:
ADHERO

SWEATSHIELD SHAPE:
On Extramützen: *Rhomboid*

MAKERS MARKS, WITH LOCATION:
No authentic example currently available.

VERIFIED PRODUCTS:

Schirmmützen:
• Army Officer Extramütze

HISTORICAL BACKGROUND:
None possible. Since the ADHERO maker label never provides the actual city or maker's name, it is not possible at this time to identify where to address an inquiry. Attempts are being made to trace the trademark name through old trademark registration records.

Confirmation Source:
Authentic examples

COMPANY: *Leonhard Paulig [LEPARO]*

BUSINESS ADDRESS: (primary)

WWII period:
Leonhard Paulig
Rothenburg-Oder

OWNER or BUSINESS LEADER:
Unknown

CURRENT STATUS:
No longer exists

TRADE NAME if any:
LEPARO

SWEATSHIELD SHAPE:
On Extramützen: *Rhomboid or Diamond*

MAKERS MARKS, WITH LOCATION:
Printed directly on the cap lining (but in some cases on the sweatshield reverse) in black or silver ink, a stylized visor cap facing to the right (as viewed) within an eye-shaped outline. Inside an upward-arcing band positioned above the eye-shape is the

LEPARO makers mark (Leonhard Paulig) on a Luftwaffe Other Ranks private purchase visor cap.

LEPARO makers mark (Leonhard Paulig) in an Army General's Old-Style Field Cap. *Courtesy of Reuben Lopez.*

trademark logo '**LEPARO**' in capitals; immediately beneath this the word "*Fabrik-marke*" [factory logo].

In a downward-arcing band below the outline, "Leonhard Paulig"; beneath this, the city: "Rothenburg-Oder". The cap size often appears as a black ink stamp above the entire logo.

VERIFIED PRODUCTS:

1. Schirmmützen:
• Army Officer/Other Ranks Extramütze
• Luftwaffe Officer Extramütze

2. *Officer's old-style field caps:* Army

COLORS:
On private purchase grade officer /Other Ranks caps :

Visor Cap Lining:	*Material:*
Army: Dark gray (officer)	Officer: Rayon or silk
Luftwaffe: Reddish-tan	(Other Ranks): Chintz

Officer's old-style field cap lining:
Army: Gray   Rayon

Sweatband colors for Visor caps:
Army: Dark tan; gray   Leather, thin leather
Luftwaffe: Reddish tan; gray

Army Officer's old-style field cap sweatband: Gray

Visor reverse Colors:
Army: Green oxide (common on private purchase grade officer /Other Ranks caps by this maker); gray; tan
Luftwaffe: Green (standard)

Army Officer's old-style field cap: Black

GRADES:
No specific grade names

PATENTS or D.R.G.M.:
None

PECULIARITIES OF DESIGN (If Any):
None noted

DISTRIBUTORS:
None noted.

HISTORICAL BACKGROUND
Leonhard Paulig could be considered the fifth most famous German cap maker of the Second World War. Though caps bearing the familiar trademark logo LEPARO (*LEonhard PAulig ROthenburg-Oder*) appear far less frequently on the militaria market than those of the "Fantastic Four", *Leparo* caps in good condition are always marked

by excellent craftsmanship and sharp appearance. Unfortunately, little is known about the company itself since no documentation has (yet) been located. This is primarily due to the fact that the territory where the small town of Rothenburg–Oder was situated became a part of Poland after the war and no longer exists as a German entity. Research efforts continue.

Confirmation Source:
Authentic examples

Rudolf Ruf makers mark on an Army field police officer's cap.

SAARBRÜCKEN

COMPANY: *Rudolf Ruf*

BUSINESS ADDRESS:

WWII period:
Rudolf Ruf
Saarbrücken, Adolf Hitlerstr. 48 (Ecke Futterstr.)*
* *[at the] corner of Futterstrasse*

Postwar:
1. **Rudolf Ruf**
Saarbrücken, Dudweilerstr. 6

2. **Rudolf Ruf**
Saarbrücken, Bahnhofstr. 73

OWNER or BUSINESS LEADER:
Rudolf Ruf

CURRENT STATUS:
No longer exists; the business closed in 1963.

FOUNDED:
1906, but located in Saarbrücken only from the end of 1928.

TRADE NAME if any:
None

SWEATSHIELD SHAPE:
On Extramützen: *Rhomboid*

MAKERS MARKS, WITH LOCATION:
Printed directly on the cap lining in black ink, in Fraktur font (German gothic):

First line, in an upward curving arc: "Sonderausführung"
Beneath this, a double-line bordered box with rounded corners. Inside the box at the top, a full-bodied dog or cat walking toward the viewer's left; below the animal, the capital letters *RR* in script. Outside the box, midway down: on the left: 19; on the right: 06.
Below the box in large letters (second line): "**Rudolf Ruf**"
Third line: "Saarbrücken"

On some caps, there may be an additional line with: "D.R.G.M. Nr. 1015092"

VERIFIED PRODUCTS:

Schirmmützen:
• Luftwaffe Other Ranks Extramütze

COLORS:
On private purchase grade officer/Other Ranks caps:

Visor Cap Lining:	*Material:*
Army: Pale gold (champagne)	Rayon (officer and Other Ranks)
Luftwaffe: Gray	Rayon or cotton

Sweatband colors for Visor caps:
Army: Reddish tan (light)
Luftwaffe: Gray

Visor reverse Colors:
Army: Reddish tan
Luftwaffe: Green (standard)

GRADES:
No specific grade names, only "Sonderklasse"

PATENTS or D.R.G.M.:
Uncertain; some Rudolf Ruf caps appear with a D.R.G.M. number, however what this number refers to is unclear. It has not yet proven traceable at the Bundes Patentamt [Federal Patent Office].

PECULIARITIES OF DESIGN (If Any):
None

HISTORICAL BACKGROUND:
Rudolf Ruf opened in 1906, according to the founding date he included with his maker information on cap linings. The original business location is not known. The man and the business first appear in Saarbrücken records at the end of 1928, listed in the 1929/30 edition of the city's Address Book as: "Ruf, Rudolf, Kürschnermeister*, fur goods, hats, caps, men's articles, Bahnhofstr. 37' (corner of Futterstr.)." *Master certified furrier*

By 1935 the company's street address had changed to "Adolf Hitlerstraße 48", though it is not clear if the company actually moved, or if (as frequently happened) the street was simply renamed and the buildings re-numbered. Available records indicate the firm was fairly small, however Ruf preferred to use the term "Mützenfabrikation" [cap manufacture] on his maker label, which implied perhaps more grandeur than was actually deserved. In fact, it is not certain whether he actually did make caps at his business premises – or simply sold another maker's caps under his own label.

Rudolf Ruf survived the end of the war in Saarbrücken, and the company appears postwar with its business premises located first at Dudweiler Str. 6, then later once again on Bahnfofstraße 73 – almost the same address where it had begun thirty-five years earlier (the number of the original site was No. 37). The company closed its doors for the last time in 1963.

Confirmation Source:
1. Stadtarchiv Saarbrücken, *Herr Schmitt, Archive employee*
2. Authentic examples

SCHWEINFURT

COMPANY: *Jean Drescher Uniformfabrik*

BUSINESS ADDRESS: (primary)

WWII period:
J. Drescher Uniformfabrik
Schweinfurt (street address unknown)

Postwar:
J. Drescher
Schweinfurt (street address unknown)

OWNER or BUSINESS LEADER:
1. Jean Drescher
2. Otto Drescher, merchant and Master furrier (from 1967), Hans-Georg Drescher, Master furrier
3. Hans-Georg Drescher (sole) from April 1971

CURRENT STATUS:
Unknown; may still exist

FOUNDED:
The date of the company's founding is unknown, but was probably a few years earlier than its first listing in the Schweinfurt Handelsregister, which is dated to December 16, 1904.

TRADE NAME if any:
None

SWEATSHIELD SHAPE:
On Extramützen: *Rhomboid*

MAKERS MARKS, WITH LOCATION:
An example of an authentic Jean Drescher makers mark is not currently available.

VERIFIED PRODUCTS:

Schirmmützen:
• Army Officer Extramütze

COLORS:

Visor Cap Lining:	*Material:*
Army: Pale gold (Champagne)	Rayon or silk

GRADES:
No specific grade names

PATENTS or D.R.G.M.:
None

PECULIARITIES OF DESIGN (If Any):
None noted

HISTORICAL BACKGROUND:
The Jean Drescher Uniformfabrik opened for business in Schweinfurt at some time around 1900. This is confirmed by a note in the company's Handelsregister file which mentions the "day of first entry on December 16, 1904." Unfortunately, no records have yet been found for the half-century period prior to 1967.

The current Handelsregister record file for Jean Drescher logged its first entry in March of 1968, that entry being notification of a change in business form to an *offene Handelsgesellschaft* [general partnership company], and listing the company owners as Otto and Hans-Georg Drescher. The company name in this 1968 entry makes no mention of *Uniformfabrik*, nor does it specify the firm's present line of business. Prior to and during the Second World War, Jean Drescher apparently manufactured uniforms, as the company's full name indicates. With the increase in military cap production in the mid to late 1930s, it was common for large companies (such as Almi of Koblenz, for example) that manufactured uniforms to include visor caps in their product line; these were, after all, part of the full uniform. In a fashion similar to many of the private tailors – C. Louis Weber comes to mind – the owner of the uniform company would hire a cap maker or Master cap maker and the necessary seamstresses to produce caps for the company. This is most likely what the Jean Drescher Uniformfabrik did, as well.

In the postwar era there was no call for military uniforms until the Bundeswehr was created, and it is not certain just what the Jean Drescher company did to fill the gap; there has not been any activity in the Drescher Handelsregister file since the last entry made on April 30, 1971 announcing the departure of Otto Drescher from the company, and the firm's continuation, unchanged, under Hans-Georg Drescher. The Handelsregister file has never been officially closed despite the apparent inactivity, and it is not certain whether the company, in fact, is still in operation. Research efforts continue.

Confirmation Source:
1. Amtsgericht Schweinfurt, *Herr Geßner, Rechtspfleger* [*judicial administrator/registrar*]
2. Authentic example

ST. WENDEL (SAARLAND)

COMPANY: *Karl Colling Mützenfabrik*

BUSINESS ADDRESS:

WWII period:
Karl Colling Mützenfabrik
St. Wendel, Luisenstr. 15

Postwar:
Karl Colling
Herrenmoden – Hüte
Luisenstr. 15
D-66606 St. Wendel

CURRENT STATUS:
Still active at the same address (over 150 years).

OWNER or BUSINESS LEADER:
Founder Karl Colling
(current) Frau Martina Eckert (née Colling) and Frau Jutta Krauser (née Colling)

FOUNDED:
1849

TRADE NAME if any:
None

SWEATSHIELD SHAPE:
On Extramützen: *Rhomboid*

MAKERS MARKS, WITH LOCATION:
In silver ink printed directly on the cap lining:
First line in script: "***Karl Colling***"
Second line: "Mützenfabrik"
Third line: "St. Wendel"

VERIFIED PRODUCTS:

Schirmmützen:
• Army Officer Extramütze

COLORS:

Visor Cap Lining:
Army: Pale gold (Champagne)

Material:
Rayon or silk

Sweatband colors for Visor caps:
Army: Reddish tan   Leather

Visor reverse Colors:
Army: Reddish tan

GRADES:
No specific grade names

PATENTS or D.R.G.M.:
None

PECULIARITIES OF DESIGN (If Any):
None

Karl Colling makers mark on an Army officer's Extramütze.

HISTORICAL BACKGROUND:

First opened by Kappenmacher [cap maker] Karl Colling in 1849, the Karl Colling business at Luisenstr. 15 has been run by six generations of Collings over a period of one hundred fifty-two years.

By 1900, the company had proudly taken on the title of "Uniform Cap Factory" – but despite the impressive sounding name, it remained, in fact, only a small business run by a cap maker with several seamstresses producing uniform caps "made to measure" (i.e. made to order). The company did a comfortable business servicing the needs of military personnel from the local area, perhaps making civilian headwear as well. It weathered the difficult decade of the 1920s and with German remilitarization in the mid-1930s, began producing high quality Extramützen for local Wehrmacht personnel. No doubt, those native sons from the St. Wendel area who could withstand the desire to purchase an EREL, Schellenberg, Clemens Wagner or Peküro cap probably favored their local cap maker with their business, instead. In this, they made a good choice: authentic Karl Colling military caps thus far observed have all been characterized by both excellent materials and first-class workmanship.

The end of World War II and with it, Germany's Wehrmacht, also put an immediate end to Colling's military cap business. The company quickly shifted its efforts to headwear for gentlemen, then branched into retail sales of men's fashion clothing. As near as can be determined without direct input from the firm (repeated inquiries to the company went unanswered), the company in fact, moved away from direct in-house manufacturing of caps and hats and concentrated instead on retail sales of men's items: hats, caps, fashion clothing and accessories.

The Karl Colling Herrenmoden – Hüte business is run today by Ms. Martina Eckert (née Colling) and her sister, Ms. Jutta Krauser (née Colling). Karl Colling's long history in business continues, at 152 years and counting.

Confirmation Source:
1. Stadtarchiv St. Wendel, *Herr Schnurr, Bibliotheksassistent* [*library assistant*]
2. Amtsgericht St. Wendel (Handelsregister)
3. Authentic examples

STRAUBING

COMPANY: *Otto Schlientz*

BUSINESS ADDRESS:

WWII period:
Otto Schlientz
Straubing (street address unknown)

CURRENT STATUS:
Unknown

OWNER or BUSINESS LEADER:
Unknown

FOUNDED:
Unknown

TRADE NAME if any:
None

SWEATSHIELD SHAPE:
On Extramützen: *Diamond*

MAKERS MARKS, WITH LOCATION:
In silver ink printed directly on the cap lining:
A stylized visor cap, facing right. The cap's center band is formed from the squared-off
letters *O* and *S*
Forming a complete circle around the stylized cap appears:
"OTTO • SCHLIENTZ • UNIFORMMÜTZEN • STRAUBING"

VERIFIED PRODUCTS:

Schirmmützen:
• Caps for *non-military* government organizations

Note: To the author's knowledge, no authentic *military* caps with an Otto Schlientz
makers mark have been identified – though the assumption that the company would
have made military caps as well seems quite logical. Schlientz is included here primarily due to the fact that its original makers mark dies may currently be in the possession
of the Janke Tailoring firm and are commonly used on Janke's reproduction *military*
caps as well as on its non-military headgear.

PATENTS or D.R.G.M.:
None

PECULIARITIES OF DESIGN (If Any):
None

HISTORICAL BACKGROUND:
No historical information or company documentation is yet available from the city of
Straubing for this company.

Confirmation Source:
Authentic examples (non-military)

The Otto Schlientz makers mark on an authentic pre-1945 non-military cap. *Courtesy of Bill Shea.*

Cap interior. This picture provides a good indication of the relative *size* of the Schlientz makers mark. *Courtesy of Bill Shea.*

WIEN (Vienna)

COMPANY: *Fillers*

BUSINESS ADDRESS:

WWII period:
Fillers
Wien (street address unknown)

CURRENT STATUS:
Unknown

FOUNDED:
Unknown

TRADE NAME if any:
None

MAKERS MARKS, WITH LOCATION:
Printed on the cap lining:
The name "**Fillers**"
Second line: "Wien"

VERIFIED PRODUCTS:

Schirmmützen:
• Army Other Ranks Extramütze (Panzer)

HISTORICAL BACKGROUND:
No information yet available for this Austrian company.

Confirmation Source:
Authentic example

WILHELMSHAFEN

COMPANY: *Mützen Scherff*

BUSINESS ADDRESS:

WWII period:
Mützen Scherff
Wilhelmshafen (street address unknown)

CURRENT STATUS:
Unknown

FOUNDED:
Unknown

TRADE NAME if any:
None

MAKERS MARKS, WITH LOCATION:
In black ink printed directly on the cap lining:

"**Mützen** *Scherff*" with "Scherff" in script.
Second line: "Wilhelmshafen"

No authentic example of the Scherff makers mark is currently available

VERIFIED PRODUCTS:

Schlöffelmützen
(Naval "Donald Duck" caps)

Note: No other type of cap has yet been seen bearing the Scherff makers mark.

HISTORICAL BACKGROUND:
No historical information or company data has yet been received from Wilhelmshafen.

Confirmation Source:
Authentic example

WUPPERTAL-RONSDORF

COMPANY: *Albert Kempf Uniform-Mützenfabrik* [*'Alkero'*]

BUSINESS ADDRESS:

WWII Period:
Albert Kempf Uniform-Mützenfabrik
Wuppertal-Ronsdorf

Postwar (final):
Albert Kempf GmbH & Co. KG
Uniformmützenfabriken
Odmiesbach 23
D-92552 Teunz

CURRENT STATUS:
Still active; Albert Kempf GmbH & CO. KG holds the distinction of being one of the very few former wartime manufacturers of military caps still active. In fact, it is the only *major* manufacturer of military caps remaining in Germany.

OWNER or BUSINESS LEADER:
1. Albert Kempf, Kürschnermeister [Master (certified) furrier]
2. Artur Kempf (son), from 1946
3. Bernd Albert Kempf (son of Artur), since 1968 (entered in the Handelsregister in 1971)

FOUNDED:
Some time prior to first entry in the Wuppertal Handelsregister (March 10, 1936).

TRADE NAME if any:
Alkero

SWEATSHIELD SHAPE:
On Extramützen: *Rhomboid* (diamond with one end rounded)

MAKERS MARKS, WITH LOCATION:
In silver ink, printed on the reverse of the sweatshield:
"Alkero" logo in script, within a canted oval, the oval held by an eagle with outstretched wings (wingtips curving down and slightly inward). Beneath the logo mark in elaborate script:

Gepolstertes Schweißleder – padded sweatband (see Peculiarities of Design note below).

On field caps, prior to the widespread use of RB numbers, the company's mark appears printed in dark blue ink as "Albert Kempf Mützenfabrik W. – Ronsdorf" [with the cap model], inside a box. Beneath and outside the box is the cap size. Not all examples of Kempf makers marks include the cap model, however.

Luftwaffe Other Ranks Government Issue cap. Most makers did not use any lining makers mark on their contract caps (normal was the ink stamp on the sweatband reverse). Albert Kempf on the other hand, apparently wanted to be certain that the company was easily recognized and used "Albert Kempf" in gold lettering with the town, "Ronsdorf", on the lining of his mass-produced contract caps. *Lieferant* means contractor.

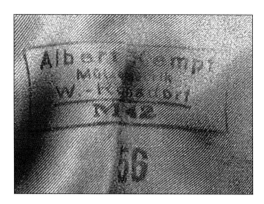

Army M42 cap with an Albert Kempf ink stamp. In this case, the makers mark includes the cap model: M42. *Courtesy of Bill Shea.*

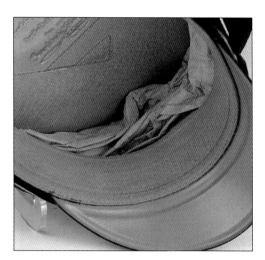

Close up of the padded sweatband. *Courtesy of Gerard Stezelberger of Relic Hunter.*

VERIFIED PRODUCTS

1) *Schirmmützen:*
• Army Officer/Other Ranks Extramütze
• Luftwaffe Other Ranks Extramütze/Contract caps (Government issue)

2) *Feldmützen:* • Army M42 caps
(field caps)

COLORS

Visor Cap Lining:
Army: Light reddish brown
Luftwaffe: Orange brown

Material:
Officer: Silk or rayon
Other Ranks: Cotton

Field Cap Lining:
Army: Gray

Sweatband colors
Army: Yellowish tan
Luftwaffe: Tan

Visor reverse Colors:
Army: Yellowish tan
Luftwaffe: Green (standard)

GRADES:
No specific grade names noted.

PATENTS or D.R.G.M.:
Though not yet confirmed, Kempf likely held either a *Reichspatent* or a D.R.G.M. for the upholstered (padded) sweatband concept. The company also holds a modern Gebrauchsmuster, under D.B.G.M. Nr. **297 23 337.8**.

PECULIARITIES OF DESIGN (If Any):
Gepolstertes Schweißleder [upholstered (padded) sweatband]. A thin strip of foam-like, pinkish padding glued around the bottom of the sweatband in order to offer the wearer more comfort, was a system apparently introduced by Albert Kempf.

This company sometimes used a partial (forehead area only) black velvet insert between the sweatband and bottom edge of the cap, as well.

HISTORICAL BACKGROUND:
Alkero – Albert Kempf – is not one of the more well-known makers of the Second World War. The company did produce fine caps, however, as existing examples verify. The Albert Kempf business started with furs and similar items. Cap manufacturing was initially only a side business, growing parallel to its main activity as a furrier. Eventually, however, the company's cap-making operations completely replaced the fur business. The company was entered in the city's Handelsregister in 1936, and was filling government cap contracts by 1939, if not before. Kempf also manufactured high-end Extramützen, and may have introduced the "padded" or "upholstered" sweatband concept. Although it has not yet been possible to locate a copy of the actual patent (or D.R.G.M.) number for this invention, it seems certain that Albert Kempf held the rights to it.

Albert Kempf was long overshadowed by its much larger competitor in Ronsdorf, the firm of Peter Küpper. Ironically, however, Albert Kempf would have the last laugh.

May of 1945 found the company still – if barely – in business, and still located in Wuppertal-Ronsdorf. Artur Kempf, Albert's son, officially took over as leader of the business in January of 1946, though it is likely that he was actually running the company already when this fact was entered in the Handelsregister (January 4, 1946). In the early postwar period the company took the opportunity to relocate from Ronsdorf to Vohwinkel, another part of the city of Wuppertal (the Vohwinkel office remains active today). Times were extremely difficult in the fledgling post-war economy, and with Küpper in perpetual competition for the available business, the Albert Kempf company eventually decided to shift some of its efforts into the production of sportswear. With the creation of the Bundeswehr (around 1955), some cap makers opted to pursue military cap production once more – Albert Kempf among them. Shortly after this, in 1958, the company founder, Albert Kempf, passed away.

By 1968, the company's sportswear manufacturing operation had become its primary business. This part of the company relocated to a larger facility in Teunz, a town in the area of Germany known as the *Oberpfalz*. Cap production (including military caps for the Bundeswehr) continued for a time in Wuppertal, (presumably) at the Vohwinkel facility.

Wartime Luftwaffe Other Ranks service visor cap for paratroop or flight personnel, manufactured by Albert Kempf. Both the cap crown and the cap band (top and bottom) are piped in Waffenfarbe.

Meanwhile, Kempf's competitors found themselves forced, one after another, to abandon the military cap market altogether and concentrate their efforts elsewhere. For some, even this was not enough, or else came too late to be of any lasting help. Albert Kempf, too, was forced by economic necessity to make a major operational decision. With rising raw material prices and labor costs, the manufacture of military caps was quickly becoming a no-win proposition and radical action was necessary. There were only two viable solutions: cease manufacturing caps (particularly military caps) altogether, or move cap manufacturing operations to a less costly locale. The company chose the second option – and the unusual step of moving most of its cap production operations to a foreign country, where labor costs were more favorable (which country or geographical area is not specified, but it is probably somewhere in Asia). This decision allowed the company to continue producing military caps, while most of its competitors were forced out of the race. At the finish line, Albert Kempf's wise business decision had left it the only major company still in the running.

The main business office of the Bundeswehr Kleiderkasse confirms that Albert Kempf is, in essence, the sole remaining major supplier of caps to the Bundeswehr and the Bundesgrenzschutz [Federal Border Patrol].

Modern (postwar) Bundeswehr Luftwaffe Other Ranks service visor cap, also manufactured by Albert Kempf. The Waffenfarbe (branch color) again is for paratroop or flight personnel. Note that only the cap *crown* is piped in Waffenfarbe – the cap band is *not* piped. Also noteworthy is the positioning of the Federal cockade (gold/red/black) where the Third Reich national emblem (eagle) normally appears on pre-1945 vintage caps.

Confirmation Source:
1. Albert Kempf GmbH & Co. KG (personal correspondence)
2. Industrie- und Handelskammer Wuppertal [Chamber of Commerce and Industry], *Herr Heinz Beier*
3. Amtsgericht Wuppertal (Handelsregister)
4. Authentic examples

COMPANY: *Peter Küpper (Peküro)*

BUSINESS ADDRESS:

WWII period:
Peter Küpper
Wuppertal-Ronsdorf

Postwar:
Peter Küpper "Codeba" GmbH & Co.

Standard Peter Küpper visor cap makers mark, in black. This piece may have also borne a rather fancy distributor's mark; unfortunately too much of it is worn off to be able to identify the company. Army officer's visor cap. *Courtesy of Gerard Stezelberger of Relic Hunter.*

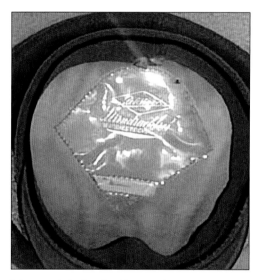

Peküro mark in silver ink. Note the full material insert of black velvet (or satin), that circles the entire cap between the sweatband bottom and the cap's base cloth. The visor underside on this cap is painted oxide green – not a common color for Peküro caps.

In der Krim 30
D-42369 Wuppertal

OWNER or BUSINESS LEADER:
1. Peter Küpper
2. Berhardine Zimmerman
3. Heinz-Joachim Zimmermann-Baum
4. Claudia Kolter (née Zimmermann-Baum)

CURRENT STATUS:
The company's registration in the Handelsregister (commercial/trade registry) of the Wuppertal Amtsgericht was officially closed in September 2000 after an insolvency investigation and subsequent proceedings were opened against the company by the local government. The firm was 'dissolved' in September of 2000, and is now closed.

FOUNDED:
Some time between 1885 and 1894. First entry in the Wuppertal Handelsregister on April 13, 1894.

TRADE NAME if any:
'Peküro' in a horizontal diamond

SWEATSHIELD SHAPE:
On Extramützen: *Rhomboid*

MARKS, WITH LOCATION:
Peter Küpper makers mark:

Company logo '**Peküro**', canted upward from left to right and underlined, within a horizontal diamond. The mark may be printed directly on the lining material, or on the sweatshield reverse. Usual ink colors: Silver or black. Beneath the logo in script: "Stirndruckfrei." Second line: "DEUTSCHES REICHSPATENT" in capitals.

VERIFIED PRODUCTS:

1) Schirmmützen:
• Army officer /Other Ranks Extramütze
• Luftwaffe officer/Other Ranks Extramütze
• Non-military caps for fire departments, police and other government agencies

2) *Private Purchase field caps* – often with a *diamond-shaped* sweatshield and makers mark in M43 caps, or an ink stamped 'Peter-Küpper'; Bergmütze

Note: Many collectors and dealers alike find makers marked private purchase field caps questionable. The reasoning is that these caps in general, frequently present enough deviation from the regulation format as it is – without any mark. A makers mark can be added to improve the believability of a questionable item.

COLORS:
On private purchase grade officer /Other Ranks caps :

Visor Cap Lining: *Material:*
Army: Reddish brown (rust); gold; light gray   Rayon, silk (officer), or cotton (Other Ranks)

Sweatband colors for Visor caps:
Army: Reddish brown; tan

Visor reverse Colors:
Army: Reddish brown; tan; oxide green (rare)

GRADES:
No specific grade names

PATENTS or D.R.G.M.:
D.R.P. No **538 943** issued in 1931

• Deutsches Reichspatent (German National Patent)

PECULIARITIES OF DESIGN (If Any):
Some caps have a black satin or velvet insert below the sweatband.

COMPANY DATA:
No specific data available.

DISTRIBUTORS:
1. Hermann Schellhorn, Offenbach am Main
2. Unidentified distributor in Aachen

HISTORICAL BACKGROUND
In 1885, the 25 year old Peter Küpper received 300 Talers (an old German monetary unit) from his father, the Eberfeld Master cap maker Peter Küpper (senior). The money was given together with a request for the youngster "not to bother him any longer with pompous plans." Young Peter had, in fact, just returned from overseas and was full of new ideas and determination. He quickly bought up a stake in an existing cap making business in Ronsdorf (a section of the city of Wuppertal), named *Krauskopf*. His diligence quickly brought the rest of the Krauskopf company into his hands, and with that success, the *Peter Küpper* firm was launched. Küpper patented one of his own unique and popular designs (a cap that folded together), and as a pioneer in the field was the first to shift production from strictly handwork to machines. To this end, he himself developed the machine designs which were then manufactured for him by the Singer sewing machine company in England. Some of these machines were eventually sold worldwide.

In 1905 Küpper managed to secure a piece of land in the Crimea where he built a factory that went operational in 1907. The eventual fate of this factory is not known, but it was probably lost during the First World War or most certainly thereafter, when Lenin took control of Imperial Russia from the Czar (Nikolas) and created the Soviet Union.

The company survived the difficult years after World War I despite the floundering economy and the shaky political management of the Weimar Republic government. With Hitler's ascendancy and the birth of the Third Reich, Peter Küpper passed away (in 1934) at the age of 74 years.

He left the company to his offspring, and to his long-serving chief company clerk, Berhardine Zimmermann. Frau Zimmermann together with the Küpper heirs steered the company through the heady years of intense military buildup in the late 1930s and on through the violence of the Second World War. Küpper had already held a long-standing reputation for high quality caps, though exactly when it began producing military headgear is not certain. By 1936, with Robert Lubstein proclaiming his company as "*Berlin's largest Uniform Cap Factory*", Peter Küpper hauled out the big guns and labeled itself (in its own advertisements) as "**Germany's** largest uniform cap factory."

Peküro makers label on a postwar (circa 1968) officer's cap for the Bundesgrenzschutz. Note the interior color combination. Early (prior to the 1970s) BGS caps like this one sometimes used thin Vulkanfiber visors similar to those on wartime Luftwaffe caps). Vulkanfiber was soon replaced completely by a heavier, plastic visor on post-1970 caps.

Peter Küpper advertisement from the July 15, 1938 issue of the *Uniformen-Markt.* The ad text reads:

Germany's largest uniform cap factory
PETER KÜPPER
Wuppertal-Ronsdorf * Delivers all kinds of UNIFORM CAPS
Free of forehead pressure
Secure visor
[Logo]
Every
<u>Peküro cap</u>
embodies quality and
advance

An unsolved mystery from the pre-war period that remains unsolved, is exactly when the company first began using its trademark acronym Peküro (formed from the *Pe* of Peter, the *Kü* of Küpper, and the *Ro* of Ronsdorf); certain is only that it was in use by 1935. During this period (or earlier), the Peter Küpper company also developed relationships with several smaller distributors, including Hermann Schellhorn of Offenbach am Main.

The end of the war in 1945 was also the end of Peter Küpper's largest customer, the Wehrmacht. With no German military force in existence, the company turned to government agencies, supplying organizations such as police and fire departments, the postal service and so on – as did many of its competitors. Some sort of new direction was needed.

Frau Zimmermann, meanwhile, had remained childless and so she adopted her young nephew Heinz-Joachim Baum. This twenty-four year-old textile engineer took the combined family name Zimmermann-Baum and joined the firm in 1954. He slowly began to convert the company from uniform cap manufacture to fashion and sport headgear; the process was very gradual. When the new Bundeswehr was created, Peter Küpper wasted no time starting production of uniform caps for its familiar old customer, and later also for the Bundesgrenzschutz. Peküro caps made for the BGS, in fact, began to appear almost immediately after the Berlin firm of Carl Halfar had been forced to cancel its exclusive contract with that organization. [See the Carl Halfar (Berlin) entry for more details.]

Peter Küpper would have been better served, perhaps, if it had made a complete break with its past and pursued only new avenues of business rather than wasting time returning to military cap production. As it was, the steadily improving German economy – this was the period of *die Wirtschaftswunder* – the *economic miracle* as the Germans refer to it, also brought rapidly increasing production and overhead costs. Peter Küpper was eventually forced to suspend its military cap production. Precious time had been lost that could have been put to better use; Herr Zimmermann-Baum worked hard to make a new name for the firm in the field of sporting goods – specifically in sport headgear. The company acted as the outfitter for and the sponsor of many events (such as the Winter Olympics). It also served as the outfitter for many "expeditions", which also offered the added benefit of serving as test beds for new Küpper products. The Peter Küpper firm, under its new trade name CODEBA, became a leading specialist in sport headgear.

After Zimmermann-Baum's death, his daughter Claudia Kolter took charge of the company. It's official line of business as listed in the Wuppertal IHK [Chamber of Commerce and Industry] company file was *"Manufacture of sport headwear."*

Nor was the popularity of the personal computer and the Internet lost on CODEBA; the company moved along with the high-tech age and even posted its own Internet web site. Despite all these successes however, the heyday of the once great Peter Küpper company had passed. Perhaps it was unable to keep pace with a rapidly changing and fickle business environment; whatever the case, its financial condition deteriorated to a point that the local Wuppertal justice ministry opened an *Insolvenz Verfahren* (insolvency proceedings) against the company on February 1, 2000. Such proceedings include assignment of an insolvency administrator; the company, as debtor, is forbidden to take any actions without the administrator's approval. The investigation then led to the actual proceedings. Such proceedings automatically result in several simultaneous actions: The firm's Handelsregister file is officially closed (September 5, 2000), and the company itself in its current form at that time is "dissolved." It is possible that, after absolving its debts and reorganizing, Peter Küpper may have continued in business, but only as a mere shadow of its former self, and never again under its original company name. (Once a company closes due to insolvency, it can never use its original name again).

The company's Internet website is no longer online, and all indications are that the famous Peter Küpper – perhaps at one time *"Germany's largest uniform cap maker"* – has closed forever. The last of the "Fantastic Four" has joined its peers as a historical note in a book.

Today, the only major wartime German military cap maker that actively produces uniform caps is Albert Kempf & Co. GmbH. Like Peter Küpper originally of Wuppertal, Albert Kempf (Alkero) was always in the shadow of its larger competitor during the Second World War; but, in the end, Kempf had the last laugh.

Confirmation Source:
1. Amtsgericht Wuppertal
2. IHK Wuppertal, *Herr Heinz Beier*
3. *Uniformen-Markt*
4. Authentic examples

WÜRZBURG

COMPANY: *Josef Rom Mützenfabrikation*

BUSINESS ADDRESS:

WWII Period:
Josef Rom
Würzburg, Leistenstrasse 3

Post-War
Josef Rom
Würzburg, Leistenstrasse 3

OWNER or BUSINESS LEADER:
1. a. Josef Rom (born December 8, 1907), Kürschnermeister (Master furrier)
b. Robert Rom, Kürschnermeister (as a soldier, killed in Greece on May 28, 1944)
c. Anna Rom (died in January of 1939)
2. Josef Rom, Anita Rom (née Thomas, born December 8, 1911), widow of Robert
3. Bernd Rom, Kürschnermeister (born December 8, 1944), from January 1, 1974, with Josef Rom and Anita Rom as limited partners.

FOUNDED:
July 15, 1937 by Josef, Robert and Anna Rom,

TRADE NAME if any:
None

Josef Rom makers mark on this badly damaged Army officer's cap. Note the three spikes at the top and bottom of the mark, and the design located between the top word and the name. The Waffenfarbe appears to be a faded medical service color.

CERTIFICATIONS:
None noted

PATENTS or D.R.G.M.:
None noted

SWEATSHIELD TYPE:
On Extramützen: *Diamond*

MARKS, WITH LOCATION:
"Mützenfabrikation" (cap manufacturing) in an arc over the name "Josef Rom" (horizontal line); below this the city, "Würzburg"(horizontal) – all printed in white ink on a black, onion-shaped background with a white border (three jagged points at each end). The logo placement should fall within the upper half of the sweat shield diamond, with the lower half blank.

VERIFIED PRODUCTS:

Schirmmützen:
• Army Officer Extramütze

COLORS:
On private purchase grade officer /Other Ranks caps

Visor Cap Lining:
Army: Gold

Material:
Rayon, or fine cotton (linen weave)

Sweatband Colors
for visor caps
Army: Tan

Visor Reverse Colors:
Army: Tan

GRADES:
None noted

HISTORICAL BACKGROUND:
Josef Rom opened the company bearing his name together with Anna and Robert Rom, in 1937. It is not clear what relation the three were to each other, perhaps sister and brothers. Anna passed away less than two years later in 1939, however, with no indication given of the cause. Although Josef and Robert were certified master furriers by trade, it was not unusual for individuals in this trade to get involved in cap manufacture; Erfurt's "Kurt Triebel Mützenfabrik" is a similar example. The Josef Rom company manufactured caps for Wehrmacht personnel, and all authentic examples of Rom caps thus far seen have been Extramützen; whether the firm also mass produced caps for government contracts has not been confirmed. The passing of Anna Rom left Josef and Robert as leaders of the business. Josef's wife, Helene, served as the company's head clerk/deputy (and would for many years).

Robert's date of birth is not given in the company records, so his age can not be determined. Josef was only 32 years old at the outbreak of World War II in 1939; Robert may have been younger. He qualified for military service at any rate, and served as a soldier.

The Rom company had meanwhile gained registration in the Würzburg Handelsregister. The date can no longer be determined, however, since the original register was at some point completely destroyed (perhaps in March of 1945), and this included all of Rom's wartime file entries.

In 1944, Robert Rom was killed in action somewhere in Greece. A postwar entry in the company's new Handelsregister file lists Robert as *gefallen,* which in German stands for *killed in action.* With his death, his wife and widow, Anita, inherited Robert's stake in the company and joined Josef at the helm. Anita and Robert had a daughter named Renate, who had been born in 1942; Renate also inherited part of Robert's legacy, but she was only two years old at the time and so a legal guardian was assigned to monitor her share. Until the end of the war (when the street name was changed), Anita and Renate's residence was located at Adolf Hitlerstrasse 11, in Würzburg.

By 1945 the Rom firm – like the Amtsgericht – had also lost most of its own private company records, either destroyed or damaged beyond hope of salvage in combat actions. The company thus reapplied for entry in the city's new Handelsregister in January of 1950, basing its claim for re-registration on its old status as of March 1945. Also by 1950, a problem had developed between Anita Rom and her daughter, some sort of hefty disagreement, concerning what remains unknown. Renate was now eight years old, but in the aftermath of a major war, children were forced by circumstances to grow up faster than would normally be the case. Whatever it was about, and however it was handled, Anita continued in her position at the Rom company alongside Josef, until Bernd Rom entered the company in January of 1974. At that time the company structure was converted into a limited partnership, with Josef and Anita becoming the limited partners and leaving the company direction to Bernd. It is not clear what relation Bernd was to Josef and Anita – perhaps he was Josef's son. It is interesting to note that all three individuals, Josef, Anita, and Bernd Rom, shared the same birthday of December 8th.

In January of 1993, Anita left the company, perhaps to retire. Some time prior to August of the same year, Josef Rom passed away. With this, the limited partnership was dissolved; Bernd assumed complete control of the company and announced his intention to continue the business. No further entries for the company appear in the Handelsregister file, nor was the file ever officially closed out. Thus, it is not clear whether or not the company still remains in business today. Efforts to confirm the company's continued existence – or closure – continue.

Confirmation Source:
1. Amtsgericht Würzburg
2. Authentic examples

MISCELLANEOUS FIRMS
(Alphabetical by company)

[Not yet confirmed, unknown, unable to identify city/company,
or information was received late]

Anton Ertle
ARGO
E.F. Kruger
Franz Ritter
Herbert Herbst

Hutrika
J. Lettel
Karl Rienäcker
Ludwig Vögele
Marke ODD
Peter Haubrich
Schmid & Menner
Spohn & Klaiber
Th. Wormanns
VIRO

UNKNOWN (city)
COMPANY: *Anton Ertle*
The Anton Ertle name was observed on an authentic Waffen-SS Officer's old-style field cap, however the mark did not include the city name and it is therefore impossible to trace the firm at this time.

UNKNOWN (city)
COMPANY: *ARGO* (maker's actual name is unknown)
This makers mark was observed on an authentic Army Officer's service visor cap. The mark included the phrase *Deutsche Wertarbeit* as a top line; beneath this, the name logo "ARGO". Third line: "Qualitäts-Marke" [quality brand]. No city or address appears, nor any specific company name. From the ARGO mark alone, the maker can not be identified unless a trademark name search through old records is conducted.

STENDAL (Sachsen)
COMPANY: *E.F. Kruger*
The E.F. Kruger makers mark appeared on an authentic Army Administrative Official's (officer Rank) service visor cap. Inquiries were made in one city with the name Stendal, but no records were found. There are, however, other cities with this name which have not yet been contacted.

The maker's mark consists only of the name: "E.F. Kruger", and the city: "Stendal (Sachsen)"

[Saxony was a province in pre-war Germany, under the Federal system it is a *Bundesland* (Federal State)].

MINDEN (Westfalen)
COMPANY: *Franz Ritter*
This makers mark has been seen in two separate examples, each with a different city. The first example was an original Army M34 with an ink stamped makers mark:

"Franz Ritter
vorm. H.K. Flander
Minden (Westf.)"
[vorm. is short for *vormalige* – previously].

Franz Ritter ink-stamped makers mark on the lining of a re-*production* Army officer's M38 cap manufactured by the Janke Tailoring firm. Note also the different name following vorm. – in this case Carl Roth, rather than H.K. Flander. *Courtesy of Bill Bureau.*

The second example was on a reproduction cap made by Janke Tailoring. The inked makers mark was the same, however the city was identified as "Dettingen" (i.e. not Minden). No documentation has been received yet from Minden to confirm the company's location in that city; if the company were entered in the city's Handelsregister, it would likely include the fact of a branch office in Dettingen (if there actually was one). Until confirmation is received, caps claimed to be authentic and bearing the Franz Ritter makers mark require careful scrutiny. Minden appears to be the correct city.

BERLIN
COMPANY: *Herbert Herbst Uniform-Militäreffekten*

BUSINESS ADDRESS:

WWII Period:
Herbert Herbst
Berlin W. 8, Friedrichstr. 180

This mark was seen on an authentic Luftwaffe Other Ranks Extramütze, with a light gray rayon lining. The makers mark includes the name: "Herbert Herbst" in blue ink. Second line: "Uniformen-Militäreffekten" Third line: Berlin W. 8, Friedrichstr. 180

SWEATSHIELD SHAPE:
On Extramützen: *Rhomboid*

COLORS:
On private purchase grade officer cap:

Visor cap lining:
Army: Light bluish gray

Sweatband:
Army: Reddish tan

The Herbert Herbst makers mark on an Army Officer's visor cap. The cap is authentic pre-1945 vintage, however no information on the company has yet been found; it is possible the company did not manufacture its own caps.

An initial records search in Berlin did not turn up any records for this firm. It was probably never listed in any Handelsregister, and if reported only at the Gewerbeamt [Business Bureau], the records may not exist anymore. The company's title is *Uniformen-Militäreffekten* – which means basically military goods/accouterments for uniforms. This may indicate that caps sold by Herbst were actually manufactured by another company, but under Herbst's own company label; there is no evidence at this time to either support or deny this conclusion.

UNKNOWN (city)
COMPANY: *HUTRIKA* (maker's actual name is unknown)
The Hutrika mark appeared on an authentic Army cavalry officer's visor cap. The cap lining was 'natural' color rayon (pale cream), with a dark brown sweatband. The Hutrika makers mark consists of several lines, printed in black:

First line: "Schutzmarke" (protected trade mark/registered trademark)
Second line: '**HUTRIKA**'
Third line: "Sonderklasse" in Fraktur (German gothic) font
Fourth line: "Stirndruckfrei"

As the trademark name provides no information that can be used to identify either the actual maker's name or the city in which the company was located, there is no way to research the company other than a search through old Patentamt trademark files.

HANNOVER
COMPANY: *J. Lettel*

This makers mark was seen on an authentic Army general staff officer's visor cap, and also on several reproductions made by the Janke Tailoring company (which may possess the original dies for the Lettel company's makers mark). The authentic cap was lined in golden yellow rayon.

The Lettel makers mark consisted of only three lines:
First line: "**J. Lettel**"
Second line: "Hannover"
Third line: "Nikolai Str. 18"

No other information was provided with the makers mark; inquiries made to the Hannover Amtsgericht office (where the Handelsregister is kept) and Gewerbeamt [Business Bureau] have *not* turned up any documentation on the company.

Note: This company's maker mark is used by the Janke Tailoring firm, which manufactures reproduction caps.

STADE
COMPANY: *Karl Rienäcker*

BUSINESS ADDRESS:

WWII Period:
Karl Rienäcker
Stade, Holzstr. 18

The makers mark for Karl Rienäcker, printed on the reverse of the sweatshield. Note the base of the cap is the correct fliegerblau (only the cap top was white).

Although the Karl Rienäcker makers mark provides a full address and name, the mark was seen at too late a date for a search to be conducted in the city of Stade for the purposes of this book. Any information that may be located will be included in a subsequent edition.

Karl Rienäcker's maker mark was observed in an off-white, rayon-lined Luftwaffe Other Ranks visor cap. The cap had a light tan colored sweatband. The makers mark, printed directly on the lining in black ink, consisted of:

First line: "die schneidige Uniformmütze"
Second line: "von"*
Third line: "*Karl Rienäcker*" in a large-sized, cursive script
Fourth line: "Kürschnermstr"
Fifth line: "**Stade** Holzstr. 18"

* *'the dashing uniform cap... from... Karl Reinäcker... master furrier'*

KARLSRUHE
COMPANY: *Ludwig Vögele Mützenfabrik*

BUSINESS ADDRESS:

WWII Period:
Ludwig Vögele Mützenfabrik
Karlsruhe, Blücherstr. 18

Orignal Ludwig Vögele invoice from the 1930s. *Courtesy of Herr Dieter Vögele*

Postwar:

Ludwig Vögele Mützenfabrik
Inh. Dieter Vögele
Stundentenartikel
Blücherstr. 18
D-76185 Karlsruhe

OWNER or BUSINESS LEADER:
1. Herr Appert [first name unknown]
2. Ludwig Vögele [son-in-law of Mr. Appert]
3. Ludwig Vögele Jr.
4. Dieter Vögele

CURRENT STATUS:
Still an active cap maker, and as far as can be determined, it is the *only* firm dating prior to World War Two that *still manufactures its headgear in Germany.*

FOUNDED:
1878

TRADE NAME if any:
None

SWEATSHIELD SHAPE:
On Extramützen: *Rhomboid*

MARKS, WITH LOCATION
The only Vögele cap encountered so far has been an Other Ranks Luftwaffe issue cap produced on a government contract. The maker mark thus appeared only as an ink stamp on the sweatband reverse, *Ludwig Vögele Mützenfabrik, Karlsruhe, Blücherst. 18,* together with the year of manufacture or delivery (1936) in black ink.

VERIFIED PRODUCTS

Schirmmützen: Luftwaffe Other Ranks *Lieferungsmütze* [contract/issue cap]
Field caps

COLORS
Unknown

GRADES
No specific grades

PATENTS or D.R.G.M.:
None

PECULIARITIES OF DESIGN (If Any):
None noted.

COMPANY DATA:
No specific data available.

HISTORICAL BACKGROUND

In 1878 an industrious gentleman named Herr Appert opened a cap manufacturing company in Karlsruhe located at Waldstraße. His initial products were uniform caps, and gloves. Twenty-two years later in 1901, the firm was entered in the Handwerk register at the Karlsruhe Handwerkskammer as a handcraft business. At the same time, Ludwig Vögele, Herr Appert's son-in-law, took charge of the company and apparently changed the name to *Ludwig Vögele Mützenfabrik*, which it remains to this day.

In 1908 the business was moved to a newly built housing area in the western part of Karlsruhe [Karlsruher Weststadt], at a corner building called the Eckhaus (at Ludwig-Marum-Straße 1). As of the year 2001, this corner address was home to an Indian restaurant. Prior to this construction this had been an undeveloped area, mostly forest. There was, however, a large military base and horse stable located opposite the new construction, and this circumstance proved to be fortuitous for Vögele in the future.

In 1911, the Vögele company acquired the house at Blücherstraße 18 (also built in 1908), located a mere two doors away, and in 1914 moved into a new facility constructed within the inner courtyard of that address. At the beginning of the 1930s, a shop was built into the ground floor of Blücherstraße 18, facing the street (and the military base). From this shop the company did a brisk business during the Third Reich period, selling besides caps and gloves also accessories such as sabers, swords and daggers. The company filled at least some government supply contracts with *Lieferungsmützen*, though the scope of this business is not certain.

Surviving World War Two and its immediate aftermath, the company doggedly continued in its business of making caps. It seems likely that the headgear produced was no longer strictly military but rather for merchant marine personnel and for government agencies such as police and fire departments, the Bundespost and similar organizations, at least until the Bundeswehr was founded. At that point, Vögele began selling caps to Bundeswehr personnel. Vögele's glove production line was shut down at some time prior to 1955, however.

In 1948, Ludwig Vögele Jr. (born in 1906) took charge of the business from his aging father. Ludwig Jr. had learned the cap-making trade and business first under his father Ludwig Sr., and then for a time in Wuppertal in training with the firm of Albert Kempf, maker of Alkero caps. By the early 1950s, Dieter Vögele, the son of Ludwig Jr., was already well into his own training in the Handwerk of cap making. As his father had done before him, Dieter Vögele finished his apprenticeship under the tutelage of a different cap maker: *Mützen Dommer*. A fellow survivor from the pre-war days, the Mützen Dommer company was located in Stuttgart (no information received in time for an entry in this book). Also during the 1950s, the Carl Isken firm of Cologne [Köln] tried to buy out the Vögele firm, but Ludwig Jr. refused the Isken offer.

After completing his training there, young Dieter returned to Karlsruhe as a journeyman, where he joined the family business in 1959. He received his Master certification in 1965 (there were difficulties in finding enough Master Cap Makers to assemble a qualified certification board). Now, as a Mützenmachermeister, he eventually took over the reins of the company from his father in 1980. The Vögele factory has never ceased to produce uniform caps of all kinds (visor caps, ski caps, garrison caps) for the Bundeswehr and for many official agencies such as the Police, customs, railway, fire departments, and so on. The Vögele factory also supplies the needs of private security firms.

Ironically enough, the Carl Isken company was once again to cross paths with Ludwig Vögele Mützenfabrik. In 1985, economic difficulties had forced Carl Isken to close out its own cap production operations. Still the Carl Isken company continued selling caps under its own name up until 1998 – but these Isken caps were all manufactured by Ludwig Vögele in Karlsruhe, the same company Isken had once tried to buy out!

In 2001, the Ludwig Vögele Mützenfabrik is the only existing pre-1945 founded company that still offers caps "*made in Germany*" – and in all likelihood, it is also the last (remember, that Albert Kempf caps are all made *outside* of Germany). When Herr Dieter Vögele retires at age 65 in the year 2005, he finds it doubtful that there will be anyone interested in taking over his business. For that matter, it is possible that there may not be anyone qualified to take over the business, as no cap makers have been trained, nor Master cap makers certified in Germany for more than a decade. Cap making for that matter, was removed from the official list of German handcraft trades around the year 2000. With the loss of the Ludwig Vögele Mützenfabrik, the final curtain will fall on the fascinating tradition of German cap making. *Die Deutsche Qualitätsmütze* ... will take its final bow.

Confirmation Source:
1. Authentic examples
2. Ludwig Vögele Mützenfabrik, owner *Herr Dieter Vögele*
3. Handwerkskammer Karslruhe

Records have been requested from Karlsruhe city offices, but these were not received in time for publication.

UNKNOWN
COMPANY: *Marke ODD*

This makers mark was seen on an authentic Army Other Ranks infantry visor cap printed on the lining in black ink.

The Marke ODD makers mark consisted of:
"Beste Qualität" (best quality)
"Marke ODD" ('**ODD** Brand')
"*Olympia Klasse*"

No other information was provided with this mark. The *Olympia Klasse* implies a grading system of some sort – it seems logical that Olympia (i.e. close to the heavens) was likely one of the company's highest grades. The meaning of ODD is not known, and with the absence of any actual manufacturer's name or city, only a trademark check at the Patentamt [patent office] offers any hope of identifying the firm. There is no "Ges.

Gesch." (registered) or other similar statement accompanying the "Marke ODD" name, however, and thus it may not have been an officially registered trademark.

The Peter Haubrich makers mark as it appears on a reproduction Army officer's cap manufactured by the Janke Tailoring firm. *Courtesy of Bill Bureau.*

KÖLN (Cologne)
COMPANY: *Peter Haubrich*

The Peter Haubrich makers mark has been observed on reproduction Third Reich military caps manufactured by the Janke Tailoring firm, which may possess the company's original dies. Whether the Haubrich firm actually existed prior to or during World War Two remains as yet unconfirmed. Inquiries made to the Cologne Amtsgericht regarding this company turned up nothing – the company is not listed in the city's Handelsregister at least. Confirmation is now being pursued through the city's Gewerbeamt [Business Bureau] files, but thus far without positive results. With the appearance of this mark confirmed on reproduction pieces, any cap offered as an authentic original and bearing the silver ink Peter Haubrich makers mark should be inspected very carefully prior to making any decision to purchase.

UNKNOWN
COMPANY: *Schmid & Menner*

To date, this company's black, ink-stamped makers mark *has not been seen on any headgear but Waffen-SS* tropical and field gray field caps (authentic). The company seems to have confined itself to supplying the Waffen-SS exclusively. Unfortunately, the stamped makers mark never provides the name of the city in which the company was located, and this makes it essentially impossible to trace the firm. For an example of this mark see Chapter 9 (tropical cap section).

ULM a.D. (am Donau, perhaps? 'on the Donau')
COMPANY: *Spohn & Klaiber*

BUSINESS ADDRESS:

Spohn & Klaiber
Ulm a. D., Syrlinstr. 18

This maker label was seen on a Kriegsmarine officer's white top visor cap. The German abbreviation "**a.D.**" suggests a river such as the Donau – this is the usual format for cities identified by a nearby river: Frankfurt a. M. "Frankfurt am Main" – "on the Main" [river] is an example. An inquiry regarding this company was therefore sent to the Amtsgericht of the city Ulm am Donau – with negative results: the company has never been entered on the city's Handelsregister roll. The Amtsgericht thus made a call to the city's Gewerbeamt office [Business Bureau] – with the same results. The company's existence thus remains unconfirmed at this time.

Note: This maker mark is also frequently used by the Janke Tailoring company on its World War II reproduction caps.

KÖLN (Cologne)
COMPANY: *Th. Wormanns*

The **Th. Wormanns** makers mark was seen on an authentic Army officer's visor cap for Signals troops (Nachrichtentruppen). The mark included only the name "Th. Wormanns" and the city "Köln" (Cologne).

Inquiries submitted to the local Amtsgericht office regarding the Wormanns company did not turn up any information, however; apparently the company never qualified for registration in the city's Handelsregister roll (not unusual).

Inquiries have also been made to the Gewerbeamt [Business Bureau], and the local Handwerkskammer (which registers small to medium-sized business within the handcraft classification), but thus far no results have been received.

UNKNOWN (city)
COMPANY: *VIRO* (maker's actual name is unknown)

BUSINESS ADDRESS:

WWII Period:
City and address unknown

OWNER or BUSINESS LEADER:
Unknown

FOUNDED:
Unknown

TRADE NAME if any:
VIRO

PATENTS or D.R.G.M.:
Registered trademark (name)

The VIRO makers mark.

SWEATSHIELD TYPE:
On Extramützen: *Diamond*

MARKS, WITH LOCATION:
"FABRIK MARKE" [factory trademark] in capitals and a small font, forming an arc over the top of a central symbol composed of angled ovals. The ovals form a circle (the ends of each oval meet one another). The open center area is in the general shape of a diamond (the space enclosed within the confines of the four ovals, that is). Centered in this space appears the "VIRO" mark, in black ink. In an arc below the symbol, again in small-sized capitals, the words: "GESETZL. GESCHÜTZT" (registered trademark). The entire mark is on a white material backing, applied separately to the cap lining.

VERIFIED PRODUCTS:

Schirmmützen:
• Army Officer Extramütze

COLORS:
On private purchase grade officer /Other Ranks caps

Visor Cap Lining: *Material:*
Army: Pale reddish gold   Rayon

Sweatband Colors
for visor caps:
Army: Reddish brown   Composition (Ersatzleder)

Visor Reverse Colors:
Army: Yellowish tan

HISTORICAL BACKGROUND:
The makers logo "VIRO" does not provide enough information to be able to identify this company. The *Vi* and the *Ro* could stand for syllables from the owner's name, or the owner's last name and the first syllable of the town (perhaps Rothenburg?). No other information is provided on the label and the trademark name has not yet been traced.

Confirmation Source:
Authentic examples

UNKNOWN (city)
COMPANY: *BeHa* (maker's actual name is unknown)

BUSINESS ADDRESS:

WWII Period:
City and address unknown

TRADE NAME if any:
BeHa

PATENTS or D.R.G.M.:
Unknown; the logo may have been registered trademark.

SWEATSHIELD TYPE:
On Extramützen: *Rhomboid*

MARKS, WITH LOCATION:
In gold ink within a double, broken-line oval: 'Be Ha' in script, at an upward cant from left to right and underlined. Beneath this, slightly offset to the right, the word 'Prima' (prime/top grade). Above the oval logo, on Luftwaffe caps, is the '*die Verkaufsabteilung der Luftwaffe*' etiquette, with the Puttkamer street address.

VERIFIED PRODUCTS:

Schirmmützen:
• Luftwaffe Officer Extramütze

HISTORICAL BACKGROUND:
The '**BeHa**' maker logo alone is not sufficient to identify the company, or the city of origin.

Confirmation Source:
Authentic examples

Collecting is a fascinating hobby for thousands of people the world over. The choice of exactly *what* to collect depends, of course, on the particular interests of each individual. Whether it be coins and stamps (two of the most traditional forms of collecting), dolls, antiques or militaria, there is something to suit every interest.

The number of people interested in militaria collecting (of any kind) has grown to a point where this field must now be classed within the top ten on any list of favorites. Wars and armies have always been an integral part of human history and for better or worse, this will likely always be true. Interest in military history is natural, and parallels this aspect of the human experience. Any of the many armies that have existed throughout history – some to a greater, some to a lesser degree – are of potential interest to a militaria collector. It does not matter what political idealogy the army/armies stood for – which side it fought for (the "Bad Guys" or the "Good Guys"). The politics is usually of no particular interest to collectors: we are *not* National Socialists simply because we like collecting German militaria – just as modelers who like building German vehicles or aircraft from plastic model kits cannot be accused of being Nazi sympathizers. Still, choosing this area of collecting unfortunately invites a certain risk of being so labeled.

I do not personally know a single collector or dealer of Soviet militaria who admires the Communist regime, but they admire the armed forces that the Soviet Union fielded. Nor do I know any collector or dealer of Third Reich era German militaria who admires the philosophy of Nazism or espouses its ideas (nor, for that matter, will they deal with anyone who does). The primary criteria used by people in selecting a country and a time period of interest for collecting, is very simple: did the military force in question *have a reputation for excellence and prowess in combat?* A positive answer is what makes that country's militaria collectible, irregardless of which "side" that army fought for. The reverse is also true: militaria from an army with a *poor* combat reputation awakens little to no interest in most collectors – you often can't even give the stuff away, no matter how sharp the uniforms might look. What is there to discuss about soldiers and an army that could not fight?

That the German military holds an excellent combat reputation, on the other hand, cannot be denied and has always been the case (despite having lost wars). More than this, the Wehrmacht has gained a certain mystique – a fascination of sorts – based on its aggressiveness and determination, its prowess and élan in combat, as well as in its amazing technical excellence in the development of military equipment and strategies. Among these can be counted the:

- First use of shaped charge [anti-tank] weapons (e.g. the Panzerfaust)
- First operational jet fighter (Me-262)
- First operational cruise missile (V-1)
- First operational ballistic missile (V-2)
- First service-wide use of camouflage paint on vehicles
- First service-wide use of camouflage uniforms (i.e. by units larger than commando size)
- First tactical use of tanks in mobile spearheads designed to punch through the enemy front and deep into rear areas

This is the reputation which leads so many collectors to choose German militaria as their focus of interest (whether Imperial forces or the Wehrmacht) – not the political idealogy behind the national emblem (the eagle and swastika).

Despite the brutal regime which it fought for, the German Wehrmacht of the Third Reich era was one of the most professional, dynamic, tenacious (and sharply dressed) military forces that has existed since the heyday of the Roman Legions. Its discipline and combat prowess was tested and proven many times over (particularly in conditions of adversity). This in no way detracts from the recognition due other armies, including the U.S. Army, which also exhibited determination, innovation (evidence the incredibly fast development and fielding of the North American P-51 Mustang) and perseverance in adversity (Bastogne, for example). Unfortunately, whether fairly or unfairly, the overwhelming American materiel superiority in France and the Allies' total control of the skies over Europe in general prevents the development of the same sense of mystique for U.S. military forces during World War II (European Theater) that many people hold for the vastly overtaxed, yet doggedly tenacious Wehrmacht.

Collectors of German militaria, whether it be uniforms, headgear, field gear, etc., are all too often falsely accused of being a "Nazi sympathizer" or worse, and it is unlikely that will ever change. Do not let this prevent you from enjoying this fascinating hobby, however! Get out and visit militaria shows, develop relationships with other headgear collectors and dealers, share information! My goals for this book were to flesh out more of the background behind German cloth military headgear and its manufacturers, as well as to help collectors improve their technical knowledge and so avoid pitfalls. I hope you have found the book useful and informative.

As a former tanker, I am personally interested in Armor-related militaria from many nations and all periods: Wehrmacht, U.S., British, Soviet, Bundeswehr, NVA, World War Two and postwar etc., as well as in German headgear from the Third Reich period. As I mentioned at the beginning of the book, this hobby is *interactive*. In keeping with this ideal, I welcome your comments and suggestions concerning the material presented here (please direct them courtesy of the publisher). Good luck, and best regards!

Gary Wilkins
Boston, MA

Appendices

Quick Reference Lists for Identifying Fakes

List I: Visor Caps, Officer Old-Style Field Caps
List II: Field Cap M34, Schiffchen, Fliegermütze, Bordmütze, Tropical
List III: Field Cap M42
List IV: Field Cap M43/Bergmütze

For collectors new to the hobby and without much experience, the following Quick Reference Lists may be the most important part of this book. Know your enemy! Study the lists well, so that you can spot and identify any discrepancies in a cap. The information must be so familiar to you that it becomes second nature.

How to use the lists:
The lists are, in fact, summaries of points already covered in individual chapters. They are designed to help you to quickly recognize modified caps and reproductions being offered as originals. The information contained in the lists is simplified, and offers:

> • A listing of items to check for which, if found, would indicate the cap being inspected is either a modified piece (contemporary cap, or pre-1945 model), or a complete reproduction. Conversely, the *absence* of these items may help confirm an authentic, original cap.
> • The kind of comparisons that you need to make between authentic pieces and bogus examples.

It is important to note that the questions presented in the *Quick Lists* are intended to determine if something is *not* correct about a given type of cap. Thus, many questions are posed *in the negative* and a "yes" response would indicate a problem with the cap under consideration. Even questions in the positive seek to determine if there is something negative about the cap – a flaw as it were. After each question, possible conclusions are suggested in brackets [ ]. The conclusions *do not go into significant detail* since that is not their purpose. For specifics, or to review, refer to the associated main chapter (indicated beneath each heading) for that type of headgear.

Some questions with a theme that applies in general to <u>all</u> caps are presented only once under the Army headings rather than being repeated again for the caps of other military services in the same model category. In such a case, a comment to this effect is included within the brackets, for example: [...*Also applies to Waffen-SS and Luftwaffe models.*], or [...*applies to caps from all military services.*].

QUICK REFERENCE LIST I
VISOR CAPS

ARMY VISOR CAPS

1. **Is the national emblem a machine-woven** (BEVo) **eagle**? [*Possible tampering – BEVo insignia was not in accord with regulations; Chapter 3]*

2. Officer cap with bullion eagle: **Is the eagle machine-applied**? [*Not usual on an original cap (bullion insignia was usually applied by hand); possible reapplication. Chapter 3. Check the eagle's proportions. See Appendix D.*]

3. **Is the insignia finish mismatched** (uneven aging, polish, etc.)? This applies to metal insignia only. [*Possibly replaced, may be reproduction; Chapter 3/Chapter 14. Applies to all Services*]

4. **Is there any indication of excess insignia holes**? [*Insignia may have been replaced. Applies to Army, Waffen-SS and Luftwaffe caps.*]

5. Hand-embroidered bullion insignia only: **Does the oak leaf wreath extend well above the top edge of the cap band**? [*Possible postwar BGS wreath; Chapter 3/Appendix D*]

6. **Is the cap interior completely gray** (that is, gray sweatband, lining, and visor reverse)? [*Indicative of postwar manufacture; Chapter 3. Applies to all the armed services.*]

7. **Are there any air vent holes under the cap overhang** (under the cap sides above the cap band)? [*Early postwar police caps; Chapter 3*]

8. **Is the crown piping noticeably uneven (lumpy)?** [*Piping may have been replaced; Chapter 2/Chapter 3*]

9. **Is the cap's piping color completely different at the very base from the color visible on the surface**?
[*Indicative of having been re-dyed; Chapter 2. Applies to Waffen-SS caps as well.*]

10. **Is the sweatband made of thin polyurethane, or vinyl**? [*Postwar; Applies to caps from all services; Chapter 2*]

11. **Is the sweatshield made of a soft, flexible material** (e.g. vinyl)? [*Postwar; Applies to caps from all services.*]

12. **Is the cap visor's brim ridge excessively wide**? [*Not an original visor, possibly reproduction; Chapter 14*]

13. **Is the cap visor made of plastic**? [*Typical of postwar Bundeswehr and East German visors; Chapter 3/Chapter 4*]

14. **Is there a strip of cream or yellowed polyurethane foam padding installed behind the sweatband at the forehead area**? [*Postwar manufacture; polyurethane did not exist prior to or during the Second World War; applies to all military services.*]

15. **Does the cap lining show any indication of tampering** (e.g. stitching where it does not normally appear, or visible stitching along the interior crown seam, etc.)? [*Possible indicator of a modification, such as lining replacement; Chapter 2. Apples to all services.*]

16. **Is the pink Panzer piping noticeably violet tinted**? [*Indication of contemporary German Bundeswehr Panzer piping color. Applies to Army and Waffen-SS caps.*]

17. **Is the cap piped in aluminum mesh cord**? [*Postwar Bundeswehr officer cap.*]

18. **Is the cap band piped only around the top edge**? [*Indicative of a modified East German NVA visor cap. Modifications of East German caps usually simulate Waffen-SS headgear, but some may be posed as Army caps.*]

19. **If the cap bears an "EREL-***Sonderklasse***" mark and a paper tag, is the tag printed completely in black ink**? If yes, check behind the sweatband (forehead area) to confirm there is no actual air vent hole through the cap band. [*Reproduction, particularly if the vent hole is missing; Chapter 14/Chapter 16*]

20. If the cap has a maker mark, **is it a confirmed maker** (confirmed existence prior to 1945)? [*Note: In and of itself this is not a conclusive indicator for, or against any given cap, but it is an "ad-up" factor.*]

21. **Is the interior cap band** (behind the sweatband) **made of plastic**? [*Modified postwar cap or a reproduction; plastic was not a material used in Third Reich era cap manufacture. Chapter 2*]

22. **Does the cap give no impression of significant age**? Does it smell of recent use (sweat, etc.)? [*Possible indication of an "artificially aged" reproduction; Applies to all services.*]

WAFFEN-SS VISOR CAPS

For detailed explanations, see Chapter 4 unless otherwise noted.

1. **Is the national emblem a machine-woven (BEVo) eagle**? [*BEVo insignia was not the correct form of insignia for Waffen-SS service visor caps.*]

2. Metal insignia: **Are the eagle and Totenkopf mismatched**? (That is, they do not appear to be an evenly matched/aged set.) [*The insignia may not be original to the cap, i.e. replaced, or, perhaps reproduction.*]

3. Metal insignia: **Is the insignia detail poor, with an uneven finish, or is there 'flash' on the edges**? **Are the proportions out of balance**? [*Indicative of a poorly cast reproduction*].

4. **Is there any indication of excess insignia holes in the cap cloth** (i.e. under the eagle)? [*Insignia may have been replaced.*]

5. For a cap claimed as that of an officer, **is the cap band black *felt* instead of the correct velvet**? [*Indicates the cap is actually for an enlisted soldier or NCO, not an officer. Exception: a cap belonging to an officer candidate (Junker/Offizieranwarter).*]

6. Using light finger pressure, **can the cap band cloth be easily moved against the underlying pasteboard backing**? Is the cap band lumpy or heavily wrinkled? [*Band may have been replaced.*]

7. Using thumbs to test, **can another set of holes be felt under the cap band cloth on each side of the Totenkopf insignia**? [*Some other insignia (e.g. an Army wreath) might have originally been on the cap, the original cap band replaced or covered. Check also for extra holes through the interior pasteboard material, behind the sweatband.*]

8. **Is the cap piped in unevenly colored Waffenfarbe**? Check whether the color at the piping base differs significantly from the color visible on the surface (allowing for normal fading). [*If it differs, the original piping may have been dyed.*]

9. **Does the cap have an RZM tag**? [*Original Waffen-SS (field gray) visor caps did not have any RZM tag.*]

10. **Does the cap lining have a gold circle mark, with SS runes**? [*This kind of mark is not found on authentic Waffen-SS service visor caps; it does appear on Allgemeine-SS caps, however.*]

11. **Is the cap band without piping along the bottom edge**? [*Indicative of a modified East German cap. East German caps are piped only around the crown and the top of the cap band.*]

12. **Is the sweatshield in the form of a diamond-shape, attached to the lining without any thread, and marked NVA**? [*Modified East German cap.*]

13. **Is the cap being offered at a price significantly lower than the current market value**? As unfortunate as it is, cheap-priced "deals" are not <u>usually</u> to the buyer's advantage. [*No honest dealer of long standing will offer an original cap at a major discount. It simply makes no economic sense to do so. Ask other individuals for a background of the dealer in question.*]

14. If a police cap, **is the color different from the normal police green**? [*Possible modification.*]

15. If a Schutzpolizei (combat police) cap, **is the cap band a color other than brown**? Or, is the brown shade lighter than normal? [*Possible modification/reproduction, remember that the band should be felt, not velvet.*]

16. **Is the police-style wreathed eagle insignia positioned very high on the cap band**? [*May have been tampered with. Check for reproduction insignia.*]

LUFTWAFFE VISOR CAPS

For detailed explanations, see Chapter 5 unless otherwise noted.

1. **Is the vertical seam in the cap band located anywhere *other* than at the front center of the cap** (i.e. it is not positioned *behind* the winged wreath insignia)? [*If yes, the cap is incorrect. The cap band seam is **always** at the front.*]

2. If an officer cap, **is the winged wreath sewn very tightly to the cap band** around the entire circumference of the insignia? [*Indicates the insignia may have been reapplied.*]

3. If an officer cap, **are both the national emblem (eagle) and the wreath in metal**? [*Possibly reapplied; all-metal insignia on an officer cap was not in accord with regulations, or with usual practice. Though metal eagles were sometimes used, an Other Ranks wreath/wings insignia (in metal) as well, is far more unusual.*]

4. **Does the cap lack internal padding, particularly in the sides** (i.e. the overhanging sides feel very thin, if squeezed)? Is the cap very lightweight? [*Little or no internal side padding – thus a low weight – is indicative of lower quality reproduction caps.*]

5. Other Ranks cap: **Is the cap piped in Waffenfarbe only around the crown** (i.e. there is no Waffenfarbe piping around either top or bottom of the cap band)? [*Indicative of a Bundeswehr*

*Luftwaffe Other Ranks cap – Bundeswehr Heer (Army) and Luftwaffe cap bands are not piped. Bundeswehr Luftwaffe **officer** cap bands, however, <u>are</u> piped.*]

6. White-top summer cap: **is the bottom edge of the cap** (below the bottom of the cap band) **a color *other* than Luftwaffe fliegerblau** (blue gray)? [*If yes, the cap is not correct. The bottom edge of the cap body should always be the normal fliegerblau.*]

7. If the cap appears to be a Government Issue grade piece*: **Are the lining material color and the sweatband color mismatched**? That is, are the colors something *other* than matching reddish tan for both sweatband and lining? [*Incorrect; possible reproduction.*]

*(i.e. made of Trikot material, no maker label.)

KRIEGSMARINE (NAVY) VISOR CAPS

For detailed explanations, see Chapter 6 unless otherwise noted.

1. White top cover: **Are the rib lines in the cover material very wide/thick**? [*Postwar, early model Bundesmarine.*]

2. White top cover: **Is there a white-enameled grommet hole at the front center**, hidden behind the eagle? [*Postwar Bundesmarine; the grommet hole is for the shaft of one version (screwback) of the modern German federal cockade insignia.*]

3. White top cap: **Are there two narrowly spaced slits through the cover cloth hidden behind the eagle**? [*Likely postwar Bundesmarine: one form of the federal cockade is attached with prongs; the spacing between the prongs is much narrower however, than for prongs from a World War II national emblem (eagle).*]

4. White top cap: **Is the cap band piped in a branch color**?
[*Kriegsmarine white top caps were never piped in any kind of branch color. Perhaps a modified Luftwaffe Other Ranks cap, or a modified cap from a non-military organization.*]

5. **Is there a wide, *flat*, enamel-coated (metal) cap spring inside the cap crown**? [*Postwar manufacture; flat, enamel-coated bands were not used during the war.*]

6. **Is the cap interior completely unlined**? [*Current model Bundesmarine caps are unlined; the flat cap spring band passes through loops at each end of the sweatshield, suspending it over the middle of the cap. The cap cover is taught, and thus does not sag onto the sweatshield or the wearer's head.*]

7. If a blue top (or white top) cap and identified as an officer model, **is the gilt wire scallop or oak leaf embroidery missing from the visor**? [*Incorrect; possibly a modified NCO cap. Note however, that pre-1936 model officer caps are an exception: these do not have visor ornamentation. Check the year of manufacture, if the stamp is present on the lining.*]

8. **Is the visor brim edge wrapping quite wide, with thick** (ungainly-looking) **stitches**? [*Indicative of postwar Bundesmarine caps.*]

OFFICER OLD-STYLE FIELD CAPS

1. **Is there a thickened rim on the front edge of the visor**? [*Authentic caps have no brim ridge of any kind on the visor.*]

2. **Is the visor made of plastic**? [*No pre-1945 vintage cap was manufactured with a plastic visor.*]

3. **Is the oak leaf wreath insignia narrower than the cap band** (i.e., it does not fill the band vertically from top to bottom)? [*Reapplied. Authentic machine-woven insignia always filled the vertical with of the cap band completely.*]

4. **Is the national emblem or oak leaf wreath made of Leichtmetall** (aluminum alloy)? [*Incorrect. These caps especially, seem to have had no exceptions to the rule: the oak leaf wreath was always the regulation machine-woven (BEVo) cloth insignia, and the eagle as well. No attachment stitching should be visible around the oak leaf wreath*]

QUICK LIST REFERENCE II
FELDMÜTZEN – [GARRISON-STYLE] FIELD CAPS

FIELD CAPS: ARMY M34

For detailed explanations, see Chapter 8 unless otherwise noted.

1. **Does the cap have a pronounced curve to the top**? [*Possible modified postwar Bundeswehr cap. The M34 top was a relatively straight cut.*]

2. **Does the cap interior lack a central** (front to rear) **seam in the top panel**? [*Modified postwar Bundeswehr or other cap.*]

3. **Are the side skirts an integral part of the cap** (that is, not movable)? [*Modified Bundeswehr warm-weather version cap.*]

4. **Do the side skirts** (which have no buttons) **pull down completely over the cheeks** (i.e. like the M42 and M43)? [*Bundeswehr Army cold-weather version cap, modified. Authentic M34 cap skirts were not double width, thus could not be pulled down around the face.*]

5. **Is the eagle insignia applied in such a way that the wreath is sewn over the edge of the front scallop**? [*Incorrect application; at a minimum, the cap has been tampered with; possibly a modified contemporary piece.*]

6. With machine-embroidered insignia: **Is the oak leaf wreath made of individual leaves with tips that curve outward**? [*Reproduction; this is one of the more identifiable flaws on poor quality reproductions. Nearly all machine-woven wreaths are made with two rings of leaves, not one.*]

7. **Is the rayon insignia backing a grassy green**? [*Dark green (bluish green) is the correct color. Reproduction insignia.*]

8. **Is the eagle insignia machine-_embroidered_**? [*Possibly reapplied. Standard insignia was the machine-_woven_ (BEVo) style; such a cap warrants closer inspection.*]

9. **Are the raw edges of the insignia backing cloth exposed** (i.e. not tucked under)? [*Insignia may be reapplied, possibly a reproduction, as well. This point applies in general to garrison caps for all services*]

10. **Is the insignia stitching visible inside the cap** (passes through the lining)? [*Indicative of incorrectly applied insignia – thus the cap may have been tampered with.*]

11. **Is the insignia hand-applied**? [*M34 insignia was normally machine-applied; caps with hand-applied insignia deserve careful scrutiny.*]

12. **Is the cap *without* grommets, or, does it have more than ONE grommet per side**? [*Incorrect; possible modified contemporary cap, or a faulty reproduction.*]

13. **If a soutache is present, do the chevron ends extend to the bottom edge of the cap**? [*Incorrect application; the cap has been tampered with.*]

14. **Is the soutache applied with stitching that does not run down the centerline of the soutache**? [*The cap has been tampered with or may be a reproduction.*]

15. Panzer Cap: Is the black color uneven **– that is, the color varies around the cap** (variances not attributable to normal fading)? [*Indicative of a cap with a different original wool color that has been re-dyed black to simulate Panzer. Applies to all military services that used black wool Panzer models.*]

16. **Is there a white size tag sewn into the lower hem of the lining**? [*Indicative of reproductions. On authentic caps an ink stamp on the lining usually indicated the size.*]

17. Tropical: **Is the material texture very smooth** (i.e. not cotton twill)? [*Modified contemporary cap, or reproduction using incorrect materials.*]

18. Tropical: **Are there <u>two</u> grommets per cap side**? [*Authentic caps had only one grommet. Most likely a modified contemporary piece.*]

19. Tropical: **Is the national emblem** (eagle) **woven in *white* thread**, or is it larger than the normal size for this type of headgear? [*Indicates that incorrect insignia was applied by an individual with poor knowledge of the subject.*]

20. Tropical Other Ranks cap: **Is the cap lining a color other than red**? [*Correct linings are normally red; khaki or light gray linings occur, but are very rare. No other colors are acceptable.*]

FIELD CAPS: WAFFEN-SS SCHIFFCHEN

For detailed explanations, see Chapter 9 unless otherwise noted.

1. **Is the cap outline straight across the top**? [*Possible modification of a contemporary cap.*]

2. **Is the insignia used on the cap a trapezoid, or metal**? [*Modification, or faulty reproduction. The insignia should be the two-piece type on black rayon for this type of cap. Exceptions, such as a metal Totenkopf, are extremely rare and cannot be proven.*]

3. **Is the stitching readily visible along the edges of the insignia** (amateurish appearance)? [*Insignia on most Waffen-SS field caps was hand-applied in a manner that hides the stitches.*]

4. **Is the head of the eagle excessively high, or excessively short** (below the top edge of the wings)? **Is the beak/head poorly shaped, or the wings, swastika, etc. poorly proportioned**? [*All indicative of reproductions.*]

5. **Is there any sign of previously applied insignia (e.g. a difference in wool c**olor around the eagle)? Are the sides of the cap very high, with simulated side skirts only? Is the cap color a brownish-gray, with a lining marked NVA in white ink? [*All points are indications of a modified East German NVA cap.*]

6. Other Ranks cap: **Does it have an Army-style eagle**? [*Rarely found on an authentic Other Ranks Schiffchen, and therefore suspect.*]

FIELD CAPS: LUFTWAFFE FLIEGERMÜTZE

For detailed explanations, see Chapter 10 unless otherwise noted.

1. **Are the top edges of the side skirts piped in Waffenfarbe**? [*Probably Bundeswehr; Other Ranks caps are piped in Waffenfarbe along the skirt edge. Piped Fliegermützen for Other Ranks personnel never went into general use prior to or during World War II.*]

2. **Is the material color a very deep blue**? [*Modified Bundeswehr cap.*]

3. **Does the material feel very thin** (the cap weighs very little)? [*Bundeswehr, modified*]

4. **Is the distance between the top edge of the front skirt and the top of the cap itself narrow**? [*Possibly Bundeswehr; since these caps have only a cockade on the front of the cap body (above the skirt), there is no need for any excess space.*]

5. **Are the side skirts very high and simulated only, with the cap color very dark blue**? [*Possible modified East German Navy cap; Chapters 10 and 11.*]

6. **Is the insignia machine-woven (BEVo) in white thread on a thin, blue rayon underlay**? [*Extremely rare form of insignia, not usually found on the Fliegermütze. Inspect very carefully – the insignia may be a reproduction. Chapters 10 and 13.*]

7. **Does the eagle have any proportional flaws** (unequal wing lengths, warped swastika, etc.)? **Is the cockade sewn to the cap with the backing material and stitching exposed**? [*Possible reproduction or reapplied insignia, which puts the whole cap in question. Applies to all cap models.*]

8. **Are interior lining markings in white ink**? [*Possibly a modified postwar cap; Bundeswehr markings are always in white ink on a black or dark colored lining. Remember that lining markings are easily faked.*]

9. **Is the cap without a central interior ventral** (front to back) **pinched ridgeline**? [*Possible modified contemporary cap. Applies to all models*]

10. Tropical: **Is the eagle insignia a trapezoid shape that also includes the cockade**? [*The tropical Luftwaffe trapezoid-shaped eagle did not include the cockade.*]

11. Officer's tropical cap: **Is the national emblem (eagle) insignia machine-woven or embroidered in a stiff, aluminum thread on khaki cloth**? [*Probable reproduction insignia, and the cap may be, as well.*]

12. Tropical caps: **Is the exterior cap material obviously green?** [*Possible reproduction.*]

13. Government Issue Tropical caps: **Is the cap lining something other than the standard red**? [*Possible reproduction, or modified contemporary cap; the only exceptions from red were tan or light gray, and these are both rare and attributable to specific makers.*]

FIELD CAPS: KRIEGSMARINE BORDMÜTZE

For detailed explanations, see Chapter 11 unless otherwise noted.

1. Blue Bordmütze: **Are the cap sides and side skirts higher than normal, with the skirts simulated only**? [*Possible modified East German People's Navy cap. Very high cap sides are the hallmarks of Soviet military design.*]

2. Blue Bordmütze/Field gray Bordmütze for Other Ranks (enlisted grades): **Are the side skirts piped**? [*Possible modified contemporary cap. Authentic wartime Other Ranks Bordmütze caps were never piped.*]

3. **Is the cap material very thin** (lightweight, with the sheen common to a synthetic such as polyester)? [*Possible modified Bundesmarine or East German cap.*]

4. If an officer cap: **Is the skirt edge piped with aluminum cord**? [*Incorrect: Kriegsmarine officer Bordmütze piping is always in gold (bullion or celleon mesh cord). Possibly a modified Bundeswehr (Luftwaffe) garrison cap.*]

5. **Is the national emblem machine-woven in white** (Other Ranks) **or aluminum** (officer) **thread**? [*Incorrect – the insignia has been replaced, or the cap is a reproduction. Kriegsmarine eagles are always either in golden-yellow thread (for Other Ranks, sometimes officers) or gilt thread (for officers).*]

FIELD CAPS: ARMY M42

For detailed explanations, see Chapter 12 unless otherwise noted.

1. **Is the front skirt scallop one complete** (unbroken) **piece** (i.e. no buttons, uncut)? [*Modified Bundeswehr garrison cap, cold-weather version.*]

2. **Is the cap without a grommet on each side**? [*Modified postwar or other cap, or a faulty reproduction.*]

3. **Is the insignia a hand-applied trapezoid, or two-piece eagle and cockade**? [*T-form insignia is correct for the M42. Inspect the insignia and the cap very carefully.*]

4. **Does the lining fail to cover the entire interior** (that is, it does not reach to within 1 or 2 mm of the bottom edge)? [*Possible reproduction; applies to all service models.*]

5. **If the cap has an inked date stamp, is the year earlier than 1942?** [*If yes, the cap is questionable.*]

6. **Is the cap piped around the upper skirt edge in aluminum cord**? [*Inspect the cap very carefully. Piping is normally incorrect since it indicates officer rank, but there was no official officer version of this cap. Authentic exceptions (upgrades to officer rank) are very rare, can not be easily proven, and will not commonly be encountered.*]

QUICK REFERENCE LIST III
DIE EINHEITSFELDMÜTZE – STANDARD [VISORED] FIELD CAP

For detailed explanations, see Chapter 13, unless otherwise noted.

VISORED FIELD CAPS: ARMY M43

1. **Is the cap material a very lightweight, fine grain wool or other material**? [*Possibly a modified Bundeswehr cap; check for holes in the material below the crown, where a Bundeswehr pronged cockade and crossed sabers may have been attached. Applies to all services.*]

2. **Is the visor length relatively short** (stubby)? [*A short visor is a characteristic of a Bergmütze (mountain cap), not an M43.*]

3. **Is there a flexible shaping ring/cord sewn into the material around the entire bottom edge of the cap**? [*Modified Bundeswehr cap, or other contemporary headgear. Original caps did not have this feature. Applies to all services.*]

4. **Is the interior color arrangement a uniform gray** (gray lining and sweatband)? [*Completely gray interiors with a full sweatband are indicative of contemporary Bundeswehr caps. Applies to all services.*]

5. **Are there two or more puncture holes in the exterior cap material just below the crown**? (If a cloth WWII eagle is in place, feel for holes hidden beneath it by running a fingertip over the eagle's surface, and check inside the cap, as well). [*Holes indicate possible previous Bundeswehr insignia (affixed with prongs). The cap may be a modified contemporary piece. Applies to all services.*]

6. **Does the insignia stitching pass through the inner lining** (i.e. is the stitching visible on the inside of the cap)? [*Normally indicative of reapplied insignia. The cap may have been tampered with, even if original. Judgement call.*]

7. **Is there a white tag indicating cap size attached to the lower lining stitching**? [*Possible reproduction, or a modified contemporary cap.*]

8. If a Government Issue grade Other Ranks cap: **are there any air vent grommets**? [*Authentic Government Issue M43 caps (for Other Ranks) did not have any grommets. Applies to all services.*]

9. **Is the front** (vertical) **cap seam backed by an internal stiffener**? [*Indicative of a Bundeswehr cap that has been modified. Applies to all services.*]

10. Tropical: **Is the cap color noticeably green**? [*Modified postwar cap, or a flawed reproduction.*]

11. Tropical: **Are the air vent grommets painted with a green paint or enamel**? [*Incorrect color, the cap is suspect. Possible reproduction.*]

VISORED FIELD CAPS: WAFFEN-SS M43

1. **Is the two-piece insignia machine-applied**? [*Inspect the cap carefully for signs of a reproduction. The majority of Waffen-SS field cap insignia (except the trapezoid) was hand-applied; machine-sewn insignia therefore raises questions.*]

2. **Are the raw edges of the insignia backing cloth** (underlay) **exposed**? (That is, not folded out of sight under the insignia.) [*Correct application left no exposed (raw) underlay edges. Possible replacement (reproduction).*]

3. **Is the insignia applied without any padding beneath it**? [*The Totenkopf in particular was usually padded; lack of same invites closer inspection of the rest of the cap.*]

4. **If an officer cap, is the aluminum crown piping uneven** (that is, "lumpy")? [*Poorly installed piping is an indicator of amateur work and thus the cap has likely been tampered with in hopes of increasing value by claiming it as an officer piece.*]

5. Panzer: **Is the cap insignia in trapezoidal form, with a machine-woven eagle in** *aluminum* **thread** [BEVo]? [*No version of a Waffen-SS machine-woven trapezoid officer insignia (i.e. in aluminum thread) is known to have been produced. Reproduction.*]

6. Tropical: **Is the cap lining color on a Government Issue Other Ranks cap a color other than red**? [*Possibly a reproduction, or a modified contemporary cap. Red was the standard lining color. A very limited number of specific makers (no more than perhaps three) used tan or light gray linings.*]

VISORED FIELD CAPS: LUFTWAFFE M43

1. **Is the insignia machine-woven** (BEVo)? [*Almost certainly a case of reproduction insignia. Machine-woven Luftwaffe insignia is problematical – extremely rare, in any case. Inspect the attachment stitching and the rest of the cap very closely.*]

2. **Is the machine-embroidered cockade applied with the underlay cloth showing**? [*If so, the insignia may have been tampered with. The underlay cloth edges on machine-embroidered insignia were normally hidden.*]

3. **Is the cap color a very deep blue, and the material very thin**? [*Modified contemporary cap.*]

4. Officer cap: **Is the front edge of the scallop piped in aluminum officer cord**? [*Most likely either a reproduction, or a modified cap. In contrast to Army practice, use of officer piping along the scallop is rare on Luftwaffe officer M43 caps.*]

5. Tropical: **If a soutache is present, do the ends pass into the cap through holes cut in the material**? [*Not the normal attachment method. Inspect carefully.*]

VISORED FIELD CAPS: KRIEGSMARINE M42

1. **Is the cap color the Army/Waffen-SS shade of field gray**? [*Reproduction or modification; the Navy version of field gray was noticeably greener than the standard Army color.*]

1. **Is the national emblem insignia** [eagle] **machine-woven in off-white thread**? [*Navy insignia was golden yellow; thus, off-white or mouse gray would indicate an Army Other Ranks type, and is therefore incorrect.*]

2. Officer caps: **Is the cap piped around the crown with aluminum cord**? [*Navy officer's caps were piped in gold or golden yellow cord, not aluminum.*]

3. Tropical: **Is the cap lining red**? [*The standard lining color for Government Issue Kriegsmarine tropical visored field caps was green, not red.*]

4. Tropical: **Is the cap insignia underlay in continental style – that is, in blue wool**? [*The insignia is incorrect, and the cap therefore, is questionable. Continental insignia on wool underlay was not authorized on tropical caps.*]

5. **Does the cap have a soutache**? [*Incorrect! The Kriegsmarine visored field cap – including the tropical version – never used a soutache.*]

VISORED FIELD CAPS: BERGMÜTZE

ARMY:

1. **Is the visor quite long**? [*The cap is either (a) a modified cap (b) an incorrectly identified original, or (b) a mislabeled reproduction. The Bergmütze's visor was short and stubby, with fairly rounded sides. Long visors are characteristic of the M43.*]

2. **Is the insignia a machine-woven** (BEVo) **trapezoid**? [*Incorrect for the Army Bergmütze, especially if the trapezoid is hand-applied. Regulation insignia for this type of cap was the machine-woven T-style. A cap with trapezoidal insignia is questionable. Check for other inconsistencies.*]

WAFFEN-SS:

1. **Does the cap have an Army type metal Edelweiss on the side skirt**? [*Not regulation for SS caps; correct is the machine-embroidered (cloth) SS Edelweiss. Inspect the cap very closely for any additional inconsistencies. Judgement call.*]

2. **Is the national emblem** (eagle) **positioned together with an Edelweiss on the side skirt**? [*Incorrect positioning; if an Edelweiss insignia is present, the eagle should be located on the cap front above the Totenkopf.*]

LUFTWAFFE:

1. Officer cap: **Is there officer piping along the edge of the front scallop**? [*Questionable. Scallop piping was not common on Luftwaffe Bergmütze caps.*]

Final Note: These questions should help novice collectors zero in on the more obvious points to look for when evaluating whether a cap is not, in fact, what it is claimed to be. They will also serve to reinforce the recognition points for what is "bad" in a cap. In some cases where exceptions do (rarely) occur but are not the standard, the final decision on authenticity is a personal judgement call. Do not be afraid to ask a more experienced collector or dealer for a second opinion! Most dealers, if they don't like something about a piece, will explain to you what it is that they find objectionable – and thereby increase your knowledge.

Good luck!

Companies Arranged Alphabetically

A. Bohnsack
Adalbert Breiter
ADHERO
Albert Kempf "Alkero"
Alfred Valet Mützenfabrik
Almi Uniformen Mützen
Anton Baumgart
Anton Ertle
Anton Freiting
ARGO
August Geiger
August Mordasch
August Schellenberg Uniformmützen-Fabrik
Bayerische Mützenfabrik Wilhelm Schreiber
 "BAMÜFA"
Berolina
Biehler-Mütze
Carl Bangert
Carl Derwig Uniformenfabrik
Carl Halfar Uniform-Mützenfabrik
Carl Isken
Carl Lippold
Christian Haug Mützenfabrikation
Christian Nagl & Sohn Mützenfabrikation u.
 Kürschnerei
C. Louis Weber Uniformen
Clemens Wagner Mützenfabrikation
Kurt Dallüge
E.F. Kruger
Emil Schebeler Mützen- und Pelzwaren Fabrik
Emil Wolsdorff
Erich Weiblen
Eugen Köch
F. Eckhard Mützenfabrik

Franz Ritter
F. Wendl & Sohn
Erwin Freudemann
Felix Weissbach
Fillers
Friedrich Bürger Mützenfabrik
Friedrich Methmann "Nordmark Mütze"
G.A. Hoffmann
Gebr. Alm Uniform-Mützen-Fabrik
Gebr. Statter
Georg Grote
Georg Kurz
Gustav Binner
Gustav Oelkers Mützenfabrik
H. Ahlers
Hans Schiederer
Heinrich Balcke Hut- und Mützenfabrik
Heinz Schmidt
Herbert Grell Mützenfabrikation
Herbert Herbst
Hermann Gollhofer Kappenerzeugung
Hermann Potthoff "HPC"
Hermann Schellhorn Feuerwehrausrüstungen
Holters Uniformen
Hut Kelg
Hutrika
Jean Drescher Uniformfabrik
J. Lettel
J.B. Holzinger
Johann Frey
Josef Rom Mützenfabrik
Josef Weithmann Militäreffekten
J. Sperb Mützenfabrikation
Karl Colling Mützenfabrik

Karl Naubert
Karl Rienäcker
Knut & Wiese Mützenfabrikation
Kurt Kläber
Kurt Triebel Mützenfabrik
Leonhard Paulig "LEPARO"
L.G. Schwind
Ludwig Vögele Mützenfabrik
Marke ODD
M. Drecksler Mützenfabrik
Michael Keck Militäreffekten
Mützen Scherff
Mützenfabrikation August Müller
Opolka & Müller
Ostland Mützen u. Bekleidungsindustrie
Otto Schlientz
Paul Kap
Paul Rienäcker
Paul Wagenmann
Peter Haubrich
Peter Küpper "Peküro" [postwar: "Codeba"]
Robert Lubstein "EREL-*Sonderklasse*"
Rudolf Ruf "RR"
Schmid & Menner
Spohn & Klaiber
Sport Metzger
Steinmetz & Hehl Mützenfabrikation
Th. Wormanns
VIRO
Wilhelm Schwarte Mützenfabrikation
Wilh. Stellrecht Mützenfabrikation
William Günther Sächsische Militär-Effekten-
 Fabrik
Willy Sprengpfeil Mützenfabrik

The Difference Between the D.R.G.M. and the D.R.P.

- Deutsches Reichsgebrauchsmuster - D.R.G.M.
- Deutsches Bundesgebrauchsmuster - D.B.G.M.
- Deutsches Reichspatent - D.R.P./D.R.P. angem.
- Gesetzlich Geschützt - "Ges. Gesch."

German patent law prior to and during the Second World War (and today, as well) offered two main levels of protection for inventions submitted by individuals or companies. One was the Gebrauchsmuster, (registered design) often called a "mini patent", for which there is no equivalent in U.S. patent law. The other was the full Patent. Both provided protection for technical inventions; both required that the invention be "new" and that it evidence a degree of "inventive achievement". The patent however, was very strict in both its submission format and in its requirement for proof of originality.

With the Gebrauchsmuster, the primary concern of the Patentamt was that the invention fall within an acceptable category for this form of protection; some items, such as a technical process, did not qualify. Absolute proof of originality, however, was not required, and the Gebrauchsmuster, therefore, was considered an "unproven" form of protection. As a result, the responsibility for defending an invention's originality against potential rival claims rested solely with the applicant. The chances for possible contention could be reduced, however, if a search for like inventions already recorded in the Gebrauchsmuster files was conducted prior to submitting any application. When different companies or individuals filed applications for similar inventions, the date of submission established priority.

A Gebrauchsmuster or Patent allowed the company to use a protection notice printed somewhere on or within the registered item. Such notices appear as follows: D.R.G.M. or D.B.G.M. (for a Gebrauchsmuster), or, if for a patent, D.R.P. or D.B.P. (see explanation below). Alternatively, the abbreviation Ges. Gesch. could also be used, though this was more often associated with a trademark registration.

D.R.G.M. / D.B.G.M. – The GEBRAUCHSMUSTER

A Gebrauchsmuster offered the following advantages:

1. Simplified application process: Required was a specific title for the invention along with its purpose, and a simple but precise description of which aspects of the item were to be protected (*Schutzanspruche*). An explanation of the design and how it functioned, and a brief declaration describing why the item was unique. The application usually included a basic diagram of the invention and possibly a sample.

2. Quick turnaround: An invention could be registered and a Gebrauchsmuster awarded within a month. The day the application was submitted served as the basis for assigning precedence should more than one individual or company submit an application for a similar invention.

3. Although the invention had to be "new", the only real requirement was that it could not be something that the average technician in the field could devise based on the current technical level within the industry – that is, it had to represent an *inventive advance.*

4. Inexpensive application and renewal fees.

5. The D.R.G.M. was sufficient to ensure that any other firm wishing to make use of the invention would have to secure a license from the D.R.G.M. holder.

The principal disadvantage to a D.R.G.M., as opposed to an actual patent, was the shorter length of time that the D.R.G.M. protection remained in force. Protection against infringement was only three years, at the end of which time (for a fee) the rights could be renewed for an additional three-year term. After the sixth year, two further renewals were possible, each for a two- year period. The limit on protection afforded by a D.R.G.M. was thus limited to a maximum of ten years, after which time the invention became public domain, available to all. Of course, if the required renewal fee was not paid, the Gebrauchsmuster lapsed at the end of the current period and the invention reverted to public domain and free access.

Gebrauchsmuster inventions falling within the category of caps were further sub-divided into a class. There were several primary classes, the most significant of which – for our purposes – appear to have been Class 41b and Class 41c.

It is important to note that the terminology used by German government offices was changed after the war to reflect the new Bundesrepublik [Federal Republic]. The terms Deutsches Reichspatentamt [German National Patent Office], Deutsches Reichsgebrauchsmuster and Deutsches Reichspatent were too well associated with Nazism.

The new titles were Deutsches Patent- und Markenamt [German Patent and Trademark Office], Deutsches Bundesgebrauchsmuster [also Bundes Gebrauchsmuster], and Deutsches Bundespatent, respectively. Any cap therefore, with a makers mark that includes D.B.G.M., B.G.M., or D.B.P. (that is, *B* for Bundes, instead of *R* for Reichs), represents a postwar item.

D.R.P. / D.R.P. angem. – The PATENT

The second – and most extensive – level of protection was an actual patent. A patent was (and still is) very difficult to secure. It required a very specific submission format with painstakingly detailed supporting documentation and an exhaustive examination of all previously recorded inventions within the same class, in order to prove beyond any doubt the absolute originality of the invention or process to be patented.

The time from submission of the patent application and supporting documents to the actual award might take from two, to two and a half years (or more). The period of protection offered was much greater than the ten years covered by the Gebrauchsmuster, and patent protection was also open to certain categories not available to the "mini-patent". Initial patent protection was for three years, but the maximum renewal limit totaled twenty years.

During the lengthy period between patent submission and actual approval and award by the Patentamt, German companies before and during the Second World War often used the phrase *D.R.P. angem.* as a warning notice, printed on a product that included or made use of the invention in question. An abbreviation for *Deutches Reichspatent angemeldet*, this phrase translates to the English "patent pending". On *EREL-Sonderklasse* caps, for example, this abbreviation is commonly found on the left temple area of the sweatband.

GES. GESCH. – REGISTERED (TRADEMARK)

In some cases, a company might prefer to use only a general notice, such as "Ges. Gesch.", short for "Gesetzlich Geschützt" , meaning literally "legally protected" and the phrase frequently appears with patent or D.R.G.M. numbers in *Uniformen-Markt* articles. In general, however, the term was used in this sense only when the company did not have a trademarked logo name, since it was normally intended to mean "registered" – equivalent to the English ® – in the sense of a registered trademark. Thus, Ges. Gesch. is commonly found somewhere near the logo on caps made by companies which had a legally registered trademark, or name (actually, there were not many companies that had a trademark logo).

Interestingly enough, Gebrüder Gloerfeld, a company whose Gebrauchsmuster is shown on page 398, was a competitor of the famous F.W. Assmann & Söhne firm, which was also located in Ludenscheid. An issue of *Uniformen-Markt* offers an entry in the recurring industry news column "Chronicle" which notes the unexpected passing of Herr Rudolf Aßmann on New Years day of 1942. At the time, he was 44 years old; the cause of his sudden death was identified as resulting from wounds received years before during World War One. Rudolf Aßmann held the Iron Cross first class, and the Wound Badge in silver from the Great War.

Acknowledgement: All the patent documents and illustrations used in this Appendix were supplied courtesy of the Deutsches Patent- und Markenamt Technische Informationszentrum (TIZ).

DEUTSCHES REICH

AUSGEGEBEN AM 3. AUGUST 1940

REICHSPATENTAMT
PATENTSCHRIFT
№ 694 529
KLASSE **41 b** GRUPPE 2
L 91394 VII/41 b

✳ **Robert Lubstein in Berlin** ✳

ist als Erfinder genannt worden.

Robert Lubstein in Berlin

Kopfbedeckung mit Schweißband

Patentiert im Deutschen Reiche vom 5. November 1936 ab

Patenterteilung bekanntgemacht am 4. Juli 1940

Um bei Kopfbedeckungen, insbesondere bei festen Kopfbedeckungen, den durch den Rand der Kopfbedeckung ausgeübten Druck zu be-
seitigen, ist schon vorgeschlagen worden, zwi-
5 schen dem Schweißleder und der Innenfläche der Kopfbedeckung ein Luftkissen zu schaffen, indem zwischen der Unterkante des Leders und dem Rand der Kopfbedeckung ein ge-
gebenenfalls auch durchbrochener Gewebe-
10 streifen eingesetzt wird, dessen mit der Kopf-
bedeckung verbundene Kante, z. B. infolge Verziehens beim Einnähen, länger ist als die mit dem Hutleder verbundene Gegenkante. Ein in dieser oder ähnlicher Weise gebildetes
15 Luftpolster ist jedoch nur während sehr kur-
zer Zeit in dem beabsichtigten Sinne wirk-
sam, da sich beim Gebrauch der Kopfbe-
deckung unter der Einwirkung von Wärme und Feuchtigkeit das aus Leder oder Leder-
20 ersatz bestehende Schweißband dehnt.

Es ist auch schon vorgeschlagen worden, auf einen perforierten Stoffstreifen längs einer Kante ein schmales Band aufzunähen, um die Elastizität des Stoffstreifens auszuschalten und durch die Naht gleichzeitig eine Verkürzung 25 der einen Kante herbeizuführen. Beim Einnähen des Schweißleders mit Hilfe eines so vorbereiteten Zwischenstreifens wird das Schweißleder ebenfalls schwebend getragen. Durch das aufgenähte Band wird aber ein 30 Dehnen des Leders auf dem ganzen Umfang verhindert. Das durch das schwebend ge-
tragene Leder gebildete Luftpolster ermög-
licht also nur eine Anpassung an die Kon-
turen des Kopfes, aber wegen der völligen 35 Ausschaltung der Elastizität keine Anpassung an die Kopfweite.

Erfindungsgemäß wird das Schweißband ebenfalls unter Zwischenschaltung eines die Bildung eines Luftpolsters bedingenden Zwi- 40

"Kopfbedeckung mit Schweißband" [Headgear with Sweatband]. Patent No.694529. First page of a patent awarded in 1940 to the Robert Lubstein company (maker of EREL-Sonderklasse) caps. The invention is an "air cushion" sweatband, characterized by three layers in the forehead area which were designed to trap air and thereby form a cushion which made the cap more comfortable to wear. In addition, the construction prevented the sweatband leather in the forehead area from undesirable stretching, while allowing the rest of the sweatband to conform to the head contours and width. Either this patent, or the vented cockade system is represented by the *D.R.P. angem.* mark on EREL sweatbands. Note: Blacking-out of swastikas is quite common on German official documents dating to the National Socialist period. In this case, done by the Patentamt.

Right and below: Diagrams of the Lubstein invention, submitted as part of the patent application.

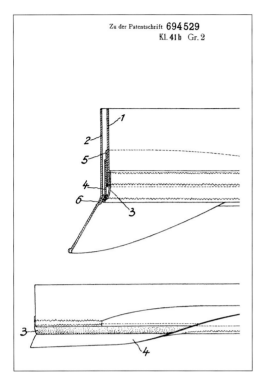

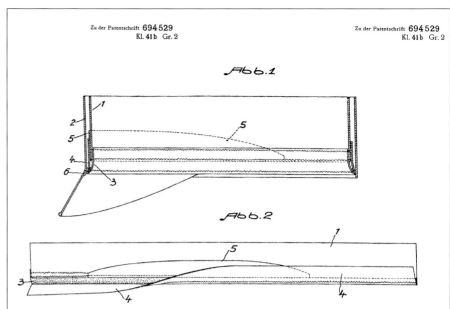

Right: August Schellenberg's application for a Gebrauchsmuster, submitted to the Reichspatentamt in October of 1937. The Gebrauchsmuster was awarded under No. 1 383 756 Class 41b. The text reads:

To the Reichspatentamt,
Berlin   22 July 1936
The undersigned, cap manufacturer August Schellenburg, Berlin C 27, Alexanderstr. 40, hereby submits the attached prototype and requests its entry in the Gebrauchsmuster Register.
The designation is: Device for the Prevention of Forehead Pressure in Caps.

Claimed as new:
With visor caps, the hard visor of Vulkanfiber or leather attached to the inside of the cap exerts an uncomfortable pressure on the wearer. This pressure is prevented if a soft intermediate layer of foam rubber, which must extend from one edge of the visor to the other, is attached to the cap band directly over the visor edge between the band and the sweatband.
   The fee of RM. 15.- is being simultaneously transferred to the Patentamt's postal checking account. Attached to this correspondence are:
   A duplicate copy of this letter, a prototype as well as a prepared receipt.

Heil Hitler!

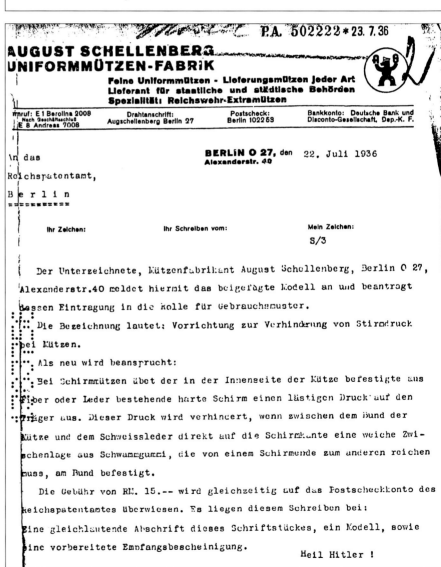

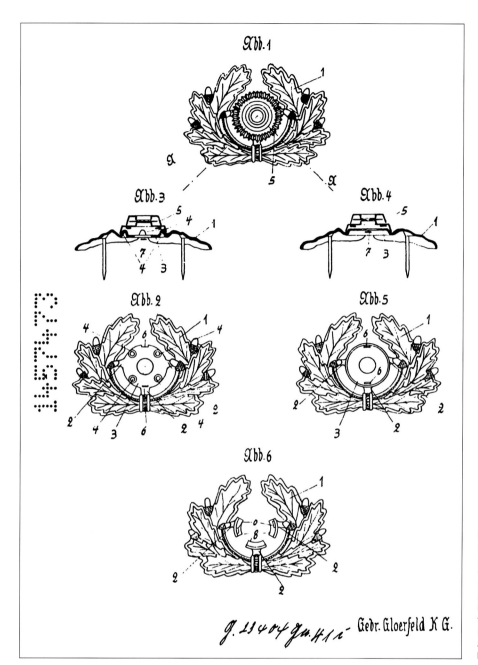

Gebrauchsmuster diagram page for an oak leaf wreath and cockade arrangement registered by the Gebr. Gloerfeld company of Ludenscheid under D.R.G.M. No. 1457 473, Class 41c "Wehrmachts-Mützenkranz". The special cockade base plate in Figure 2 [Abb.2] has raised nubs (Item No.4) which press against the interior edge of the cockade and thereby block it from moving in any direction. Figure 5 shows an alternate variation in which the edges of the plate are raised into a sectioned rim rather than nubs; the rim sections press against the inner, bottom edge of the cockade base and thus keep it correctly positioned within the wreath.

Insignia and Accessories

INSIGNIA & BUTTONS

Machine-woven (BEVo) insignia: Reproduction/Original
Following are two examples of reproduction machine-woven (i.e. flat) insignia, and an original. All have been cut from large rolls, and the background material is therefore rectangular. The accuracy of some current reproductions – particularly those manufactured in South Korea – is of extremely high caliber; no visual inconsistencies or proportional errors may be evident. The thread pattern is also excellent, both on the front, and rear of the insignia. Thus, the only means

Front side of three types of machine-woven insignia. The first example is a reproduction Waffen-SS Panzer [officer's] trapezoid, in aluminum thread on black rayon; however, while very well done, no authentic example of such an insignia has ever been seen. Indeed it has never been confirmed that such an officer's insignia was, in fact, ever produced. The Army example in the center is authentic, machine-woven in white thread (early form) on a black rayon backing, for use on a Panzer M34, or M43 cap. The third (bottom) insignia is a reproduction Army officer's national emblem in aluminum thread on a dark green rayon backing. The green backing is slightly off from the correct shade. Both reproductions were manufactured in South Korea.

Rear view. The "negative" pattern made by the thread on the insignia reverse is nearly identical between reproduction and original.

of detecting such reproductions may be flaws in the method and quality of the application – i.e. the sewing.

The study of cloth insignia is a field of its own, and something every collector needs to take time to review. When attending militaria shows, use every opportunity available to study the insignia attachment on original caps. Uncut original insignia examples should also be viewed whenever possible.

Hand-embroidered Bullion Insignia - Reproduction

Handgestickte [hand-embroidered] bullion insignia is one of the few areas where German manufacturers made occasional errors. While metal and machine-woven (BEVo) insignia are usually flawless (with examples of the latter showing almost no variation within a production run), each and every piece of bullion insignia was individually hand made. Variations in the bullion wire used, slight errors that occurred from time to time during the actual embroidery operation, and the embroiderer's own style all contributed to differences from one insignia to the next. Most examples are very well made, but now and again spacing between oak leaves might be a tad off, acorn stems perhaps a little crooked, swastika arms a little "crowded" within the wreath…the head perhaps a bit stubby. These individual variations are very beneficial to reproduction insignia makers, since they allow any errors in a reproduction to be passed off as "idiosyncrasies" in an original insignia. Despite minor flaws, however, the German embroiderers were for the most part well trained, with varying degrees of experience. The large numbers of insignia produced on a daily basis at German insignia factories built experience quickly – and any major mistakes were almost certainly trashed at the factory. Small discrepancies appear to have been acceptable to German manufacturers, and an insignia with minor flaws may thus be explained away; major flaws, however, can not be excused on the grounds of authentic production mistakes.

The very same difficulties the Germans faced in producing hand-embroidered bullion insignia exists today among reproduction makers, who produce far smaller quantities of insignia (no major demand), and therefore have far less experience in producing pieces of consistently high quality. Errors in proportion are usually the key indicators of reproductions, as evidenced in the two reproduction examples that follow.

The eagle in this case is a very poor quality reproduction. Aside from the fact that only <u>one</u> type of wire is used (there should be crinkley wire highlights to the top edges of the wings and the inner feathers). The proportions of the circular wreath and swastika are completely incorrect: the wreath ring is much too small, the swastika arms truncated since they would not otherwise fit into the little remaining space. A close inspection of this eagle will also reveal poor quality backing cloth, not the badge cloth wool normally used in German production. The application is also poor. This wreath is a postwar item. Its design is completely incorrect for any pre-1945 German Army wreath in shape, excessive height, indistinct jumble of oak leaves, and green (not bluish "bottle" green) backing cloth.

Another example of a reproduction hand-embroidered Army eagle. In this case, the oak leaf wreath is too large; the swastika arms are once again out of balance (and proportion). Though the correct crinkley wire is used on the inner feathers, it is used for the entire inner wing area, which is not correct, either. The eagle's head is recessed well below the upper edge of the wings (completely wrong), the body sides taper down and outward instead of being straight, and the leg area is formed as an ugly, squat, inverted V shape (wrong). The cap sporting this insignia was identified as an original— but the national emblem, at least, is easily identifiable as a fake and with that the entire cap is thrown into question.

Left: An original cap eagle and oak leaf wreath for comparison. This is a matching set. The wreath includes the high-profile (3D), all bullion cockade. The insignia was produced at about optimum level – with no significant production idiosyncrasies of note on the eagle. The head is at the usual (standard) height: Slightly above the top edge of the wings'. The swastika is the correct size, canted at the correct angle, and well formed (as is the wreath ring itself. The inner feathers and the top vein of the wing are embroidered with the correct crinkley wire; also correct are the non-crinkley highlight wires between each feather.

Standard pebbled aluminum Army, Waffen-SS (including generals) or Luftwaffe officer's cap cord button, with the correct twisted aluminum cord loop. Note the positioning of the button on the cap band in relation to the end of the visor, and the lower cap band piping. This example is from a Waffen-SS cap.

Kriegsmarine officer's chinstrap button, with the end of the patent leather chinstrap.

BUTTONS, CORD LOOPS AND KNOTS

Authentic buttons and cord loops used for Army, Waffen-SS and Luftwaffe caps, and leather cap band and button for a Kriegsmarine officer cap are shown at left.

More on the EREL Air Vent and EREL Maker Marks

The ventilation system developed and marketed by Robert Lubstein was an optional item offered on EREL-*Sonderklasse* caps. As far as is known, this system was never offered with military caps from the Robert Lubstein line (eventually replaced in any case by the EREL-*Sonderklasse* cap line).

The foil or paper tag used to advertise the ventilating system has already been introduced in the brief discussion of this topic in Chapter 16 under the Lubstein maker entry. Examples of the various insignia types found with the vented cockade screen are presented below. It should be noted, however, that some authentic caps do appear with the cockade and the original advertising tag – but without the actual vent hole through the cap band and interior pasteboard band. Why this is so, is not known. In general, lack of the actual vent is normally an indication of a reproduction, particularly if the paper tag is printed in black and white.

Cap interior showing the maker mark and the ventilation system advertisement tag, this example in foil. *Courtesy of Klaus-Peter Merta and the Deutsches Historisches Museum, Berlin.*

Left: Pre-1943 general's cap (EREL-Sonderklasse) made by Robert Lubstein, with the vented cockade. Gold bullion piping and chincords. This cap is from the historical military collection of the Deutsches Historisches Museum, Berlin. *Courtesy of Klaus-Peter Merta and the Deutsches Historisches Museum, Berlin.*

Army Leichtmetall (aluminum alloy) wreath and metal cockade with ventilation mesh. *Courtesy of Gerard Stezelberger of Relic Hunter.*

Right: Luftwaffe officer's hand-embroidered bullion wreath and cockade, with ventilation mesh (compare with the vented gold celleon general's cockade in Chapter 5). *Courtesy of Gerard Stezelberger of Relic Hunter*

The cap dates from around 1934 or earlier and has a very rarely seen circular form Offizier Kleiderkasse etiquette—though this appears as usual, above the EREL diamond logo mark. Note that the sweatshield shape is itself a diamond, not the rhomboid usually found on military caps of later manufacture made by Robert Lubstein. *Courtesy of Gerard Stezelberger of Relic Hunter*

The text of the advertisement tag states that ventilation is achieved by means of a "special cockade or insignia" (spezial Kokarde oder Abzeichen). Waffen-SS caps with ventilation appear to have used the normal Leichtmetall Totenkopf insignia (the exterior form is unchanged), with the eyeholes serving as the air inlet.

Army gilt aluminum wreath (for general's caps after January 1943), metal cockade with ventilation mesh. *Courtesy of Gerard Stezelberger of Relic Hunter..*

Army hand-embroidered bullion wreath and metal cockade with ventilation mesh. *Courtesy of Gerard Stezelberger of Relic Hunter.*

ADDITIONAL ROBERT LUBSTEIN MAKER MARKS:

As mentioned earlier under the Robert Lubstein maker entry, Reichswehr and early Third Reich period EREL-Sonderklasse caps often used a diamond-shaped sweatshield instead of a rhomboid. Following is an example of an EREL-Sonderklasse cap offered through the [Heeres] Offizier Kleiderkasse, Berlin.

Important Institutions

- Handwerk
- Gewerbeamt
- Handelskammer
- Industrie und Handelskammer
- Amtsgericht and the Handelsregister

The German business classification system was and still is, relatively complicated. Under this very old system, any small to medium-sized business may be classified as a Handwerk [handcraft/tradecraft], provided the following three conditions are met:

1. The company is rated in the small to medium-sized business category;
2. The business activity or product falls into one of the fields classified as a Handwerk;
3. The business does not qualify for registration at the Chamber of Commerce and Industry.

Handwerk categories cover some sixty-five different fields and include fashion clothing and textile manufacture, furrier work, cap-making and many others. It is the largest single business segment in the modern German economy, and back in the 1800's, provided the economic backbone for each of the independent German kingdoms until these were unified into a single nation under Prussian leadership, by Bismark.

Small or medium-sized companies in the Handwerk category normally registered with the local Handwerkskammer (in effect, a form of small business trade chamber based on skilled handcraft professions). Such registration was on a voluntary basis. On the other hand, every business – regardless of type, size or other category (with few exceptions) – was required to register with the local city Gewerbeamt [Business Bureau] before it could legally open its doors to the public.

If the business was rated too large (ergo, an industrial rating) to qualify for registration with the Handwerkskammer, then a company could move up a notch to the local Industrie and Handelskammer (IHK) branch (Chamber of Commerce and Industry). Again, this registration was voluntary, but the IHK offered useful assistance to companies by conducting business inspections and offering training courses. More importantly, it also submitted official application letters to the city Amtsgericht [Justice Ministry] office on a firm's behalf, recommending the business for entry in the Handelsregister rolls [Commercial Registry] and certified that the firm met the necessary financial and business qualifications. The Amtsgericht offices, which actually maintain the Handelsregister records, have cooperative relationships with their local IHK branches.

Thus, all of these agencies, Handwerkskammer, Gewerbeamt, IHK and the Amtsgericht, serve as important sources for historical documentation. The primary requirement for entry in the city's Handelsregister rolls was that a company must be run in a fully "mercantile" fashion [vollkaufmännisch geführt], with a specific level of sales and a specific form of bookkeeping and inventory control. A document to this effect, including annual sales figures, monthly projections and company net worth, had to be submitted to the Amtsgericht together with the Handelsregister application.

The company's form of business structure, be it a limited partnership company [Kommanditgesellschaft], a general partnership company [offene Handelsgesellschaft], or some other form, was generally inconsequential with the exception of a limited liability company, *Gesellschaft mit Beschränkter Haftung* or "GmbH" [similar to English L.L.C.]. This particular business form was (and still is) legally required to register for entry with the Handelsregister.

Once actually listed, a business had to continue to satisfy the original qualification requirements in order to maintain its position. If it failed to do so, the Amtsgericht usually closed out the company's active registry file and removed the firm from the register (though the file documentation remained on hand). The IHK generally followed suit. This did not necessarily mean that the company had physically gone out of business (though this was usually the case). Sometimes, though no longer being actively registered with either institution, the company may still have existed and remained in business. On the other hand, sometimes a firm quietly went out of business with this fact apparently escaping the notice of the Amtsgericht, in which case the file remained open and was never officially closed out. The end result of these different scenarios is that it is sometimes quite difficult to determine exactly when a business really did close its doors permanently.

Perhaps the greatest power of the Amtsgericht over companies listed in the Handelregister is its authority to initiate an insolvency investigation for any of these firms when petitioned by creditors, and then to conduct subsequent legal proceedings if such are warranted. The opening of an official investigation is officially noted in the company's Handelsregister file. With the start of actual legal proceedings, the firm is then stricken from the Handelsregister roll and officially terminated as a viable, legal business entity. If the company somehow rebuilds and starts the business anew at a later date (after fully servicing all of its debts), it is nonetheless legally barred from reusing its former name. Thus, for all intents and purposes, the original company is, in fact, terminated forever. In the case of the famous maker Clemens Wagner (Braunschweig), for example, the company may eventually have reopened after being found insolvent – assuming it paid off all of its debts; however, the famous Clemens Wagner name could never be used again.

If a firm was never listed on the Handelregister, and if the Gewerbeamt records no longer exist, then it becomes impossible to determine the year the firm was founded, or to even prove its existence. This state of affairs is entirely possible, since many Gewerbeamt offices regularly destroy old records. Only three possible sources then remain: old Handwerkskammer records (but remember that registration at the HWK or IHK was on a voluntary basis); old address books from the city archive, or some commercial source – an advertisement in Uniformen-Markt, for example. The possibility always remains, however, that no physical record will be found for a company having ever existed, should these last sources also turn up blank.

The Heeres [Offizier] Kleiderklasse Catalog

The Army Kleiderkasse published a yearly catalog of available goods ranging from caps, to officer's daggers, to uniform cloth to storage trunks. Catalogs were published through 1944, though it is not known with certainty whether a 1945 issue was ever released.

In general, the next annual catalog was released in August or September of the current year, with the new prices going into effect as of October. Thus, to give an example, the 1936 catalog was released prior to October of 1935, with the new catalog prices effective as of 1 October 1935. The 1936 catalog exists in a reprint published many years ago, apparently by a European auction house; the availability even of the reprints is at best rare. They are extremely difficult to find. Even more difficult to locate of course, are authentic catalogs (from any year, but in particular those published during the war).

In order to gain a good sense of just how much a private purchase [Extramütze] cap could cost an officer, it is very useful to compare prices between a range of items offered from the catalog, rather than merely read the price of a cap from the price list. No specific manufacturers are identified in the item descriptions; this may have been a policy with the Kleiderkasse. For example, with caps, only the model is given, identified as *Sonderklasse*; but since we know that Robert Lubstein had an exclusive contract to supply caps to this institution, we can assume that this probably meant "EREL-*Sonderklasse*" (i.e. caps made by Lubstein). Average grade, generic officer caps manufactured for government stocks might also be meant, but these were more likely to have been offered only through military divisional supply depots, not the Kleiderkasse. No, it is almost certain that such *Sonderklasse* caps were indeed ERELs.

Thus, the Kleiderkasse catalog provides us with an idea of the cost for Lubstein caps sold (at discounted prices) directly through the Kleiderkasse. No doubt, at a haberdasher or military goods store, the price for the same EREL-*Sonderklasse* cap model would have been perhaps ten to fifteen (or more) Reichs Marks higher.

Note that 100 Pfennigs equaled 1 Reichs Mark, and that with numbers, Germans use a comma where Americans use a period. Here then, is a list of various items that an officer would need to purchase at one time or another – including caps of course. Each entry is taken from the Heeres Kleiderkasse 1936/37 Price List [catalog].

MÜTZEN [*CAPS*]

Dienstmütze [service visor cap] with insignia, no cords; made of the finest Eskimo, "Sonderklasse" model, top workmanship with silk lining, elegant form.
Price: RM 6,90 (per item) [i.e. Six Reichs Mark and ninety Pfennigs]

Dienstmütze for Generals with insignia, no cords; made of the finest Eskimo, "Sonderklasse" brand, top workmanship with silk lining, elegant form.
Price: RM 11,—

Feldmütze [Old-style field cap] in doeskin, with soft leather visor and silk lining
Price: RM 5,75

The same, but for Generals.
Price: RM 9,75

Eichenlaub [*Oak Leaf Wreath*]

[Hand-] embroidered, with metal national cockade: *RM 2,30 per piece*

[Hand-] embroidered, with embroidered cockade: *RM 2,90 per piece*

In metal [Leichtmetall – aluminum]: *RM 0,40 [forty Pfennigs] per piece*

Machine-woven, for [old-style] field cap: *RM 0,15 [fifteen Pfennigs] per piece*

Cap cords, gold: *RM 5,—*

Cap cords, aluminum: *RM 1,75*

Cap case, oval, covered with black artificial leather, sufficient for packing two caps or helmet and one cap: *RM 7,50*

Cap trunk: black autoduck; nickel lock, moire lining; sufficient for packing two caps: *RM 14,50*

SHOULDERBOARDS:
For the Waffenrock, without insignia, from aluminum thread
For Generals: *RM 7,75 per pair*
For Lieutenants, etc.: *RM 1,80 per pair*

Matte, for the Field Blouse, without insignia
For Staff Officers: *RM 4,40*
For Lieutenants: *RM 1,60*
For Officials, with Rank of Lieutenant to Captain: *RM: 5,75*

COLLAR TABS:
for the Waffenrock
For Generals, gold-embroidered: *RM 12, – per pair*
For officers of all branches, hand-embroidered in aluminum: *RM: 7, – per pair*

SIDE ARMS
Standard Officer's Saber with double gilted hilt: *RM 21,—*
Officer's Dagger in best quality: *RM 13,80*

OFFICER'S [Storage] **TRUNKS**
Leutnant's [Lieutenant's] Trunk with waterproof cover, metal fittings; length 60 cm, width 38,5 cm, height 36 cm, Quality level II: *RM 30,—*

The same, but Quality Level I: *RM 36,—*

GLOVES

Pigskin Gloves, imitation, gray; one button or slide [*closure*]: *RM 6,—*

Pigskin Gloves, natural leather, gray; one button or slide: *RM 10,50*

Lined Nappa Gloves, clasp closure, sheepskin lining, gray: *RM 11,50*

SHIRTS

Sport shirts

Silk sport shirt *"Herold"*, raw silk, tan color: *RM 13,75*

Silk sport shirt *"Senator"*, multi colored, pure Japanese silk : *RM 14,—*

Riding and Service Shirt *"Hubertus"*, oxford material, green speckled: *RM 7,20*

Duty shirt for Panzer troops, grayish black poplin, with two collars: *RM 8,50*

In looking over the various prices, it is interesting to note that a pair of officer's hand-embroidered collar tabs cost more than a visor cap; a civilian style "Senator" sport shirt cost twice as much, while the top model *Handschuh* [gloves] was more than a third higher. An officer's dagger cost twice as much as a visor cap, and *more* than twice the cost of an old-style field cap. A hand-embroidered [officer's] wreath and embroidered cockade cost nearly half of the price for the cap itself. The cap cords (and buttons), which had to be purchased separately, cost the same as a pair of shoulderboards. Every model of shoe and boot listed in the catalog cost over ten Reichsmarks, with the highest coming in at RM 28,50.

A white summer uniform tunic (without shoulderboards) cost only RM 14,50 for in-stock sizes (44-52), about the same price as a general's cap with cap cords. A tunic ordered in non-stock sizes (i.e. custom) would cost an additional two marks (delivery time: 14 days).

Caps purchased through the Kleiderkasse were clearly not cheap; on the other hand, they were nowhere near the most expensive uniform item that an officer needed to purchase (despite being made by Robert Lubstein)!

More research needs to be done with the aim of ascertaining – if at all possible – the actual average market price for such a cap at local shops, as opposed to the purchasing power of military pay at the time. Only this will help provide a truly accurate picture of the real cost of an *Extramütze*: how much of a dent, so to speak, such a purchase put in the buyer's bank account. Would that these caps were still available today at those prices!

Care and Maintenance of Headgear

PRESERVATION

A cap collection represents a significant monetary investment, the value of which will increase over time. Interest in militaria in general has never waned over the fifty-five years since the war ended, nor does it appear likely to in the future. Prices have increased dramatically over the past fifteen years, and all indications are that they will continue to climb. A Luftwaffe officer's visor cap that cost $475 in 1985 was valued at $925 in the year 2000. Cap collecting has clearly become a very expensive hobby. With this in mind, care of the caps in your collection is an important consideration, together with convenience in displaying them.

For cleaning caps of pet (or human) hairs and lint, scotch tape is recommended. The adhesive material on packaging tape and other heavy duty tapes (including cloth backed tape) is too sticky, and may pull the cap band felt or damage a doeskin surface if used for cleaning loose matter from the cap material. Any stains present on the cap cover or elsewhere should be left untouched; they add to the history of the cap, and attempts at removal may discolor or damage the cap material. If a cap in top condition (i.e. without stains) is more to your taste, then this is what you should purchase in the first place.

Preventing Damage to Sweatbands

Sweatbands are fragile and are often damaged, or are already damaged. Care in handling is absolutely vital, particularly when pulling down the sweatband to get a look at the reverse. Perforated sweatbands are particularly susceptible to ripping, since the perforations reduce the strength of the leather surrounding them. There are a number of leather preservatives on the market, which will return a degree of moisture to the leather without damaging it in other ways. This reduces the potential for cracking or tearing, however it should be noted that many dealers and headgear collectors object to the use of such preservatives and feel that the value of a cap may be adversely affected by any such treatment. In the end, the decision to proceed lies with you, the cap owner. For those who wish to use such leather preservatives, several products are listed at the end of this appendix. For sweatbands already exhibiting tears, some collectors advocate repairs in order to prevent damage from worsening; others once again feel that any such repairs lessen the value of the cap, since they are not original. This is another judgment call, but if done, then repairs can be made with cotton thread in a matching color, or a piece of cloth tape applied to the reverse side of the sweatband over the area of the tear. Ersatz leather sweatbands that have begun to flake can not be repaired, though the rate of additional flaking can perhaps be at least slowed somewhat by refraining from handling the band except when absolutely necessary.

As the moisture leaves the leather, the sweatband will tighten up to a degree. For this reason, it is very important to keep your caps on display heads (Styrofoam or hard molded) in order to keep the sweatband stretched to the proper position. In some cases, a cap size may be too small to fit the standard head; find something smaller to use – never overstretch a sweatband. On the other hand, some cap sizes will prove too large to fit correctly on the standard display head. In this case, you can fashion a handmade headband or two out of folded paper in order to take up the slack; make sure however, that the thickness of the paper band(s) is the same throughout its length. Once these strips are inserted between the head and the sweatband, this will prevent any potential warping of the leather. Also, make sure that you use acid free paper (paper for artists, for example, is often acid free: look for such a statement on the pad). This is important, as some acids commonly used to bleach and process paper may also leach out and damage the sweatband leather over time. Even when the cap size fit is correct, a single sheet thickness of paper should be used between the head and the sweatband in order to prevent any chemical crossover from the Styrofoam itself. Styrofoam is actually a product name that over time, has become a generic term for EPS, or expandable polystyrene. EPS is a styrenic resin, with a principal raw ingredient of styrene monomer – a volatile liquid derivative of petroleum distillation and cracking. While EPS products that come into contact with food are strictly regulated in terms of the amount of styrene monomer that may leach out, other products have few or no such limitations.

Preventing Visor Damage

Another primary reason for using display heads for your caps is to protect the stitching used to attach the visor. Setting a cap directly on a shelf puts a strain at each corner of the visor where it joins the cap body (this part of the cap is not resting on the shelf). The ends of the visor are often the first place where the visor attachment threads break. Once one visor end is loose, the rest will eventually follow. Mounting a cap on a display head completely removes any damaging pressure.

Caps should not be displayed in a location that receives direct sunlight. As well as the fading of the piping color that sunlight causes, periods of prolonged exposure with significant changes in temperature can also start or increase the rate of fine surface cracking in the visor's lacquer coating, known as *crazing*.

Moth Damage

Moth damage is a common problem with headgear collectibles made from cloth in general, and wool in particular. "Tracking" is the most minor form of damage unless it takes up a very large area, while "nips" are more serious. There are degrees of tracking, with the lowest degree (and most common) being a slight change in material color where a bit of the wool surface has been nibbled away in the manner of sheep grazing on a grassy field. Such damage tends to occur either in patches, or in lines – hence the term "tracking". Sometimes, moth larvae will focus on one particular area (something about the wool used for piping, for example, seems to hold a special allure), and actually eat completely through the material leaving either a hole, or if the target is piping, leave large chunks missing. The question of whether to make repairs (not much can be done with piping damage) or not rests with the owner; again, any repairs made may decrease the value of a cap. If the damage is so severe that the cap has little value as is, however, then there is sometimes no good reason not to make such repairs, particularly if you never plan to re-sell the cap in the future. At least the appearance can be improved to some degree. This is particularly true with visor caps if there are major moth holes *under* the side overhangs, or under the overhang at the rear. Skillfully done sewing repairs can pull the edges of the hole(s) together and improve the cap's appearance in an area not normally visible as well as prevent further decay. At the least, any such sewing work should be done with original German field gray, fliegerblau or navy blue thread, if this can be found (spools do still exist, though they are difficult to locate).

Display

Some collectors use expensive glass or Plexiglas cases to protect and display their headgear collections. While visually attractive and offering good protection, such cases do not allow easy removal of a cap for close-up viewing or light cleaning and may take up considerable space. Another alternative is the vinyl bag used for packaging of winter blankets. These bags usually have a zippered flap along one side, and the medium size (twin bed or larger blanket) bag will actually fit a Styrofoam head wearing a cap. The zipper panel serves as the base. These vinyl bags are very lightweight and easy to remove, while at the same time they provide for a clear view of the cap contained within (as long as the bag has not been wrinkled excessively). You can also ask relatives to turn any bags from their own blankets over to you, rather than throw them away.

These bags do not prevent moisture from entering, but they do keep dust buildup to a very low level as well as providing a fairly good level of protection from moths. If you prefer even more moth defense, place a small square of cedar wood inside the bag with the cap. Cedar is the best form of [natural] moth protection since it poisons moths and moth eggs, but will not damage your caps in any way. Note, however, that every few months you will have to lightly sand the surface of the cedar square, in order to revitalize the scent.

Mothballs and similar products are *not* recommended, since the camphor that they release is very strong and has the potential to damage materials used within the cap.

Leather Preservers:

Recommended
Pecard Antique Leather Dressing
Connolly Hide Care

Not recommended
Saddle Soap
Neatsfoot Oil (and any other leather care product based on petroleum distillates will rot the sweatband stitching and should not be used)